EARTH SCIENCE

The Physical Setting

Thomas McGuire

AMSCO SCHOOL PUBLICATIONS, INC.
315 Hudson Street, New York, N.Y. 10013

The publisher would like to thank the following teachers who reviewed the manuscript.

Dr. James R. Ebert
Professor, Earth Science Education
SUNY College at Oneonta
Oneonta, New York

Bernadette Tomaselli
Science Department Chair
Lancaster High School
Lancaster, New York

Howard Gottehrer
Former Earth Science Teacher
Martin Van Buren High School
Queens Village, New York

Gary Vorwald
Science Department Chair
Paul J. Gelinas Junior High School
Setauket, New York

Thomas Lewis
Earth Science Mentor
Monroe BOCES #2
Rochester, New York

Editor: Margaret Pearce

Text and Cover Design: Mel Haber

Composition: Northeastern Graphic, Inc.

Art: Hadel Studio

Cover Photo: Getty Images, Inc. Herbert and Bow Lakes, Banff National Park, Canada

Please visit our Web site at:
www.amscopub.com

To the Student

Earth Science: The Physical Setting, which follows the New York State Core Curriculum, is an introduction to the study of Earth Science. With this book, you can gain a firm understanding of the fundamental concepts of Earth Science—a base from which you may confidently proceed to further studies in science and enjoy a deeper appreciation of the world around you. You also will need to become familiar with the *Earth Science Reference Tables*, a document prepared by the New York State Education Department. You will find the individual tables within the appropriate chapters of this text. You can obtain a copy of the entire document from your teacher or it can be downloaded from the State Education Web site (www.nysed.gov).

This book is designed to make learning easier for you. Many special features that stimulate interest, enrich understanding, encourage you to evaluate your progress, and enable you to review the concepts are provided. These features include:

1. **Carefully selected, logically organized content.** This book offers an introductory Earth Science course stripped of unnecessary details that lead to confusion. It covers the New York State Core Curriculum for the Physical Setting—Earth Science.

2. **Clear understandable presentation.** Although you will meet many new scientific terms in this book, you will find that the language is generally clear and easy to read. Each new term is carefully defined and will soon become part of your Earth Science vocabulary. The illustrations and photographs also aid in your understanding, since they, like the rest of the content, have been carefully designed to clarify concepts. Words in **boldface** are defined

in place and in the Glossary. Words in *italics* are important science words you already should know.

3. **Introduction.** An introductory section at the beginning of each chapter sets the stage for the rest of the chapter.

4. **Step-by-step solutions to problems followed by practice.** Problem solving is presented logically, one step at a time. Sample solutions to all types of Earth Science problems are provided. These sample problems will help you approach arithmetic problems logically. To enhance your newly acquired skill, you will find practice problems following most sample problems.

5. **End-of-chapter review questions.** The Regents-style, multiple-choice questions at the end of each chapter help you to review and assess your grasp of the content. The open-ended questions provide practice in answering questions found in Part B-2 and Part C of the Regents exam.

6. **Appendices.** Appendix A introduces you to laboratory safety. In Appendix B, you will be presented with a format to follow when preparing laboratory reports. Appendix C reviews the International System of Units. Appendix D lists the physical constants important to Earth Science. Appendix E explores the use of graph in science.

7. **Glossary.** This section contains all the **boldfaced** words found in the text along with their definitions.

The study of Earth Science can be both stimulating and challenging, The author sincerely hopes that this book will increase your enjoyment of this science.

Contents

Chapter 1

The Science of Planet Earth

 WHAT IS SCIENCE?

Science is a way of making and using observations. The applications of science have played a central role in the advancement of civilization. The Latin origin of the word *science* (*scire*) can be translated as "to know." While some people might think of scientific conclusions as unchanging facts, our understanding is never complete. As the understanding of nature grows, old ideas that no longer seem to fit our observations are discarded. The so-called facts of science are often temporary while the methods of science (observation and analysis) are permanent.

Science often attempts to answer questions such as: Why is the sky blue? Why do we see the moon on some nights, but not on others? What causes clouds to form? Why are there violent storms, earthquakes, and volcanoes? How can people protect themselves from these disasters? How can people wisely use Earth's resources and still preserve the best features of a natural environment? Understanding Earth and how it changes is essential for human survival and prosperity. (See Figure 1-1 on page 2.)

1

Figure 1-1 Earth is our home; we must keep it livable.

Great works of art are valued, in part, because they have strong emotional impact. However, unlike works of art, scientists generally want their work to be as free of bias and individual judgments as possible. Rational thought and clear logic support the best scientific ideas. Scientists often use numbers and mathematics because mathematics is straightforward, logical, and consistent. These qualities are valued in scientific work.

Scientific discoveries need to be *verifiable*. This means that different scientists who investigate the same issues should be able to make their own observations and arrive at similar conclusions. When a climate prediction is supported by the work of many scientists or by computer models, the prediction is considered to be more reliable. In fact, the ability reproduce results or verify ideas is a significant characteristic of science.

Science at Work

Alfred Wegener proposed his theory of continental drift in the early 1900s; it was based on indirect evidence. During his lifetime, he could not find enough evidence to convince most other Earth scientists that continents move over Earth's surface. However, new evidence gathered by other scientists working 50 years later gave renewed support to his ideas. Today, plate tectonics, as the theory is now known, is supported by precise measurements of the changing positions of the continents. This is a good example of how the efforts of many scientists resulted in a new way of thinking about how our planet works.

Science can therefore be defined as a universal and continuous method of gathering, organizing, analyzing, testing, and using information about our world. Science provides a structure to investigate questions and to arrive at conclusions. The reasoning behind the conclusions is clear, and the conclusions are subject to continued evaluation and modification. The body of knowledge of science, even as presented in this book, is simply the best current understanding of how the world works.

ACTIVITY 1-1	GOOD SCIENCE AND BAD SCIENCE

Sometimes it is easier to understand science if you look at what is *not* science.

Tabloids are newspapers that emphasize entertainment. They publish questionable stories that other media do not report. Bring your teacher an article from a questionable news source that is presented as science. Your teacher will display the stories for the class to discuss. What are the qualities of these stories that make them a poor source of scientific information?

WHAT IS EARTH SCIENCE?

The natural sciences you study in school are generally divided into three branches: life science (biology), physical science (physics and chemistry), and Earth science. (See Figure 1-2.) **Earth science** generally applies the tools of the other sciences to study Earth, including the rock portion of Earth, its oceans, atmosphere, and its surroundings in space.

Earth science can be divided into several branches. **Geology** is the study of the rock portion of Earth, its interior, and surface processes. Geologists investigate the processes that shape the land, and they study Earth materials, such as minerals and rocks. (See Figure 1-3.) They also actively search for natural resources, including fossil fuels.

Meteorology is the study of the atmosphere and how it changes. Meteorologists predict weather and help us to deal with natural disasters and weather-related phenomena that affect our lives. They also investigate climatic (long-term weather) changes.

Oceanography is the study of the oceans that cover most of Earth's surface. Oceanographers investigate ocean currents, how the oceans affect weather and coastlines, and the best ways to manage marine resources.

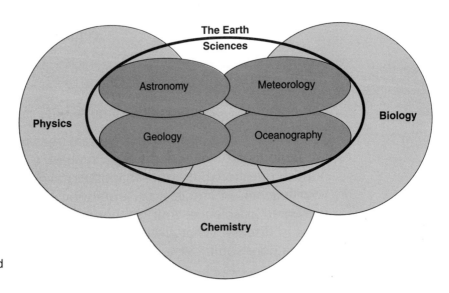

Figure 1-2 Earth sciences study the major parts of the planet by using other branches of science, such as biology, chemistry, and physics.

Figure 1-3 Earth science is an exploration of our planet to understand how it came about and how it changes. This man is exploring a slot canyon.

Astronomy is the study of Earth's motions and motions of objects beyond Earth, such as planets and stars. Astronomers consider such questions as: Is Earth unique? How big is the universe? When did the universe begin, and how will it end?

Many Earth scientists are involved in **ecology**, or environmental science, which seeks to understand how living things interact with their natural setting. They observe how the natural environment changes, how those changes are likely to affect living things, and how people can preserve the best features of the natural environment.

 ## HOW IS EARTH SCIENCE RELATED TO OTHER SCIENCES?

One important feature of Earth science is that it draws from a broad range of other sciences. This helps present an all-encompassing view of the planet and its place in the universe. Earth scientists need to understand the principles of chemistry to investigate the composition of rocks and how they form. Changes in weather are caused by the energy exchanges at the atomic level. By knowing the chemical properties of matter, scientists can investigate the composition of stars. Knowledge of biology allows Earth scientists to better interpret the information preserved in rock as fossils.

The movements of stars and planets obey the laws of physics regarding gravity and motion. Physics helps us understand how the universe came about and how stars produce such vast quantities of energy. Density currents and the circulation of fluids control the atmosphere, the oceans, and even changes deep within our planet. Nuclear physics has allowed scientists to measure the age of Earth with remarkable accuracy.

The Earth sciences also make use of the principles of biology and, in turn, support the life sciences. Organic evolution helps us understand the history of Earth. At the same time, fossils are the primary evidence for evolutionary biology. The relationships between the physical (nonliving) planet and life forms are the basis for environmental biology. Only recently have people grown to appreciate how changes in Earth and changes in life forms have occurred together throughout geologic time.

WHY STUDY EARTH SCIENCE?

Although some readers of this book may become professional geoscientists, it is more likely that you will find work in other areas. Regardless of the career you choose, Earth science will affect your life. Everyone needs to know how to prepare for changes in weather, climate, seasons, and earth movements.

Natural disasters are rare events, but when they occur they can cause devastating loss of life and property. To limit loss, people can prepare for hurricanes, tornadoes, floods, volcanic eruptions, earthquakes, and climate shifts. Humans can survive the effects of cold and drought if they plan ahead, but they need to know how likely these events are and how best to avoid their devastating consequences. How will humans be affected by general changes in climate? Can it be prevented? Will a large asteroid or comet strike Earth, and how will it affect Earth's inhabitants?

Our civilization depends on the wise use of natural resources. Freshwater, iron, and fossil fuels are among the great variety of materials that have supported a growing world economy. These resources have brought us unprecedented wealth and comfort. How much of these materials are available for use? What will happen if these materials run out? What is the environmental impact of extracting, refining, and using these resources

These issues affect all of us regardless of our profession. As citizens and consumers, we make decisions, and as citizens, we elect governments that need to consider these issues.

How can you, as one individual among millions in the United States, among billions in the world, make a difference? Environmental activists have a useful way of thinking about this, "Think globally, but act locally." If you consider broad issues as you conduct your daily life, you can contribute to solving global problems. One person conserving resources by reusing and recycling materials has a very small impact. But when all people contribute their small parts, the beneficial effects are multiplied. One person buying a more fuel-efficient car or using mass transportation has a small impact. However, when these practices become widespread through public education, they can become powerful forces.

Working with Science

Figure 1-4 Cynthia Chandley

CYNTHIA CHANDLEY: Water Rights Lawyer

Cynthia Chandley is not an Earth scientist, but she knows how important it can be to understand Earth. (See Figure 1-4.) She earned a degree in geology, and, after several years of working in the mining industry, attended law school and became an environmental lawyer. Ms. Chandley now works as a water rights litigator for a law firm. "I constantly use my geoscience background to influence the use and preservation of an essential resource. But these issues go well beyond my profession. Everyone needs to understand our planet to help determine how our resources can be most effectively managed for ourselves and for future generations."

OBSERVATIONS, MEASUREMENT, AND INFERENCES

You gather information about your surroundings through your five senses: sight, touch, smell, taste, and hearing. The processes and interpretations made by scientists depend on making use of information gathered using their senses. These pieces of information are called **observations**. Some observations are qualitative. Relative terms, such as long or short, bright or dim, hot or cold, loud or soft, red or blue, compare the values of our observations without using numbers or measurements. Other observations are quantitative. When you say that the time is 26 seconds past 10 o'clock in the morning you are being very specific. Quantitative comes from the word *quantity* meaning "how many." Therefore quantitative observations include numbers and units of measure.

Scientists use measurements to determine precise values that have the same meaning to everyone. Measurements often are made with instruments that extend our senses. Microscopes and telescopes allow the observation of things too small, too far away, or too dim to be visible without these instruments. (See Figure 1-5.) Balance scales, meter sticks,

Figure 1-5 Instruments help us make better observations.

clocks, and thermometers allow you to make more accurate observations than you could make without the use of instruments.

People accept many things even if they have not observed them directly. An inference is a conclusion based on observations. For example, if Liz meets a friend late one afternoon, and he appears tired and is carrying a baseball, bat, and glove, Liz would probably infer that her friend had been playing baseball. Although Liz never saw him playing, this inference seems reasonable. When many rocks at the bottom of a cliff are similar in composition to the rock that makes up the cliff, it is reasonable to infer the rocks probably broke away from the cliff.

Scientists often make inferences. When scientists observe geological events producing rocks in one location and they find similar rock in other locations, they make inferences about past events, although they did not witness these events. No person can see the future. Therefore all predictions are inferences. In general, scientists prefer direct observations to inferences.

 Exponential Notation

Scientists deal with data that range from the sizes of subatomic particles to the size of the universe. If you measure the universe in subatomic units you end up with a number that has about 40 zeros. How can this range of values be expressed without using numbers that are difficult to write and even more difficult to work with? Scientists use exponential numbers, sometimes called scientific notation, which uses powers of ten to express numbers that would be more difficult to write or read using standard decimal numbers.

Numbers in exponential notation take the form of $c \times 10^e$, where c is the coefficient (always a number equal to or greater than 1 but less than 10) and e is the exponent. Being able to understand and use exponential notation is very important. Any number can be changed into exponential notation in two steps.

Step 1: Change the original number to a number equal to or greater than 1 but less than 10 by moving the decimal point to the right or left.

Step 2: Assign a power of 10 (exponent) equal to the number of places that the decimal point was moved.

A good way to remember whether the power of 10 will be positive or negative is to keep in mind that positive exponents mean numbers greater than 1, usually large numbers. Negative exponents mean numbers less than 1, which are sometimes called decimal numbers. Once you get used to it, it becomes easy.

Let us see how this is done. The mass of Earth is 5,970,000,000,000,000,000,000,000 kilograms. Move the decimal 24 places to the left to get 5.97. The power of 10 is therefore 24 Expressed in exponential notation this number is 5.97×10^{24} kilograms.

SAMPLE PROBLEMS

Problem 1   The age of Earth is 4,600,000,000 years; express this number in exponential notation.

Solution

Step 1: Change the original number to a number equal to or greater than 1, but less than 10 by moving the decimal point to the right or left. (Zeros that appear outside nonzero digits can be left out.) In this case, you get 4.6.

Step 2: Assign a power of 10 (exponent) equal to the number of places that the decimal point was moved. This decimal point was moved nine places. In this case the decimal point was moved left, make the power of 10 a positive number. So the age of our planet is 4.6×10^9 years.

Problem 2   Light with a wavelength of 0.00004503 centimeter (cm) appears blue. Express this value in scientific notation.

Solution

Step 1: After moving the decimal point five places to the right, the coefficient becomes 4.503. The zero before the 3 is kept because it appears between nonzero digits. This zero is needed to establish the number's value.

Step 2: When the decimal point is moved right, you make the exponent a negative number. The power of 10 is −5. The number is 4.503 × 10^{25} cm.

ACTIVITY 1-2	EXPONENTIAL NOTATION IN THE REAL WORLD

Make a list of 5 to 10 values expressed in scientific notation, document their use, and translate them into standard numbers. Your examples must come from printed or Internet sources outside your Earth science course materials.

For each example you bring, include the following:

1. The value expressed in exponential notation. (If units of measure are present, be sure to use them.)

2. What is being expressed. (For example, it might be the size of a particular kind of atom.)

3. The same value expressed as a regular number.

4. Where you found the value. Please give enough information so that another person could find it easily.

The International System of Measurement

Over the course of time, different countries developed their own systems of measurements. The inch and the pound originated in England. There were no international standards until the European nations established a system now known as the "International System of Units." This system is called "SI," based on its name in French, *System Internationale*. SI units are now used nearly everywhere in the world except the United States. SI is similar to the metric system.

In a temperature-controlled vault in France, a metal bar has been marked at exactly 1 meter. In the past, it was the precise definition of meter, and all devices used to measure length were based on that standard. Everyone knew the length of a meter and everyone's meter was the same. Today the meter is defined as a certain number of wavelengths of light emitted by krypton-86 under specific laboratory conditions. The advantage of this change is the standard length can be created anywhere and is not susceptible to natural or political events.

In everyday life, people often use a system of measures called "United States Customary Measures." Units such as the mile, the pound, and the degree Fahrenheit have been in use in this country for many years. Most Americans are familiar with them and resist change. As this country becomes part of a world economy, SI units will gradually replace the United States Customary units. Many beverages are now sold in liters. A variety of manufactured goods created for world markets are also measured in SI units. (See Figure 1-6 and Table 1-1.)

However, scientists everywhere use SI units for several reasons:

- They are universal. Scientists do not need to translate units when they communicate with their colleagues in other countries.

- Most SI units are related by factors of 10. For example, there are 10 millimeters in a centimeter and 100 cm in a meter.

- Scientific instruments on the world market are generally calibrated in SI units.

Figure 1-6 In some places, road signs with SI (metric) units are replacing signs that used United States Customary Measures.

TABLE 1-1. International System of Units

Physical Quantity	SI Unit	Symbol	U.S. Customary Measure
Length	meter	m	inch, foot, mile
Volume	liter	L	fluid ounce, quart, gallon
Mass	gram	g	ounce, pound, ton
Time	second	s	(same as SI)
Temperature	kelvin	K	degree Fahrenheit
	degree Celsius	°C	degree Fahrenheit

ACTIVITY 1-3   **MAKING ESTIMATIONS**

Estimation is a valuable skill for anyone, but especially for scientists. If you want to know whether a measurement or calculation is correct, it can be helpful to estimate the value. If your estimate and the determined value are not close, you may need to give some more thought to your procedure.

If you were to estimate the distance from your home to the nearest fast-food restaurant, you might say that you can walk there in 30 minutes. If you walk at a rate of 5 kilometers per hour (km/h), in half an hour you can walk 2.5 km. So your estimate would be 2.5 km.

Working in groups, estimate the volume of your classroom or your school building. No measuring instruments may be used. Your group must write a justification of your estimate. Please use only SI (metric) units.

USING SI UNITS   Density is an important property of matter. For example, differences in density are responsible for winds and ocean currents. Density is defined as the concentration of matter, or mass per unit volume. For example, if the mass of an object is 30 grams and its volume is 10 cubic centimeters, (cm³), then its density is 30 grams divided by 10 cm³, or 3 grams/cm³. The formula for calculating density is given in the *Earth Science Reference Tables*.

SAMPLE PROBLEM

Problem   The measurements of a rectangular block are length 5 cm, width 3 cm, and height 8 cm. Find the volume of the block.

Solution

You can calculate the volume by multiplying the length by the width by the height:

$$\text{Volume} = \text{length} \times \text{width} \times \text{height}$$
$$= 5 \text{ cm} \times 3 \text{ cm} \times 8 \text{ cm}$$
$$= 120 \text{ cm}^3$$

 Practice Problem 1
A rectangular bar of soap measures 10 cm by 2 cm by 7 cm. Find the volume of the bar of soap.

 # HOW IS DENSITY DETERMINED?

Density is the concentration of matter, or the ratio of mass to volume. Substances such as lead or gold that are very dense are heavy for their size. Materials that we consider light, such as air or Styrofoam, are relatively low in density. Objects made of the same solid material usually have about the same density. (Density does change with temperature as a substance expands or contracts.) As shown in the following problem, density can be calculated using the formula given in the *Earth Science Reference Tables*. Density is generally expressed in units of mass divided by units of volume. Note that the units are carried through the calculation, yielding the proper unit of density: grams per cubic centimeter (g/cm³).

SAMPLE PROBLEM

Problem What is the density of an object that has a volume of 20 cm³ and a mass
of 8 g?

Solution

$$\text{Density of a substance} = \frac{\text{mass}}{\text{volume}}$$

$$= \frac{8 \text{ g}}{20 \text{ cm}^3} = 0.4 \text{ g/cm}^3$$

 Practice Problem 2
A 105-g sphere has a volume of 35 cm³, what is its density?

Water, with a density of 1 g/cm³, is often used as a standard of density. Therefore, the process of flotation can be used to estimate density. If an object is less dense than water, the object will float in water. If the object is more dense than water, the object will sink. Most wood floats in water because it is less dense than water. Iron, glass, and most rocks sink because they are more dense than water. The idea of density will come up many times in Earth science and it will be discussed as it is applied in later chapters.

The instrument shown in Figure 1-7 is called a Galileo thermometer. It is named for the Italian scientist who invented it. This thermometer is based on the principle that the density of water changes slightly with changes in temperature. As the water in the column becomes warmer and less dense, more of the glass spheres inside the tube sink to the bottom. Therefore, the number of weighted spheres that float depends on the temperature of the water. Reading the number attached to the lowest sphere that floats gives the temperature.

A demonstration of the relative density of liquids can be made by first pouring corn syrup, then water, followed by

Figure 1-7 In a Galileo thermometer, as the water inside the tube becomes warmer and less dense, more of the weighted glass spheres sink to the bottom. The tag on the lowest sphere that floats indicates the approximate temperature.

cooking oil, and finally alcohol into a glass cylinder. Care must be taken not to mix the liquids. They will remain layered in order of density as shown in Figure 1-8. If a rubber stopper with a density of 1.2 g/cm³ were added, it would sink through the water layer. The stopper would remain suspended between the water and the corn syrup. Rubber is more dense than water, so it sinks in water. Corn syrup is more dense than rubber. Therefore, the rubber stopper would float on top of the corn syrup layer.

Errors in Measurement

No matter how carefully a measurement is made, it is likely that there will be some error. Using measuring instruments more carefully or using more precise instruments can reduce error, but error can never be eliminated. In general, errors are reduced to the point that they are not important or that it is not worth the effort to make them smaller. Sometimes measurements are used in calculations, such as the determination of density. In these cases any errors in measurement will result in errors in the calculated value.

Percent error is a useful way to compare the size of the error with the size of the value being measured. For example, an error of 1 cm in the size of this book is a large error. But an error of 1 cm in the distance to the moon would be a very small error. They are both errors of 1 cm. However, because the book is so much smaller, a 1-cm error is far more significant.

Within the chapters of this book you will find the components of the *Earth Science Reference Tables*: charts, maps, physical values, and mathematical equations that you will need throughout this course. You do not need to memorize any of the information in the Reference Tables because this document will always be available to you for classroom work, labs, and tests. However, you should become familiar with the Reference Tables so you know when to use them.

In the *Earth Science Reference Tables* is an equation called "Percent Deviation from Accepted Value." This is a more precise term for percent error. The term "accepted value" is used because no measured value is known with complete accuracy.

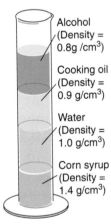

Alcohol
(Density =
0.8g /cm³)

Cooking oil
(Density =
0.9 g/cm³)

Water
(Density =
1.0 g/cm³)

Corn syrup
(Density =
1.4 g/cm³)

Figure 1-8 These four liquids will remain in place in order of density unless they mix or evaporate. The most dense liquids sink to the bottom and the least dense liquids remain on top.

LAB 1-1: Densities of Solids

Density can be used to identify different substances. In general, no matter how much you have of a certain substance, its density is the same. Rather than measuring density directly, usually the mass is measured, and the volume is determined so that density (density = mass/volume) can be calculated. The equation volume = length × width × height is used to determine the volume of rectangular solids. There are equations that can be used to determine the volume other regular solids.

Your teacher will supply a variety of objects. Create a data table in which to record your data. Measure the mass and determine the volume of each sample, then calculate the density of each. Be sure to use SI (metric) measurements.

After you have calculated the density of each sample, place a star next to the name of those that will float on water. How can you tell that they will float?

In many cases, however, an expected or accepted value can be determined. The following Sample Problem will show how to use this equation.

SAMPLE PROBLEM

Problem   A student estimated the height of a tree to be 15 m. However, careful measurement showed the true height was 20 meters. What was the percent deviation?

Solution
_____

$$\text{Deviation (\%)} = \frac{\text{difference from accepted value}}{\text{accepted value}} \times 100$$

$$= \frac{5\,\text{m}}{20\,\text{m}} \times 100$$

$$= 25\%$$

Please notice the following features in this calculation:

1. The calculation starts with the complete algebraic formula. The only numbers that show in this first step are constants used in every application of the formula.

2. Values are substituted into the formula, including numbers and their associated units of measure.

3. The steps to the solution are organized so that they are easy to follow, leading to the answer at the end.

 Practice Problem 3

A student determined the density of a piece of rock to be 3.5 g/cm³. The accepted value is 3.0 g/cm³. What was the student's percent error?

 # USING GRAPHS IN SCIENCE

A graph is a visual way to organize and present data. Instead of reading paragraphs of information or studying columns of figures, a graph make comparisons between variables easier. Unlike a data table, a graph enables the reader to visualize changes in data, to understand relationships between variables within the data, and to picture trends or patterns.

 ## Line Graphs

A line graph, such as the one in Figure 1-9, shows how a measured quantity changes with respect to time, distance, or some other variable. Line graphs are constructed by plotting data on a **coordinate system**, a grid in which each location has a unique designation defined by the intersection of two lines. A coordinate system is set up on vertical and horizontal axes. The horizontal (x) axis is usually used for the independent variable. It usually indicates a uniform change, such as hours, years, or centimeters. Normally, the regular change expected in the independent variable is well understood. The vertical (y) axis is used for the dependent variable. It usually indicates the amount of the measured quantity being studied, such as temperature, height, or population. The values of

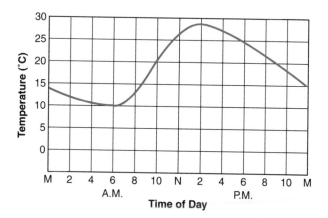

Figure 1-9 This line graph shows how the temperature changed on a summer day in central New York State. Note that on a line graph you can read the temperature at any given time.

the dependent variable are what you are trying to find. The graph shows how the dependent variable changes with respect to the independent variable.

The rise or fall of the line in Figure 1-9 on page 20 shows the increase or decrease in temperature during a typical summer day in central New York State. When the line on the graph moves upward and to the right, it represents a continuous increase. When the line on the graph moves downward and to the right, it indicates a continuous decrease. A horizontal line on the graph represents no change. The steeper the line segment rises to the right, the greater the slope of the segment, and the greater the increase in temperature. Likewise, the steeper the line segment falls to the right, the greater the decrease in temperature. Not all graphs are curved lines. Some line graphs are straight lines.

 Pie and Bar Graphs

Sometimes, a line graph is not the best kind of graph to use when organizing and presenting data. In Earth science, bar and pie graphs are often used. The bar graph is useful in comparing similar measurements at different times or in different places. For example, the bar graph in Figure 1-10 on page 20, which is based on the data in Table 1-2 on page 20, compares monthly rainfall, or precipitation (PPT), in millimeters (mm) over the period of 1 year.

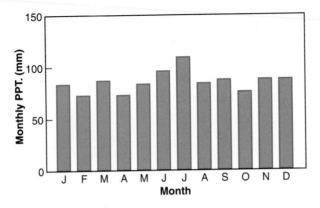

Figure 1-10 This bar graph represents the average monthly precipitation for Lake Placid, New York.

TABLE 1-2. Average Monthly Precipitation for Lake Placid, New York

Month	PPT (mm)	Month	PPT (mm)
January	81	July	107
February	71	August	84
March	86	September	86
April	71	October	74
May	81	November	86
June	94	December	86

The pie graph is used to show how a certain quantity has been divided into several parts as well as to show the comparisons among these parts. The pie graph in Figure 1-11 shows by percent the most abundant chemical elements in the rocks of Earth's crust.

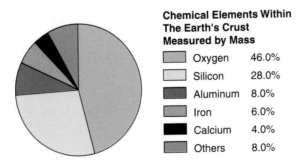

Chemical Elements Within The Earth's Crust Measured by Mass

Oxygen	46.0%
Silicon	28.0%
Aluminum	8.0%
Iron	6.0%
Calcium	4.0%
Others	8.0%

Figure 1-11 A pie graph shows how a quantity has been divided and the comparison between the divisions.

Guidelines for Making Graphs

Graphs are all around us. They are especially common in news and in advertising where it is important to convey information quickly. However, in the effort to keep the graph simple, they sometimes contain unfortunate errors. When you construct graphs in science you should take care to follow these guidelines:

- Keep in mind that the purpose of a graph is to convey information. The graph should have a title to clarify the relationships represented. All essential information should be presented as clearly and simply as possible. The axes should be labeled with both quantity and units. One axis might be time in years while the other is price in U.S. dollars per barrel. (See Figure 1-12.)

- The independent variable should be plotted on the horizontal axis. Usually, data shows how one factor changes depending on changes in the other. For example, in Figure 1-10, it is clear that the price of oil does not determine the passage of time. The price of crude oil depends on when it is purchased. In this case, time is the independent variable and the price is the dependent variable. Time (the year, month, etc.) belongs on the bottom axis.

Figure 1-12 In this graph, time is the independent variable, and the price of crude oil is the dependent variable.

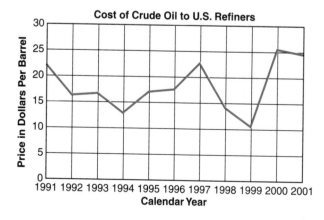

● Fit the graph to the data. Design your vertical and horizontal axes so that the data reasonably fills the graph but does not go beyond the scales on the two axes.

ACTIVITY 1-4   **MAKING A GRAPH OF THE REVOLUTION OF THE PLANETS**

Does the distance of a planet from the sun affect how long it takes to make one orbit of the sun? You can investigate this question by drawing a graph.

In the *Earth Science Reference Tables* there is a table labeled "Solar System Data." Use this data in this table to graph the relationship between the distance of a planet from the sun and its period of revolution. Label each data point with the name of the object from Mercury through Pluto. (Do not include the sun or Earth's moon.)

As a follow up, you might try graphing planetary distance and other factors in this table.

 # TECHNOLOGY IN EARTH SCIENCE

How science is "done" has always depended on the tools available. Some tools have revolutionized Earth science. Computers provide a good example. When they are attached to a variety of other devices, computers can be used for an amazing variety of applications. Computers help us analyze data, produce and edit images, and quickly access information. The first electronic computers filled entire rooms, and were so expensive that only a few research facilities could afford them. Today, a laptop computer can have computing power equal to that of a supercomputer of the 1970s.

Connecting computers in networks has progressed to the point where you can almost instantly access information stored in millions of computers all over the world. This is the World Wide Web connected by the Internet. It allows all of us to communicate faster than ever before.

| ACTIVITY 1-5 | AN INTERNET SCAVENGER HUNT |

A scavenger hunt is an activity in which the goal is to collect a variety of unrelated objects. In this case, the "objects" will be bits of information. Each example will require two responses: (1) give the answer to the question, and (2) record where on the Internet you found it. [Please, provide the Internet address (URL) and/or the name of the Internet site.] It is unlikely that you will be able to answer all these questions, so just find as many as you can.

1. What is the weather like today in Phoenix, Arizona?

2. Where and when has a major earthquake occurred in the past 6 months?

3. Other than the sun, what is the nearest star to Earth?

4. What is the human population of New York City?

5. What is the current value of gold per ounce?

6. How many sunspots were recorded in 1990?

7. What name was applied to the third tropical storm in the Atlantic Ocean last year?

8. What is the chemical composition of emeralds?

GIS and GPS are two of the most exciting, recent technological advances for the Earth sciences. The Geographic Information System (GIS) is visual resource that allows you to plot the spatial relationships of data. Because GIS is based on information in computers all over the world, a wide variety information can be retrieved and mapped. It also can be updated regularly.

Figure 1-13 Using a Global Positioning System (GPS) device, this person can determine her position.

The Global Positioning System (GPS) depends on satellites that transmit information, which can be received by a handheld device. The information enables you to determine your location with remarkable accuracy. (See Figure 1-13.) Installed in your car, a GPS unit can direct you to an unfamiliar location in real time. The GPS is so accurate that it has been used to measure the slow movement of continents over Earth's surface.

Inquiry in Science

Many people would say that an inquiring mind is the most important asset humans have. Using observations, information resources, and a variety of analytical tools, people can often make important discoveries by asking the right questions and following productive leads. As long as there is the curiosity to ask questions and the will to find the answers to them, science will help find those answers.

◹ LAB 1-2 The Thickness of Aluminum Foil

Given the following:

1. Density $= \dfrac{\text{mass}}{\text{volume}}$

2. Volume $=$ length $\times$ width $\times$ thickness

3. Density of aluminum $= 2.7$ g/cm³

Materials: a metric ruler, a kilogram scale, a small piece of aluminum foil (about 30 cm on each side).

Figure 1-14 Materials needed to determine the thickness of aluminum foil.

Problem: Determine the thickness of the aluminum foil to the nearest ten-thousandth of a centimeter (two significant figures).

Hint. Combine the two equations above into a single equation with one unknown. Then substitute measurements, to solve for thickness.

TERMS TO KNOW

astronomy	**Earth science**	**inference**	**oceanography**
coordinate system	**ecology**	**meteorology**	**percent error**
density	**geology**	**observation**	**science**

CHAPTER REVIEW QUESTIONS

1. Some scientists estimate that age of the universe is about 1.37×10^{10} years. Which choice below correctly expresses this value?

   (1) 13,700,000,000 years
   (2) 13,710,000,000 years
   (3) 1,370,000,000,000 years
   (4) 1,371,000,000,000,000 years

2. The average distance from Earth to the sun is about 149,000,000 kilometers. Which choice below correctly expresses this in scientific notation?

   (1) 1.49×10^4 km
   (2) 1.49×10^8 km
   (3) 1.49×10^{10} km
   (4) 1.49×10^{13} km

3. A student recorded information about a rock sample. Which is an observation?

   (1) If placed in water, the rock may float.
   (2) The rock has a mass of 93.5 g.
   (3) The rock is billions of years old.
   (4) The rock formed deep inside Earth.

4. The following statements are taken from a student's notes about the current weather conditions. Which statement is an inference?

   (1) The temperature 3 hours ago was 20°C.
   (2) The current air pressure is 1000.4 millibars.
   (3) The sky is completely overcast with clouds.
   (4) It is probably cooler 500 miles north of this location.

5. A certain rock has a mass of 46.5 g and a volume of 15.5 cm³. What is the density of this rock?

   (1) 0.300 g/cm³   (3) 3.50 g/cm³
   (2) 3.00 g/cm³   (4) 767.25 g/cm³

6. What is the most important reason that scientists display data in graphs?

   (1) Graphs never contain errors.
   (2) Graphs take less room than data tables.
   (3) Graphs make data easier to understand.
   (4) Graphs make papers easier to get published.

7. Which of the following would be a complete label for the vertical (y) axis of a graph?

   (1) mass of the sample
   (2) volume of the sample
   (3) degrees Celsius
   (4) number of correct responses

8. A student measured the mass of a rock as 20 g. But the actual mass of the rock was 25 g. What was the student's percent error?

   (1) 5.5%   (3) 33%
   (2) 20%   (4) 75%

9. What is the principal reason for using percent error rather than simply expressing the size of the error itself?

(1) Percent error gives more information than the value of the error itself.
(2) If there is no error, percent error makes this more clear.
(3) Percent error emphasizes the importance of errors.
(4) Sometimes the value of the error itself is not known.

10. The density of quartz is 2.7 g/cm³. If a sample of quartz has a mass of 81 g, what is its volume?

(1) 0.03 cm³   (3) 11.1 g
(2) 8.1 g   (4) 30 cm³

11. Gold has a density of 19.3 g/cm³. A prospector found a gold nugget with a volume of 10 cm³. What was the mass of the nugget?

(1) 10 g   (3) 29.3 g
(2) 19.3 g   (4) 193 g

12. Pumice is an unusual rock because it can float on water. What does this tell you about pumice?

(1) Pumice is usually found in very small pieces.
(2) Pumice is most common in high mountain locations.
(3) Pumice is less dense than water.
(4) Pumice absorbs water.

13. The density of granite is 2.7 g/cm³. If a large sample of granite is cut in half, what will be the density of each of the pieces?

(1) 1.35 g/cm³   (3) 5.4 g/cm³
(2) 2.7 g/cm³   (4) 27 g/cm³

14. If two leading scientists are investigating the same question and they reach similar conclusions, what does this show?

(1) They probably changed their results to get agreement.
(2) Their conclusions have a good chance of being correct.
(3) Their scientific work showed a lack of originality.
(4) The conclusion they both made is probably in error.

15. Which is usually considered a division of Earth science?

(1) chemistry   (3) physics
(2) biology   (4) geology

Open-Ended Questions

Base your answers to questions 16 through 19 on the information and data table below.

The snowline is the lowest elevation at which snow remains on the ground all year. The data table below shows the elevation of the snowline at different latitudes in the Northern Hemisphere.

Latitude (°N)	Elevation of Snowline (m)
0	5400
10	4900
25	3800
35	3100
50	1600
65	500
80	100
90	0

16. On a properly labeled grid, plot the latitude and elevation of the snowline for the locations in the data table. Give the graph a title. Use a dot for each point and connect the dots with a line.

17. Mt. Mitchell, in North Carolina, is located at 36°N and has a peak elevation of 2037 m. Plot the latitude and elevation of Mt. Mitchell on your graph. Use a plus sign (+) to mark this point.

18. Using your graph, determine to the *nearest whole degree*, the lowest latitude at which a peak with the same elevation as Mt. Mitchell would have permanent snow.

19. State the relationship between latitude and elevation of the snowline.

20. The diagram below shows three liquids of different density in a 100-mL cylinder. A sphere of oak wood about half the diameter of the cylinder is dropped in the cylinder without mixing the liquids. The wooden sphere has a density of 0.9 g/cm³. Where will the sphere come to rest?

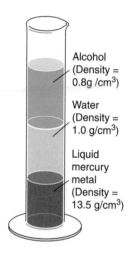

Alcohol
(Density =
0.8g /cm³)

Water
(Density =
1.0 g/cm³)

Liquid
mercury
metal
(Density =
13.5 g/cm³)

Chapter 2

Earth's Dimensions and Navigation

 ## WHAT IS EARTH'S SHAPE?

Most ancient people thought of Earth as a flat and boundless expanse. Earth is so large that a person on the surface cannot see its curvature. (See Figure 2-1.) Until people became world travelers and they invented electronic communication, the idea of a flat and endless surface was all people needed. Besides, some people reasoned that if Earth's surface were curved, gravity would pull us off the edge.

 ### Evidence of Earth's Shape

Although Earth looks flat and endless, there were some ancient scholars who believed that Earth is a gigantic sphere. The scholars came to this conclusion because they noticed that as a ship sails away to sea, it seems to disappear hull first. Ships appear to sail over and below the horizon as shown in Figure 2-2.

Another indication of Earth's shape came from observing the moon. During an eclipse of the moon, Earth's shadow

Figure 2-1 To an observer on the surface, Earth looks flat and endless.

moves over the surface of the moon. The edge of that shadow is always a uniformly curved line. Ancient Greek observers knew that the only shape that casts a uniformly curved shadow is a sphere. (See Figure 2-3 on page 32.)

You may know that if you place a telephone call to someone several hundred miles away, that person's local time will probably be different from yours. If it is noon in New York, it is only 9 A.M. for a person in California. At the same time, people in Europe are having their evening meal. For a person in central Asia or Australia, it might be midnight. When time differences over the whole planet are considered, it is clear that Earth is a gigantic sphere.

There is also evidence of Earth's shape in the observation of distant objects in the night sky. A person at the North Pole

Figure 2-2 As a ship sails over the horizon, it seems to disappear from the bottom upward.

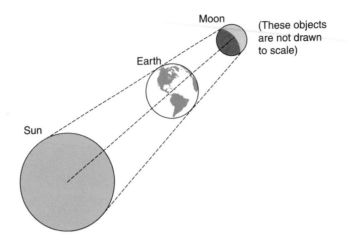

Figure 2-3 During an eclipse of the moon, the moon moves into Earth's shadow. Because the edge of Earth's shadow always shows a uniform curvature, scientists infer that Earth is a sphere.

sees the North Star, Polaris, directly overhead. To a person located farther south, Polaris appears lower in the sky. In fact, at the equator, Polaris is along the horizon. (See Figure 2-4.) The **equator** is an imaginary line that circles Earth half way between the North and South Poles. South of the equator, Polaris is not visible at all. (Observers south of the equator can see the stars of the Southern Cross, which is never visible in New York.) These observations support the idea of a spherical planet.

The exploration of space has allowed direct observations and photographs of Earth to be taken from far above its surface. The Apollo program, which explored the moon in the late 1960s, brought astronauts far enough from Earth to show that our planet is a nearly spherical object orbiting in the vastness of space.

ACTIVITY 2-1 HOW ROUND IS EARTH?

Careful measurements of Earth have shown that it is not a perfect sphere; its equatorial radius is 6378 km, and its polar radius is 6357 km. Earth's rotation on its axis causes a bulge at the equator. But how much of a bulge is there? Earth is not perfectly round. It is **oblate**, or slightly flattened at the poles.

To calculate Earth's degree of flattening, use the formula on page 34. If the result is a large number, Earth is not very round.

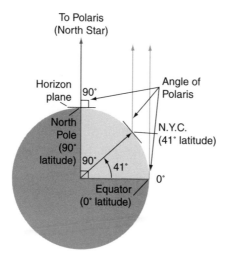

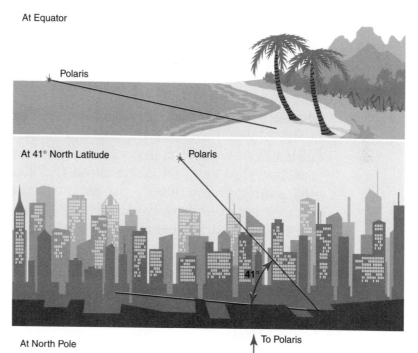

Figure 2-4 The observation of changes in the angular altitude of Polaris that are made as one travels north or south are consistent with a spherical planet.

$$\text{Degree of flattening} = \frac{\text{difference between equatorial and polar radii}}{\text{equatorial radius}}$$

Next, use a drawing compass to draw two large circles centered on the same point. Make one circle according to the polar radius and the second determined by the equatorial radius. By comparing these two circles, state how far from round Earth would appear from space.

How Large Is Earth?

A Greek scholar named Eratosthenes (era-TOSS-then-ease) made the first recorded calculation of Earth's size about 2000 years ago. He knew that on the first day of summer the noon sun was directly overhead at the town of Syene in Egypt. In Alexandria, 5000 *stadia* (approximately 800 km, or 500 mi) to the north, the sun was 7.2° from the overhead position. (*Stadia* is the plural form of *stadium*, a unit of distance used in Eratosthenes' time.) Since 7.2° is $\frac{1}{50}$ of a circle, Eratosthenes reasoned that the distance around the Earth must be 50 × 5000 *stadia*, or 250,000 *stadia*. Although the exact length of a *stadium* is not known, Eratosthenes' figure appears to be remarkably close to the more accurate measurements made today.

WHAT ARE EARTH'S PARTS?

Based on differences in composition, Earth can be divided into three parts. These parts form spheres, one inside the other, separated by differences in density. Each sphere is also a different state of matter: gas, liquid, or solid.

The **atmosphere** is the outer shell of gas that surrounds Earth. The **hydrosphere** is the water of Earth. About 99 percent of this water is contained in Earth's oceans that cover about three-quarters of the planet. The **lithosphere** is the solid rock covering Earth. (The crust is the rocky outer layer the lithosphere.)

TABLE 2-1. Average Chemical Composition of Earth's Crust, Hydrosphere, and Troposphere

Element (symbol)	Crust		Hydrosphere, Percent by Volume	Troposphere, Percent by Volume
	Percent by Mass	Percent by Volume		
Oxygen (O)	46.40	94.04	33.0	21.0
Silicon (Si)	28.15	0.88		
Aluminum (Al)	8.23	0.48		
Iron (Fe)	5.63	0.49		
Calcium (Ca)	4.15	1.18		
Sodium (Na)	2.36	1.11		
Magnesium (Mg)	2.33	0.33		
Potassium (K)	2.09	1.42		
Nitrogen (N)				78.0
Hydrogen (H)			66.0	
Others	0.66	0.07	1.0	1.0

Table 2-1 lists the abundance of chemical elements in each sphere. Rocks in Earth's crust represent the lithosphere because these are the rocks that are found at and near the surface. (Deep inside Earth, denser elements, such as iron and magnesium, are more common than they are near the surface.) Notice that oxygen is among the most common elements in all three parts of Earth. Elements are shown rather than chemical compounds because the crust is composed of thousands of minerals, each with a different chemical composition. However, most minerals contain roughly the same elements. Most of the atmosphere is composed of elements in the form of gases. Only the hydrosphere is made mostly of a single compound: water. Water is composed of two parts hydrogen to one part oxygen.

ACTIVITY 2-2 **PIE GRAPHS OF EARTH'S SPHERES**

Use the data in Table 2-1 to make a pie graph of the chemical composition of each of the following: the crust, hydrosphere, and lithosphere.

 The Atmosphere

A thin layer of gas, the atmosphere, surrounds the solid Earth and oceans. Most of the mass of the atmosphere, clouds, and weather changes occur in the troposphere, the lowest layer of the atmosphere. (See Figure 2-5.) Although the atmosphere accounts for a tiny part of the total mass and volume of the planet, it is in this changing environment that people and most other life-forms live.

Air is a mixture of gases composed of about 78 percent nitrogen (N_2), a stable gas that does not readily react with other elements or compounds. About 21 percent of the atmosphere

Figure 2-5 The atmosphere is the layer of air that surrounds us.

is oxygen (O_2), which combines with many other elements in the processes of oxidation, combustion, and cellular respiration. Living things depend on cellular respiration to make use of the energy stored in food. The inert gas argon, which almost never reacts with other elements or compounds, makes up about 1 percent of the atmosphere.

The proportions of other gases in the atmosphere are variable. The amount of water vapor, water in the form of a gas, can vary from as high as several percent in warm, tropical locations to a tiny fraction of a percent in deserts and cold areas. Carbon dioxide, a common product of respiration and the burning of fossil fuels, makes up far less than 1 percent. However, carbon dioxide needed by plants for photosynthesis, and it may play an important role in climate change, which will be explored in Chapter 25.

The paragraph above describes the composition of the atmosphere's lowest layer, the **troposphere**. Note that the names of the layers of the atmosphere end in –*sphere* because this is their shape around Earth. The names of the boundaries between layers end in –*pause* as in stopping. Therefore, the tropopause is the place where the troposphere ends.

The atmosphere is divided into layers based on how the temperature changes with altitude. (See Figure 2-6.) Because

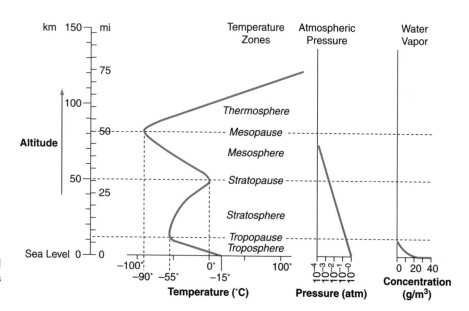

Figure 2-6 Selected properties of Earth's atmosphere.

the layers of the atmosphere are a result of density differences, the atmosphere is most dense at the bottom of the troposphere. Actually, the troposphere contains most of the mass of the atmosphere even through it extends only about 7 mi (12 km) above Earth's surface. Nearly all the atmosphere's water vapor, clouds, and weather events occur in this lowest layer.

Within the troposphere as altitude increases, temperature decreases. Have you ever noticed that snow lasts longer in the high mountains? The world's highest mountains extend nearly to the top of the troposphere. Above that height, the temperature change reverses and it actually becomes warmer with increasing altitude. The altitude at which the reversal occurs is known as the tropopause.

The next layer of the atmosphere is the **stratosphere**, in which the temperature increases with increasing altitude. The stratosphere extends up to the stratopause, where another change in temperature trend takes place. In the **mesosphere**, the temperature falls as altitude increases. Above the mesopause, is the highest layer, the **thermosphere**, in which the air temperature rises significantly. However, that increase in temperature speeds the motion of very few atoms. This increased agitation separates the molecules into positive and negative ions. This layer is sometimes called the ionosphere.

The lower boundary of the atmosphere is quite distinct: the surface of the land or the hydrosphere. However, because the atmosphere thins with altitude, there is no clear upper boundary of the atmosphere. The atmosphere just gets thinner and thinner as you get farther from Earth. When people refer to the atmosphere, they usually mean the troposphere, which contains about three-quarters of the atmosphere's total mass. This is layer in which we live.

ACTIVITY 2-3   **INTERPRETING REFERENCE TABLES**

You have probably ridden an elevator to the top of a building. What would you experience if you could ride in an open elevator through the atmosphere? Based on the information in Figure 2-6, write a travelers' guide to an elevator ride to a point 150 km above

Earth's surface. Describe changes in temperature, air pressure, and water vapor concentration that a traveler would encounter on the ride. In addition, describe the protective equipment that a traveler would need to survive the trip.

The Hydrosphere

Earth's oceans cover nearly three times as much of our planet as do the continents. People may think oceans are vast, featureless expanses of water; but oceans are not infinite and not featureless. The bottom of the oceans are almost as variable as the land areas. The hydrosphere is Earth's thinnest layer, averaging about 4 km in depth. Furthermore, scientists think the oceans are where life began on Earth

The liquid hydrosphere can be divided into two parts. About 99 percent of the hydrosphere is made up of the oceans, which are composed of salt water. Salt water is about 96.5 percent water and about 3.5 percent salt, mostly sodium chloride (common table salt). The remaining 1 percent of the hydrosphere is freshwater, which contains much smaller concentrations of dissolved solids. Freshwater is found in streams, rivers, and lakes. However, far more freshwater exists in the spaces within soil and rock. In fact, groundwater is estimated to be 25 times as abundant as the freshwater on Earth's surface.

The Lithosphere

This natural arch in Utah (see Figure 2-7 on page 40) is part of the lithosphere, Earth's rigid outer layer. The great bulk of Earth is the geosphere. We can define the **geosphere** as the mass of solid and molten rock that extends more than 6000 km from Earth's solid surface to its center. The lithosphere, the top 100 km of the geosphere, is the most rigid (unbending) part of the geosphere. Direct explorations in mines have taken humans to a depth of less than 4 km. The deepest drill hole is about 12 km deep. Everything we know about the

Figure 2-7 This natural arch in Utah is part of the lithosphere, the rigid outer layer of the solid Earth.

geosphere at depths greater than 12 km comes from indirect evidence, such as the increasing temperatures with depth, the passage of seismic (earthquake) waves, examination of meteorites, and from the determination of Earth's bulk properties, such as its density.

HOW IS LOCATION DETERMINED?

How can sailors far out on the ocean determine their position? With no familiar landmarks, such as roads, cities, and geographic features, they cannot describe their location in terms of surface features the way people usually do on land. Long ago, explorers solved this problem by establishing a coordinate system that covers the whole Earth.

Terrestrial Coordinates

The grid on a sheet of graph paper is a type of coordinate system. Each point on the paper can have a unique address expressed in terms of numbers along the x- and y-axes. Many cities are laid out in a coordinate system. Much of New York

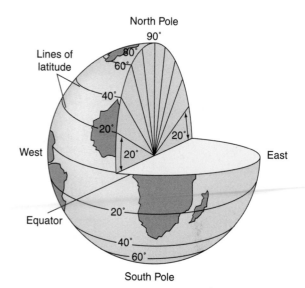

Figure 2-8 Latitude is the angular distance north or south of the equator.

City has numbered avenues that run north-south and numbered streets that run east-west. Knowing the street address of a building can help a person quickly locate it on a map or in the city itself. However, in Earth's undeveloped areas there are no roads or street signs, and there is no way to mark the oceans' surface. Because of this, explorers used their observations of the sun and stars to find their position on Earth's surface.

The coordinate system established by early sailors and explorers is Earth's system of latitude and longitude, called **terrestrial coordinates**. This system is based on the spin (rotation) of Earth on its axis. The **axis** is an imaginary line that passes through Earth's North and South Poles. Halfway between the poles is the equator, an imaginary line that circles Earth. The first terrestrial coordinate value is latitude. As shown in Figure 2-8, **latitude** is the angular distance north or south of the equator.

Lines of equal latitude are called parallels because they run east-west and, unlike longitude lines are parallel to each other. The equator is the reference line at latitude 0°. Both north and south of the equator, latitude increases to a maximum of 90° at the poles. Parallels can be drawn at any interval of latitude from the equator (0°) to the north and south poles (90°N and S).

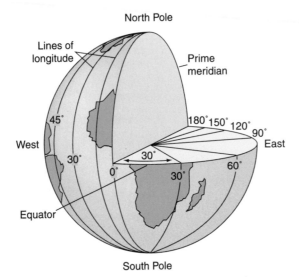

Figure 2-9 Longitude is the angular distance east or west of the prime meridian.

The second terrestrial coordinate value is longitude. As shown in Figure 2-9, **longitude** measures angular distance east and west. Unlike latitude, there is no natural or logical place to begin longitude measurements. English explorers established their reference line at the Royal Observatory in Greenwich (GREN-itch), England. Since England dominated world exploration and mapmaking, a north-south line through Greenwich became the world standard for measurements of longitude. Today, the Greenwich meridian, also known as the **prime meridian**, has become the reference line from which longitude is measured.

Lines of equal longitude are called meridians. Meridians all run from the North Pole to the South Pole. The prime meridian has a longitude of 0°. Longitude increases to the east and west to a maximum of 180°, a line that runs down the middle of the Pacific Ocean. Meridians are not parallel because they meet at the North and South Poles. As shown in Figure 2-10, the Eastern and Western Hemispheres are the two halves of Earth bounded by the prime meridian and the north-south line of 180° longitude.

Unfortunately, some people think of latitude and longitude only as lines. For example, they confuse latitude, the angular distance from the equator, with the lines on a map that show constant latitude. If your only purpose is to read the coordinates on a map, this is not a problem. But, if you want to

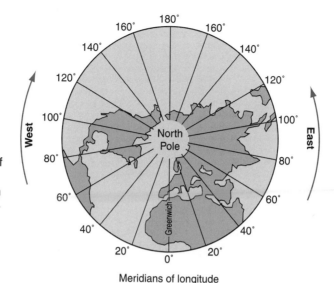

Figure 2-10 Lines of longitude meet at the North and South Poles. The 0° meridian and the 180° meridian separate the planet into the Eastern and Western Hemispheres.

Meridians of longitude

understand what latitude is and how it is determined, you need a deeper understanding.

Finding Latitude

Earlier in this chapter you read that observations of Polaris, the North Star, were used to show that Earth is a sphere. Those observations can also be used to tell how far north a person is from the equator. It takes Earth one day, 24 hours, to complete one rotation on its axis. That rotation is responsible for day and night. Although there is no scientific reason that Earth's axis should be aligned with any particular star, it is. The direction in which Earth's axis points moves through a 26,000-year cycle. However, at this time the axis lines up with a relatively bright star called Polaris, or the North Star. (The alignment is not perfect, since Polaris is a little less than 1 degree from the projection of Earth's axis.) Figure 2-11 on page 44 shows how to locate Polaris.

An observer at the North Pole sees Polaris directly overhead in the night sky. The angle from the horizon up to Polaris is therefore 90°. That observer is also located 90° north of the equator. As the latitude of the observer decreases, the altitude of Polaris also decreases. At the equator, Polaris is visible

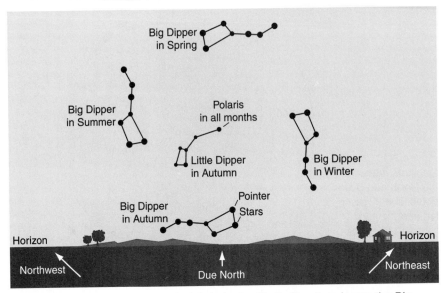

Figure 2-11 The easiest way to find Polaris, the North Star, is to locate the Big Dipper and follow the pointer stars at the end of its bowl as they point to Polaris. This diagram shows the way the Big Dipper looks in the evening sky at the middle of each season. The Little Dipper also rotates around Polaris, but it is shown only in its autumn position.

right on the northern horizon. Here, Polaris is 0° above the horizon, the latitude at Earth's equator is 0°. Therefore, for any observer in the Northern Hemisphere, latitude can be determined by observing the angle of Polaris above the horizon. The altitude of Polaris equals the latitude of the observer.

South of the equator the North Star is not visible. However, with a star map, an observer can determine the point in the night sky that is directly above the South Pole. It is near the constellation called the Southern Cross. In a procedure similar to what is done in the Northern Hemisphere, south latitude is equal to the angle of that point in the starry sky above the horizon. For people used to sighting on Polaris, it did not take long to master finding latitude in the Southern Hemisphere.

ACTIVITY 2-4   DETERMINING YOUR LATITUDE

You can construct an instrument to measure your latitude using the following simple materials: a protractor, a thin string, a weight, and a sighting device such as a soda straw. This instrument is called an

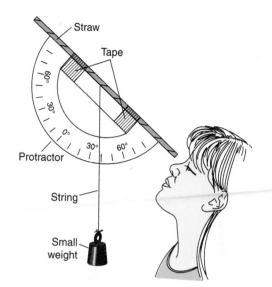

Figure 2-12 The astrolabe is used to measure the angular altitude of an object in the sky.

astrolabe. It is similar in principle to instruments used by mariners for hundreds of years. Figure 2-12 shows how to construct and use a simple astrolabe.

If you use a standard protractor when you sight along the horizon, the string will fall along the 90° line. Similarly, if you look straight up, the string will line up with 0°. In these cases, you will need to subtract your angle readings from 90° to find your latitude. Your latitude is equal to the angle of the star Polaris above a level horizon.

 Finding Longitude

Longitude can be determined by observations of the position of the sun. If it is noon where you are, it must be midnight halfway around Earth. (A full circle is 360°, so halfway around the planet is 180°.) The sun appears to move around Earth from east to west. Therefore, when it is noon where you are, in places to your east, the local time is afternoon, and in places to your west it is still morning. Because the sun appears to move around Earth in 24 hours, each hour of time difference represents $\frac{1}{24}$ of 360°, or 15°. So, each 1-hour difference in time from one location to another represents 15° of longitude.

Using the time difference of 15° per hour, you can determine the numerical value of longitude. But how can you determine whether it is east or west longitude? If local time is earlier than Greenwich time, the observer is located in the Western Hemisphere. Observers in the Eastern Hemisphere will note that local time is later than Greenwich time. To make this clearer, you can look at a globe and imagine the sun at the noon position in England. Remember that Earth spins toward the east. On your globe, most of Europe and Asia, at eastern longitudes, are in the afternoon or evening. At the same time in those places to the west of England, it is still morning.

Of course, this is based on solar time. Solar noon is the time the sun reaches its highest point in the sky. Clock time may differ from solar time by half an hour, even more if daylight savings time is in effect. If people set their clocks to the apparent motion of the sun across the sky in their location, clock time would be different from one place to another. This was done before time was standardized. In those days, towns had a clock that chimed on the town hall, so the citizens would know the local time. At that time, watches were difficult to make and too expensive for most people to own. Only places on a north-south line (at the same longitude) would have exactly the same clock time. If you wanted to meet someone in another town at a particular time, you could not use a clock set to the time in your town because you would probably show up early or late. Radio and television programs would not necessarily begin on the hour or half-hour. To standardize time, the United States is divided into four time zones: Eastern, Central, Mountain, and Pacific Time. In each time zone, all clocks are set to the same time.

ACTIVITY 2-5   **FINDING SOLAR NOON**

It is quite easy to measure local time by observations of the sun. To determine the time of solar noon, you will need to be at a location where a tall, vertical object, such as a flagpole or the high corner of a tall building, casts a shadow onto a level surface. Throughout the middle of the day, mark the exact position of the point of the shadow, and label the positions with the accurate clock time. (To avoid making permanent marks, use a substance such as chalk that will wash away in the rain.) Call these marks the time points. Connecting the points will form a curved line north of the object casting the shadow.

The next step is find where the curved shadow line comes closest to the base of the shadow object. (You will probably need to use a long metal tape measure to measure the distance.) Mark this point "Solar Noon." Finding the clock time of solar noon will probably require you to estimate between the marked time points to establish the precise time of "Solar Noon."

An added benefit of this procedure is that it provides a line that runs exactly north to south. The line from the vertical base of the shadow object could actually be extended to the North Pole. The English navigators dominated world exploration and mapmaking after the defeat of the Spanish Armada in 1588. These mapmakers set their clocks to observations of the sun made at the Royal Observatory in Greenwich, England. Therefore a north-south line running through Greenwich became the line from which longitude was measured. Modern clocks have become so precise they can measure small changes in Greenwich noon throughout the year. Therefore, **Greenwich Mean Time (GMT),** which evens out these small annual changes, is used as the basis of standard time throughout the world.

The method of determining longitude became quite clear. However, in practice, it was not so easy. A navigator at sea needs to know the precise time noon occurs back in Greenwich. Although ships carried the most accurate clocks available at that time, after a long sea voyage, changes in temperature, and the rocking motion of the ship caused these mechanical devices to become inaccurate. It was easy enough to observe local time by observing when the shadow of a vertical object pointed exactly north. Buy comparing local time with the time back in England depended on those mechanical clocks. Until very accurate clocks could be manufactured, measurements of longitude were poor, and maps generally showed large errors in the east-west direction.

ACTIVITY 2-6

DETERMINING YOUR LONGITUDE

Longitude equals the time difference between local solar time and Greenwich Mean Time (GMT) in hours and hundredths of an hour multiplied by 15° per hour of difference. If you performed the solar noon activity earlier in this chapter, you can use your data to determine your longitude. In that activity, you determined the difference between clock time and solar time. For example, if you determined that solar noon occurred 11:55 A.M., the difference between clock time and solar time is −5 minutes. If solar noon occurred at 12:09 P.M., the difference between clock time and solar time is +9 minutes. (Convert your solar time from hours and minutes to hours and hundredths of an hour by dividing the minutes by 60.)

For any location in New York State, you can find Greenwich Mean Time by adding 5 hours to your clock time. For example, if it is 1:15 P.M. clock time, Greenwich Mean Time is 6:15 P.M. (*Note*: If it is daylight savings time in New York, you would add only 4 hours.)

Once you have calculated your longitude experimentally, you can check your results with a map of New York State, such as the *Generalized Bedrock Geology of New York State* found on page 433 or in the *Earth Science Reference Tables*, that shows local latitude and longitude.

(*Note*: Mean Time and solar can differ by as much as 15 minutes. This could cause an error of as much as 4° of longitude. The following Web site explains this issue, known as the Equation-of-Time <http://www.analemma.com/Pages/frames Page.html>)

Today, finding the angle north or south of the equator (latitude) and the angle east or west of the prime meridian (longitude) has become simple and accurate thanks to modern technology. There are very accurate timepieces that use the precise vibrations of quartz crystals to measure time. In addition, radio and telephone communications provide Greenwich time to great accuracy. Even better, is the use of Global Positioning System (GPS) devices that analyze signals from orbiting satellites, allowing people to find latitude and longitude with great accuracy. Using a GPS device allows people to pinpoint their position to within a few meters. These devices are now small and inexpensive enough to be used by hikers and sportsmen. On land or at sea, it has become remarkably easy to find your place on the planet's terrestrial coordinates.

DIVISIONS OF ANGLES Just as meters can be divided into centimeters and millimeters, degrees of angle can be divided into smaller units. Each degree is made up of 60 minutes (60′) of angle. So, $23\frac{1}{2}$ degrees is 23 degrees and 30 minutes (23°30′). Furthermore, one minute of angle ($\frac{1}{60}$ of a degree) can further be divided into 60 seconds (60″). On Earth's surface, an accuracy of a second of latitude or longitude would establish your location to within a rectangle that measures about 30 meters on each side.

ACTIVITY 2-7	READING LATITUDE AND LONGITUDE ON MAPS

The Earth Science Reference Tables contain three maps (one New York map and two world maps) that can be used to read latitude and longitude. However, these world maps do not show cultural features, such as cities and political boundaries. To complete this activity it is best to use an atlas or a world map.

Your teacher may ask you to make a small "X" at each world location on a paper copy of the Tectonic Plates world map from the Reference Tables.

What cities are located at the following coordinates? (Please use a sheet of notebook paper. You should not mark in this book or on any reference materials.)

(1) 36°N, 122°W   (3) 35°N, 140°E
(2) 33°S, 18°E   (4) 55°N, 37°E

List the latitude and longitude coordinates of the following places on a world map. Please estimate values to the nearest degree of latitude and longitude.

(5) New York City   (7) Sydney, Australia
(6) London, England   (8) Honolulu, Hawaii

Find the terrestrial coordinates of each of these places in New York. Ask your teacher if and how you should divide degrees of latitude and longitude and how accurate your answers should be.

(9) Buffalo   (11) Mt. Marcy
(10) Albany   (12) Rochester
(13) What three lines of latitude and longitude have been used to define the political boundaries of New York State?

TERMS TO KNOW

atmosphere	hydrosphere	oblate
axis	latitude	prime meridian
equator	lithosphere	stratosphere
geosphere	longitude	terrestrial coordinate
Greenwich Mean Time	mesosphere	thermosphere

CHAPTER REVIEW QUESTIONS

1. As altitude increases within Earth's stratosphere, air temperature generally

(1) decreases only.   (3) increases, only.
(2) decreases, than increases.   (4) increases, then decreases.

2. When the time of day for a certain ship at sea is 12 noon, the time of day at the prime meridian (0° longitude) is 5 P.M. What is the ship's longitude?

(1) 45°W   (3) 75°W
(2) 45°E   (4) 75°E

3. To an observer in Buffalo, New York, the North Star, Polaris, is always located above the northern horizon at an altitude of approximately

   (1) $23\frac{1}{2}°$.

   (2) $43°$.

   (3) $66\frac{1}{2}°$.

   (4) $90°$.

4. The dashed line on the map below shows a ship's route from Long Island to Florida. As the ship travels south, the star Polaris appears lower in the sky each night.

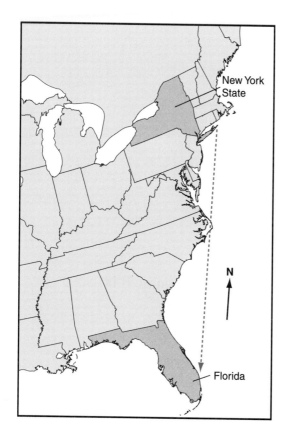

   The best explanation for this observation is that Polaris

   (1) rises and sets at different locations each day.

   (2) has an elliptical orbit around Earth.

   (3) is located directly over Earth's Equator.

   (4) is located directly over Earth's North Pole.

5. Earth's hydrosphere is best described as the

   (1) solid outer layer of Earth.
   (2) liquid outer layer of Earth.
   (3) liquid layer located below Earth's crust.
   (4) gaseous layer of Earth located above Earth's crust.

6. The diagram below shows the latitude-longitude grid on an Earth model. Points A and B are located on the surface.

   The solar time difference between point A and point B is

   (1) 1 hour.   (3) 5 hours.
   (2) 2 hours.   (4) 12 hours.

7. Earth's troposphere, hydrosphere, and lithosphere contain relatively large amounts of which element?

   (1) iron   (3) hydrogen
   (2) oxygen   (4) potassium

8. Ozone is concentrated in Earth's atmosphere at an altitude of 20 to 35 km. Which atmospheric layer contains the greatest concentration of ozone?

   (1) mesosphere   (3) troposphere
   (2) thermosphere   (4) stratosphere

9. At which location will Polaris be observed at the highest altitude?

    (1) equator

    (2) Florida

    (3) central New York State

    (4) Arctic Circle

10. Earth's shape is most similar to a

    (1) basketball.

    (2) pear.

    (3) egg.

    (4) apple.

11. Which two New York State cities experience solar noon (the time when the sun is highest in the sky) at almost exactly the same time?

    (1) Watertown and Binghamton

    (2) Buffalo and Albany

    (3) Binghamton and Elmira

    (4) Buffalo and Ithaca

12. What is the most abundant element in Earth's atmosphere?

    (1) oxygen

    (2) argon

    (3) hydrogen

    (4) nitrogen

13. An explorer rode a balloon high into Earth's atmosphere, taking a continuous record of atmospheric pressure. In which layer was the explorer most likely located when the atmospheric pressure was 10^{-2} atmospheres?

    (1) troposphere

    (2) stratosphere

    (3) mesosphere

    (4) thermosphere

14. What percent of Earth's hydrosphere is fresh water?

    (1) 99%

    (2) 75%

    (3) 23%

    (4) 1%

15. Approximately what percentage of Earth's surface is covered by water?

    (1) 100%

    (2) 75%

    (3) 50%

    (4) 25%

Open-Ended Questions

16. Give three types of evidence that are a result of Earth's spherical shape.

17. Describe a simple procedure that can be used to find latitude in the Northern Hemisphere.

The table below shows the concentration of ozone, in ozone units, in Earth's atmosphere. (One ozone unit is equal to 10^{12} molecules per cubic centimeter.)

Concentration of Ozone	
Altitude (km)	Ozone Units
0	0.7
5	0.6
10	1.1
15	3.0
20	4.9
25	4.4
30	2.6
35	1.4
40	0.6
45	0.2
50	0.1
55	0.0

18. In which zone of the atmosphere is the concentration of ozone the greatest?

19. On what basis, has Earth's atmosphere been divided into its four layers?

20. What is the relationship between the density of Earth's layers the position of each layer?

Chapter 3

Models and Maps

WHAT IS A MODEL?

A **model** is anything that is used to represent something else. A photograph helps you remember your loved ones when they cannot be near. A photograph is an example of a physical model. There are also mathematical models. The formula for density you used in Chapter 1 is a mathematical model. This formula represents the relationship between mass and volume for any object made of a uniform substance. Memories and dreams are models of real or imagined human events. In Chapter 2, you may have developed a mental model of the size and shape of planet Earth. Such models can be useful as you try to understand this planet.

Most models are a compromise between reality and convenience. Children often play with toys that look like adult objects, such a doll or a toy car. Using toys is an important way for children to learn adult skills. It would be foolish to give young children real babies or cars because they do not yet have the skills to manage the real objects safely. Models must be realistic enough to show the important characteristics of the real object, but avoid problems that would occur if the real thing were used.

Many physical models are scale models. They show the shape of the real object, but they are made smaller or larger as a matter of convenience. Toys and photographs are usually smaller than the real thing. But a diagram showing the parts of an atom would need to be much larger than the original, perhaps by a factor of 10^9 (a billion). A convenient way to express scale is as a ratio. We often show a ratio as two numbers separated by a colon. A common scale used to make a small toy automobile is 1:64. This is read as "one to sixty-four." One cm on the toy car represents 64 cm on the real car. The model atom mentioned earlier would be at a scale of 1,000,000,000:1.

| ACTIVITY 3-1 | MODELS IN DAILY LIFE |

Make a list of some of the models people use in their daily lives. Include physical models, mathematical models, and mental models. Organize your list into four columns: (1) Model (name each), (2) What It Represents (3) Why It Is Used Instead of the Real Object, and if it is a scale model, the (4) Approximate Scale. Avoid different examples of the same kind of model.

 Maps

Some maps are models of the whole Earth; other maps model just a part of its surface. Road maps help people to drive from one place to another. Political maps show the geographic limits of laws and government services. It is hard to imagine traveling without using maps. (See Figure 3-1 on page 57.)

Because Earth is spherical, its surface is curved. The only kind of map that shows Earth's surface without distortion must also have a curved surface. Therefore, a globe is the most accurate model of Earth. Directions and distances are shown without distortion. However, flat maps are easier to carry. You can fold them for storage and open them on a flat surface. For small regions, the distortion of a flat map is not significant.

When world maps are transferred to flat surfaces, they may show increasing distortion in areas far from the equator.

Figure 3-1 Person using a map to find her way.

Compare the Ocean Currents maps (Figure 3-2) with the Tectonic Plates map (Figure 3-3). Figure 3-2 shows the Scandinavian Peninsula in northern Europe as much smaller than Australia. But Figure 3-3 on page 58, the tectonic plates map, incorrectly shows these areas to be about the same size. Although the Ocean Currents map shows less size distortion, it

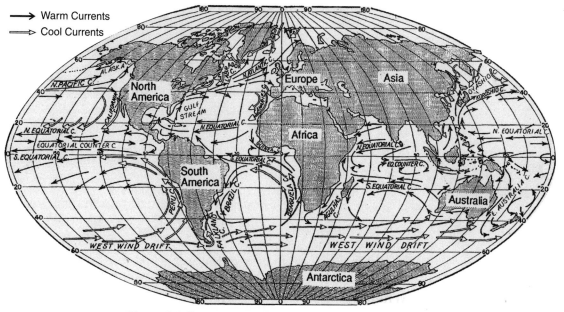

Figure 3-2 Ocean currents

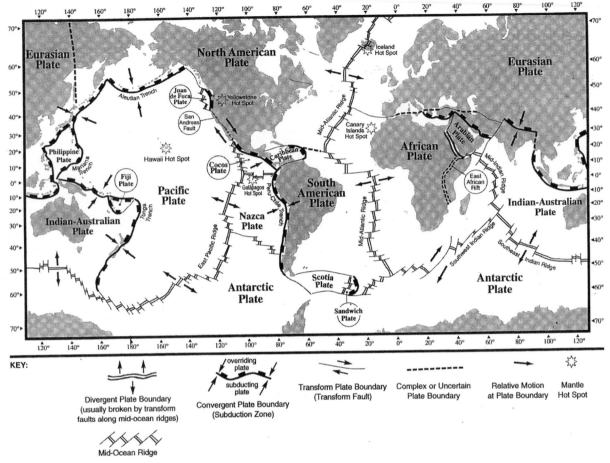

Figure 3-3 Tectonic plates.

distorts directions. Notice how North America seems to be slanted on the ocean currents map.

ACTIVITY 3-2 A MAP TO YOUR HOME

Use a sheet of $8\frac{1}{2} \times 11$ inch paper to draw a map that shows the most direct route from school to your home. You may use a computer drawing program if it is available. Show the landmarks (buildings and natural features) that would be most useful in guiding a person who is unfamiliar with your community. Also include a scale of distance.

What Are Compass Directions?

Directions are a part of your daily life. Many people can give you directions, but Earth itself can also give you directions. In general, the sun rises in the east and sets in the west. For any observer in New York State, the noon sun is in the south. If you want to travel to a cooler climate, head north. To avoid cold winters, travel south. In addition to the four principal directions shown on Figure 3-4, there are intermediate directions such as northeast and southwest. Compass direction can also be specified as an angle known, as **azimuth**, which starts at 0°, which is directly north and proceeds to east, south, west, and back to north at 360°.

Most maps are printed with north toward the top of the map. This convention helps align the map with the area it represents. However, there may be reasons to align the map in a different way. Therefore, this rule is not always followed.

Earth as a Magnet

You may have used a magnet to pick up metal objects, such as paper clips. Ancient people noticed that a large piece of the iron ore magnetite, sometimes called lodestone, attracts other iron objects. They called this attractive force magnetism. Furthermore, they noticed that a piece of lodestone floating on

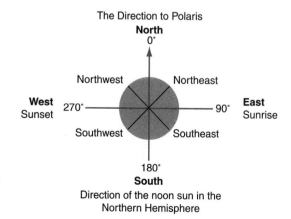

Figure 3-4 Events in your life, such as the apparent passage of the sun through the sky, help you become aware of directions.

water tends to align itself in a consistent geographic direction. (Because this kind of rock is heavier than water, it must be placed in a flotation device.) The earliest magnetic compasses were made in this way long before anyone understood magnetism.

ACTIVITY 3-3	MAKING A WATER COMPASS

You can construct a water compass by floating a strong bar magnet in a large container of water. Float the bar magnet in a shallow bowl that does not touch the sides of the large container. Be sure to try this activity well away from any magnetic metals or electric motors. Does the magnet tend to align consistently toward magnetic north?

A compass needle points north because Earth acts as a giant magnet. You may know that opposite magnetic poles attract each other. Then why does the north pole of a magnet tend to point toward Earth's North Pole? The reason is simple. The pole of a magnet that points north is labeled N because it is the "north-seeking pole." That end actually has a magnetic field like Earth's South Pole.

Why does Earth have a magnetic field? That question took a long time to answer. Scientists now think that Earth's outer core is made of molten liquid that circulates due to heat flow. The planet is very dense, so molten iron is a strong possibility for the primary substance in the outer core. Scientists think it is this circulation guided in part by Earth's rotation (spin) that keeps the currents running generally north and south.

There are two other important things to know about Earth's magnetic field. First, the field seems to reverse itself as often as every 30,000 years. This is more evidence to support the outer core circulation theory. Second, the magnetic poles do not align exactly with the geographic (spin) poles. In fact, the magnetic poles wander through the Arctic and Antarctic regions.

ACTIVITY 3-4	MAGNETIC DECLINATION

Geographic north is determined by Earth's spin axis. There are several ways to find it. In Chapter 2, when you found solar noon, the noon shadow pointed to geographic north. At night you can use the North Star, which is directly above the North Pole. You can also use a map, which may show man-made features such as roads that run north to south.

Magnetic north, which is a result of Earth's magnetic field, can be determined with a compass. Magnetic north does not always align with geographic north. Although a compass needle points in a northerly direction for most places on Earth's surface, the difference between these two directions changes from place to place. Magnetic declination is the angle between geographic north and magnetic north. This angle can be measured with a protractor.

WHAT ARE FIELDS?

A **field** as a region of space in which the same quantity can be measured everywhere. Let us look at a simple example. Have you ever noticed that when dinner is being prepared, the aroma of the food drifts through the house? As you walk into the kitchen area, the aroma increases. In this example, the house is the field, and the aroma is the field quantity that is observed. In the previous section, you read about Earth's magnetic field. All around Earth, scientists can measure the direction and strength of Earth's magnetism in its magnetic field. Gravity is another field quantity that changes over Earth's surface. Sensitive instruments can measure very small changes in Earth's gravity from place to place. Temperature is also a field quantity. Wherever you go, you can measure the temperature and notice how it changes. Scientists often make maps to show how field quantities change over a geographic area.

Understanding Isolines

An **isoline** is a line on a field map that connects places having the same field quantity value. If you have looked at a weather map of the United States that shows the daily high temperatures, you have seen isolines that cross the country from east to west. These lines are **isotherms** that connect places having the same temperature. Other weather maps use isolines to show atmospheric pressure or the amount of precipitation.

ACTIVITY 3-5   **CHARACTERISTICS OF ISOLINES.**

Based on Figure 3-5, or a similar national isoline map, discuss the following features of isolines:

1. Do isolines ever touch or cross each other?

2. Do isolines usually have sharp angles or gentle curves? (Please pick one.)

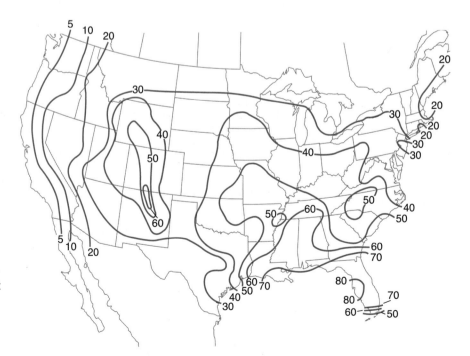

Figure 3-5 This isoline map shows that thunderstorms are especially common in Florida and New Mexico.

3. What does each point on an isoline have in common with all other points on the same line?

4. Do isolines ever end, except at the edge of the data area?

5. On a single map, is the change in value from one isoline to the next always the same?

6. Do isolines tend to run parallel as they extend around the map?

7. Does every isoline have one side where the values are higher and another side where the values are lower?

ACTIVITY 3-6   DRAWING ISOLINES

Your teacher will supply you with a diagram that shows a pattern of numbers within a boundary. Use these numbers to draw isolines of the bounded field areas at an interval of one unit.

Please draw your isolines with a pencil and an eraser. It is easy to make mistakes when drawing isolines. It is important to be able to completely erase and correct your errors.

Every isoline map is drawn at a specific interval. Many newspaper weather maps show temperatures at an interval of 10°F. That means when you move from one isotherm to the adjacent isotherm, there is a change of 10°F. A smaller interval, such as 1°F, would make the map cluttered and hard to read. A larger interval would result in a map that does not give enough information. For every isoline map, someone has to decide the most appropriate interval.

ACTIVITY 3-7   A TEMPERATURE FIELD

In science, you sometimes collect and analyze your own data. Gathering accurate field data can be difficult, but it is an excellent way to learn about science. For example, each student in the class can use a thermometer to read and record the air temperature at the same time throughout the classroom. Individuals or lab groups can write these numbers at the appropriate location on a floor plan of the classroom and then use the numbers to draw an iostherm map of the science room. This can be a very challenging activity, but it is a good way to learn about isolines. Take simultaneous temperature readings at different levels in the classroom (floor, desk level, and 1 meter above the desks).

WHAT IS A TOPOGRAPHIC MAP?

A **topographic map** is a special type of field map on which the isolines are called contour lines. These maps show a three-dimensional surface on a two-dimensional page. **Contour lines** connect places that have the same elevation (height above or below sea level). Each contour line is separated from the next by a uniform interval, called the contour interval. A common interval is 10 meters. As you move your finger up or down a slope, each time it crosses a contour line, the elevation has changed the amount equal to the map's contour interval. Where the contour lines are close together, the land is steep. Where there are broad spaces between contour lines, the land is more flat.

It may help you visualize the area represented on a topographic map if you think of each contour level as a layer on a cake. Multiple levels stack to make hills. Figure 3-6 is a topographic map of an imaginary location. Figure 3-7 is photograph of a step model constructed layer-cake style from that map. (See Activity 3-8.) A real landscape would have a more rounded shape.

Hikers and sportsmen often use topographic maps because they show the shape of the land. These maps tell them where roads, hills, lakes, and streams are as well as a wide variety of other landmarks. Topographic maps can be used to plan the best route from one location to another. The United

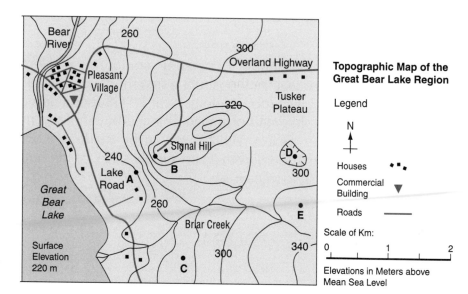

Figure 3-6 Topographic map of Great Bear Lake.

States Government publishes topographic maps at various scales. Camping, sporting goods, and hardware stores often carry local United States Geological Survey (USGS) topographic maps. These maps also can be ordered from government map distribution offices.

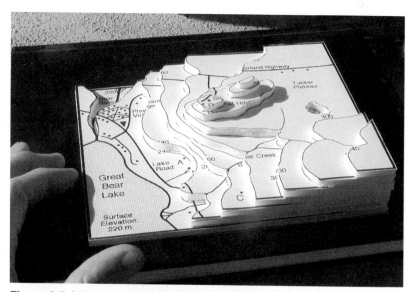

Figure 3-7 A three-dimensional model of a land surface can be constructed using a contour map and foam board from an art supply store.

| ACTIVITY 3-8 | MAKING A TOPOGRAPHIC MODEL |

From a simple topographic contour map selected or approved by your teacher, construct a step model of the map area. You may select a portion of your local USGS topographic map or the topographic map or image of a nearby landform from which to make your model. The layers can be made of corrugated cardboard, foam board from an art supply store, or the bottoms of clean foam meat/food trays.

 Common Features on Topographic Maps

Topographic maps use contour lines to show the topography, or shape of the land. In their steeper sections, small streams often cut gullies by erosion. These distinctive features show up as a sharp bend in the contour lines, which creates a V-shape that points upstream. On Figure 3-6 you can see that Briar Creek flows down to the west because the contour lines make Vs that point upstream to the east.

Reading the elevation of a point on a contour line is easy because it is the elevation of that contour line. If you want to know the elevation of a point between contour lines you estimate the elevation in reference to nearby contour lines. For example, point C on Figure 3-6 is about half way between the 260- and 280-meter (m) lines. You can therefore estimate the elevation of point C to be 270 m.

When contour lines form an enclosed shape that is something like an irregular circle, the center of the enclosed area is usually a hill. The top of the hill is within the smallest circle. By reading the elevation of the highest contour level, you can estimate the height of the hill. A useful convention is to add half of the contour interval to the elevation of the highest contour. On Figure 3-6 the highest contour line around the house on Signal Hill is 340 m. The contour interval on this map is 20 m. So your best estimate for the elevation of the house is 350 m.

Sometimes, contour lines enclose a dip in the land surface, or a closed depression. Water running into a closed depression will collect there unless it finds an underground exit, because water cannot run uphill out of a closed depression. To distinguish depressions from hills, mapmakers use contour lines with small bars that point down toward the center of the depression. Point D on Figure 3-6 is within the 300-m contour, so its elevation at the bottom of the depression is about 290 m. The first depression contour line always has the same elevation as the lower of the two adjacent contour lines.

Most maps have a legend or key printed outside the map area. This explains the meanings of various symbols shown on the map. The more complex the map the greater the variety of symbols that must be explained in the legend.

As with other physical models, every topographic map has a map scale. The scale can be expressed as a ratio, such as 1:24,000. The scale can be a translation of distances, such as 1 inch represents 2000 feet. But, you will probably find the most useful scale is a line located outside the map area that is marked with divisions of distance. Figures 3-8 to 3-10 illustrate the three steps in determining the distance between two map locations or the size of a geographic feature.

Step 1. To determine the length of Turquoise Lake, you will need to locate it on a topographic map.

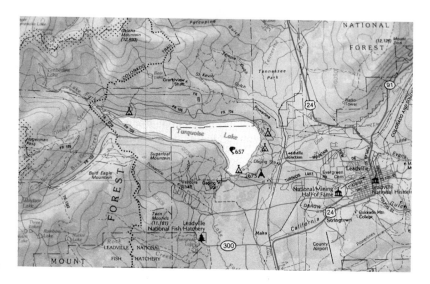

Figure 3-8 Locating Turquoise Lake on the map.

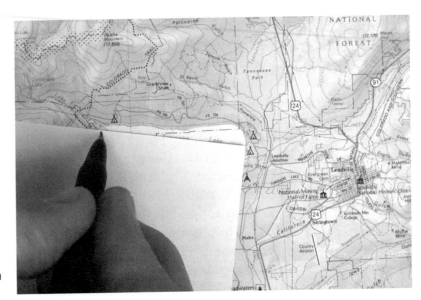

Figure 3-9 Marking the length of the lake on a clean sheet of paper.

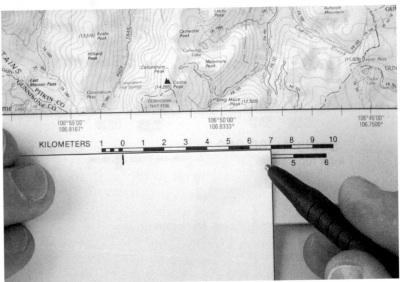

Figure 3-10 Holding the marked paper along the map's scale of kilometers, show that Turquoise Lake is about 7 km long.

Step 2. Hold the edge of a blank sheet of paper along the longest part of Turquoise Lake. Place the corner of the paper at one end of the lake. Make a mark along the edge of the paper where it touches the other end of the lake.

Step 3. Move the marked paper to the distance scale on the map. This indicates that Turquoise Lake is about 7 km long.

| ACTIVITY 3-9 | READING YOUR LOCAL TOPOGRAPHIC MAP |

Look at a copy of the United States Geological Survey topographic map that includes your town or city. Please, do not make any marks on this map unless your teacher directs you to do so. Notice the locations of familiar features such as roads, buildings, and streams. Notice how hills and valleys are represented by contour lines. Find a symbol for your house or identify its location.

1. What are the latitude and longitude coordinates of the bottom right corner of this map?

2. When was this map published? Has it been revised?

3. How many centimeters represent 1 km?

4. What is the local magnetic declination?

5. What is the contour interval of this map?

6. Locate a place where contour lines become V-shaped as they cross a stream. Do the Vs point upstream or downstream? In your notebook, make a sketch of that place.

7. What and where are the highest and lowest points on this map?

 Gradient

If a hill changes quickly in elevation, it has a steep gradient. In fact, slope is often used as a synonym for gradient. At every location, a field value has a measurable gradient. If the field value is not changing over a particular line, we say that the gradient is zero. We can therefore define **gradient** as the change in field value per unit distance. (The field value on a topographic map is land elevation.)

You can tell where gradients are the steepest by looking at a field map. The places where the isolines are closest are the places with the steepest gradient. On Figure 3-6 the gradient is steep between points A and B where the contour lines run close together.

The following formula from the *Earth Science Reference Tables* can be used to calculate gradient:

$$\text{Gradient} = \frac{\text{change in field value}}{\text{distance}}$$

SAMPLE PROBLEMS

Problem 1   The temperature at the center of a town is 20°C, but 10 km west at the river it is only 15°C. What is the temperature gradient between these two locations?

Solution

$$\text{Gradient} = \frac{\text{change in field value}}{\text{distance}}$$

$$= \frac{20°C - 15°C}{10\ km}$$

$$= \frac{5°C}{10\ km}$$

$$= 0.5°C/km$$

Problem 2   Calculate the gradient from point C to point E on Figure 3-6

Solution

Point C, half way between the 260- and 280-meter contour lines, must be at about 270 m. Similarly, point E is at about 330 m. The distance between them can be determined using the scale of kilometers in the map legend as shown in Figures 3.8 to 3.10. That distance is 2 km.

$$\text{Gradient} = \frac{\text{change in field value}}{\text{distance}}$$

$$= \frac{330\ m - 270\ m}{2\ km}$$

$$= \frac{60\ m}{2\ km}$$

$$= 30\ m/km$$

Please note the following features of both solutions.

- Each is started by writing the appropriate formula.
- Values including units of measure are substituted into the formula.

- Each step to the solution is shown.

- The units of measure are part of the solution.

In Problem 2, the units are meters per kilometer. You probably know that a kilometer is 1000 meters. So, this answer could be written 30 m/km or 0.03 m/m. Recall that the first part of these units is measured in meters in the vertical direction. The second part (meters or kilometers) is measured in the horizontal direction. So each one measures something different.

☑ Practice Problem 1

On a sunny day, the temperature at the floor of the classroom was 27°C. At the ceiling, 4 m above the floor, the temperature was 31°C. What was the temperature gradient between the floor and the ceiling?

☑ Practice Problem 2

The source of a stream is at an elevation of 600 m. The stream enters the sea 50 km from its source. What is the gradient of the stream?

 Making Topographic Profiles

A profile is a cross section or a cutaway view. If you stand between a bright light and a wall, your body will cast a shadow on the wall. When you stand with your shoulder perpendicular to the wall, your shadow will include a profile of your face with features such as your chin and nose indicated clearly. As illustrated in Figure 3-11, a topographic profile shows the elevation of the land surface along a particular route. Along

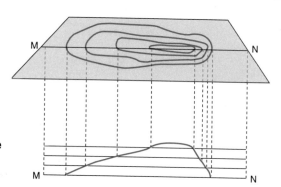

Figure 3-11 You can make a profile using a contour map.

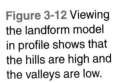

Figure 3-12 Viewing the landform model in profile shows that the hills are high and the valleys are low.

the profile route, the hills show as high places and the valleys as low places. (See Figure 3-12.)

A topographic map can be used to draw a profile along any straight-line route. You will need a sheet of paper that is marked with parallel lines, such as writing paper, and a blank strip of paper a little longer than the profile route on the map. By following the steps below, you can draw a profile from a contour map.

Step 1.   Place an edge of the blank strip along the profile route. Each time the edge of the blank strip crosses a contour line, make a mark at that point along the edge of the strip as shown in Figure 3-13.

Step 2.   Label each mark on the strip with the elevation of the contour line crossed. Take care with this step, since some marks will have the same elevation as others that cross the same contour line or cross another line at the same elevation.

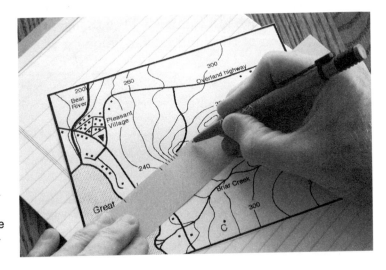

Figure 3-13 The first step in making a profile from a contour map is marking the blank strip where it crosses each contour line.

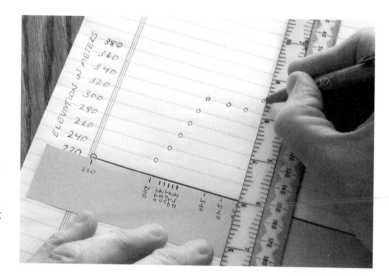

Figure 3-14 After labeling each mark with the elevation of the contour line it represents. the next step is to draw a dot at the correct elevation on the lined paper.

Step 3. Along the left side of the lined paper, label the horizontal lines with the elevations of the contour lines crossed by the blank strip as shown in Figure 3-14.

Step 4. Lay the marked blank strip along the lowest labeled horizontal line on the lined paper. Directly above the marks on the strip, make dots on the lined paper at the height indicated by the marks on the strip.

Step 5. As shown in Figure 3-15, connect the dots on the lined paper with a gently curved line. Valleys and hilltops should be rounded above the highest dots.

Step 6. Labeling features from the map line can help you visualize the curved profile. Look at Figure 3-15 again.

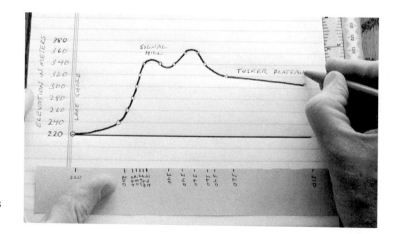

Figure 3-15 Complete the profile by connecting the dots with a smooth line.

A similar procedure can be used to make a profile from any isoline map, such as a weather map that shows local temperatures. Some people can look at a field map and visualize its profile without following this procedure. As with making mathematical estimations, visualizing a profile is a skill that can be helpful in making accurate profiles and reading any isoline map.

ACTIVITY 3-10 **A PROFILE ON A LOCAL TOPOGRAPHIC MAP**

Use the steps above to construct a profile from your local USGS contour map along a line selected by your teacher.

 Using Isoline Maps for Practical Purposes

People buy topographic maps for many uses. Developers and construction companies use maps to plan roads and the placement of buildings. Search and rescue teams use them when they look for lost or injured people and to plan rescue efforts. You, on the other hand, might want to find the best way to get to a fishing spot. Whether you will walk or travel by car will be an important consideration. Do you mind going over hills or crossing streams? How long will it take you to get to your destination? Considerations such as these can be important depending on how and when you will travel.

ACTIVITY 3-11 **PLANNING A TRIP**

Your teacher will select two points on your local topographic map. Describe the best way to get from one place to the other. Justify your route and tell what you will find along the trip. (Mention only features shown on the map.)

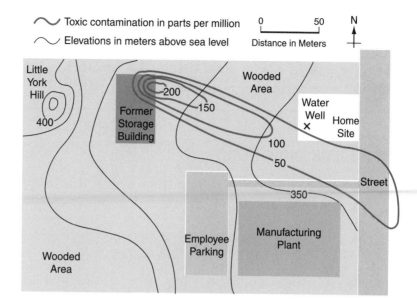

Figure 3-16
Groundwater contamination at the York Instrument Factory site.

ACTIVITY 3-12   **INTERPRETING ISOLINE MAPS**

The map in Figure 3-16 shows two field quantities. The black isolines show surface elevations, and orange isolines show the concentrations of a toxic substance in the groundwater. What is the contour interval and the pollution interval on the map? Where did the water pollution probably originate? Which way is it moving? How can you tell?

ACTIVITY 3-13   **RESCUE AND EVACUATION PLANNING**

Your teacher will indicate a location on a local topographic map where a person has been injured in a fall. Devise three different, detailed plans to evacuate the injured person to an ambulance waiting along a main road. In your planning consider the cost of the evacuation and the best route to take in transporting the victim. The plans should be written in enough detail so that the evacuation team will know how to proceed and what obstacles to avoid. Compare the planning of different groups.

TERMS TO KNOW

azimuth	field	isoline	model	topographic map
contour line	gradient	isotherm	profile	

CHAPTER REVIEW QUESTIONS

Base your answers to questions 1–3 on the topographic map below, showing locations X, Y and Z. Elevations are expressed in meters.

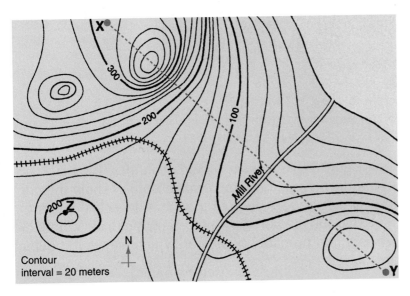

1. Which profile best represents the topography along the dashed line from point X to point Y?

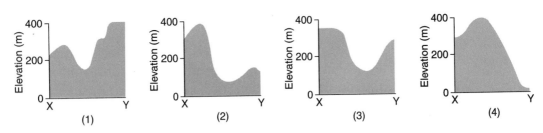

2. Mill River generally flows toward the

(1) southeast.

(2) southwest.

(3) northeast.

(4) northwest.

3. What is the elevation of point Z?

    (1) 190 m   (3) 240 m

    (2) 220 m   (4) 250 m

4. What is the average temperature gradient between two places in a classroom that are 10 m apart if one place has a temperature of 25°C and the other has a temperature of 23°C?

    (1) 0.2 °C/m   (3) 5 °C/m

    (2) 0.2 m/°C   (4) 5 m/°C

5. A map has a scale of 1 cm:12 km. What is the distance on this map between two locations that are actually 30 km apart?

    (1) 2.5 cm   (3) 18 cm

    (2) 12 cm   (4) 30 cm

6. On a topographic map, the gradient is steepest where the contour lines are

    (1) straight.   (3) far apart.

    (2) curved.   (4) close together.

7. In what general direction does sunrise occur?

    (1) north   (3) east

    (2) south   (4) west

8. On a topographic map, what feature is indicated by small, closed "circle"?

    (1) stream   (3) building

    (2) road   (4) hilltop

Base your answers to questions 9 and 10 on the map below.

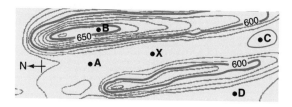

9. What is the contour interval on this map? (Elevations are in feet.)

    (1) 5 ft   (3) 20 ft

    (2) 10 ft   (4) 25 ft

10. Points X, A, B, C, and D mark places on this land surface. All are within a distance at which a person at one point would be able to see people at the other points. A student is standing at point X. Other students are standing at points A, B, C and D. The student at X would have the most difficulty seeing the person at which other point?

(1) A

(2) B

(3) C

(4) D

Open-Ended Questions

Base your answers to questions 11–15 on the topographic map below of an area of New York State. Points X, Y, and Z are marked on the map. Points X and Y are located along Squab Hollow Creek.

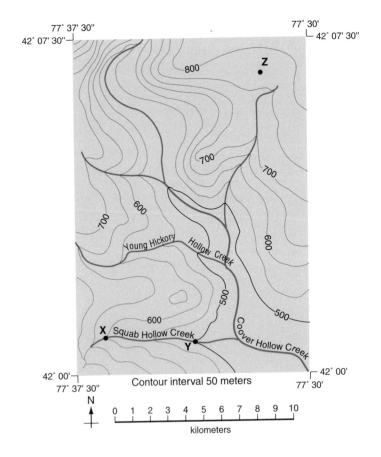

Contour interval 50 meters

11. Determine the gradient of Squab Hollow creek between point X and point Y according to the directions below.

a. Using the *Earth Science Reference Tables* write the equation used to determine gradient.

b. Substitute the values into the equation.

c. Solve the equation and label the answer with the correct units.

12. Describe one way to determine the direction in which Young Hickory Hollow Creek is flowing.

13. What is the elevation of point Z?

14. Where on this map is the gradient of the land surface steepest?

15. Based on the coordinates of latitude and longitude shown on the map, what is the distance from this map area to Buffalo, New York, in kilometers?

16. The diagram below shows time zones of the United States. If it is 9 A.M. in Dallas, what time is it in each of these three cities: New York, Denver, and San Francisco. (Be sure to list the three cities and the clock time at each.) All parts of the country are on standard time.

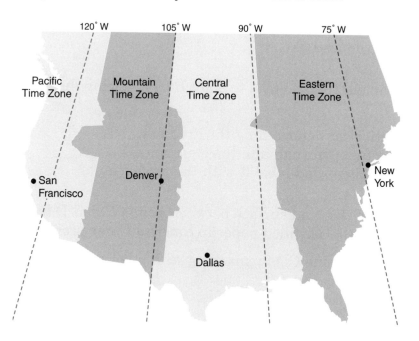

Base your answers to questions 17, 18, and 19 on the topographic map below that shows points A, B, C, and D. Elevations are in feet.

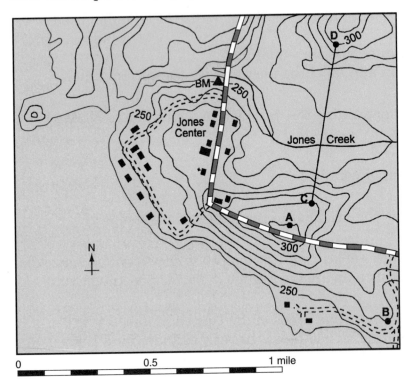

17. Explain briefly how the map can be used to determine that Jones Creek flows westward into Jones Lake.

18. Determine the gradient from point A to point B by following the directions below.

 a. Write the equation for determining the gradient.
 b. Substitute data from the map into the equation
 c. Calculate the gradient and label it with the proper units

19. Construct a profile of the land surface between points C and D by following the directions below. Use lined paper to construct your profile.

 a. Plot elevations along line CD by marking with a dot each point where an isoline is crossed by line CD.
 b. Connect the dots to complete the profile.

The map below shows a location in New York State where several students went camping. The dotted line indicates the route they took in approaching their campsite along Hidden Lake. Use this map for questions 20 through 24.

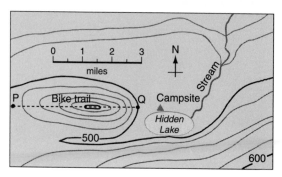

20. State the evidence shown on the map that the area directly north of Hidden Lake is relatively flat.

21. State the general compass direction in which the stream is flowing.

22. Use a sheet of lined paper to construct a profile line along the dotted line from P to Q. On your profile, show each point where a contour line is crossed with a dot on your profile line.

23. What is the contour interval on this map?

24. What is the highest elevation shown on the map?

Chapter 4

Minerals

 ## WHAT ARE MINERALS?

If you look around the natural environment, you will probably see two kinds of things: living and nonliving. Plants and animals are parts of the living environment. The living environment is sometimes called the biosphere. Nonliving things are usually the subjects of the earth sciences. Much of this part of the natural environment is composed of rock and soil. Just as a house is made of a variety of building materials, for example, wood, nails, concrete, and brick, so soil and rock are made primarily of minerals.

Some rocks such as granite are composed of crystals of different colored substances. (See Figure 4-1.) The differences in the properties of the crystals identify them as different minerals. What is a mineral? Defining minerals as the substances of which rocks are made could be acceptable in some situations, but as you will read below, there are a few exceptions to this idea. A more exact way is needed to define what a mineral is and what it is not.

Geologists have identified thousands of minerals. In fact, new minerals are discovered and named all the time. Most of the newly discovered minerals are rare and have no practical use. The wide variety of minerals makes it difficult to define

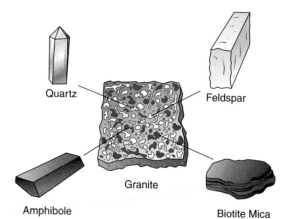

Figure 4-1 As are most rocks, granite is composed of a variety of minerals. These four are especially common.

exactly what a mineral is. However, geologists do have certain characteristics that identify minerals.

Natural, Inorganic Solids

Artificial substances such as steel or plastic are not considered minerals even though these substances have other characteristics of minerals. Man-made gems, such as synthetic sapphires and rubies, are not minerals either because they were not made by a natural process. The first property that characterizes all minerals is that they formed naturally.

All minerals are also inorganic. This means that they were made by physical processes, and are not the result of biological events. You have probably heard of and perhaps seen coal. Careful study of the origin of coal shows it is formed from layers of plant remains that have been buried and compressed by the weight of more layers above. Since coal forms organically, it is not a mineral. Coal is mined in many areas of the United States, and is often used as a fuel. If we define a rock as a natural solid part of the lithosphere (the top part of the solid Earth), coal is a rock. Many samples of coal show beautiful plant impressions.

While coal is a kind of rock, it was formed from living materials. Therefore coal is an example of a rock that is not made of minerals. (Impure coal does contain some mineral material.) Another example of a rock that is not composed of minerals is fossil limestone. This kind of limestone is the result of the accumulation of shells or coral. Coral is a colony of tiny marine animals that live in a hard external skeleton. These colonial skeletons can be several inches or more in length and take unusual shapes. Because fossil limestone is made of the hard remains of coral, shells, and other living tissue, it is not composed of minerals. Minerals must be of inorganic origin.

All minerals are solids. The water pumped from the ground is not considered a mineral because it is a liquid. Petroleum, or crude oil, which is taken from deep underground, is a geologic resource but it is not considered a mineral for two reasons: First, petroleum is a liquid; second, like coal, it is of biological origin.

Solids have a definite shape and volume. Gases, such as air, can be compressed. When enough pressure is applied, a sample of a gas can be compressed to a very small volume. When the pressure is released, the gas expands nearly without limit. Therefore, gases do not have a definite volume. Gases and liquids take the shape of the container in which they are placed. Liquids and gases are called fluids. In general, a fluid is a substance that can flow and take the shape of its container. A solid can be taken out of its container and it will keep its shape. You can measure the dimensions of a solid without the object being in a container. This is not true for gases and liquids. Figure 4-2 illustrates the three states of matter.

Solids maintain their shape and volume because of their atomic structure. Large crystals, such as those you might see

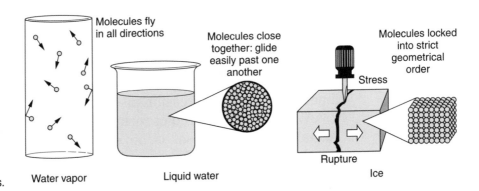

Figure 4-2 Matter, such as water, can exist in three states.

Molecules fly in all directions

Molecules close together: glide easily past one another

Molecules locked into strict geometrical order

Stress

Rupture

Water vapor Liquid water Ice

in a museum or a mineral collection, usually have sharp corners and flat faces. Careful examination with powerful electron microscopes shows that these crystal shapes are a result of the way their atoms are arranged. An atom is the smallest part of an element. In crystalline substances, the atoms generally have a regular arrangement in rows and layers. The distance between the atoms in a solid changes very little with changes in temperature and pressure. Furthermore, the atoms in a solid cannot move over or around one another. Sometimes, enough atoms line up to make crystals that are visible, such as those in some kinds of rock. In rare occurrences, crystals can grow to the size seen in museums. The largest natural rock crystals can be several meters long. It is the internal arrangement of atoms that gives crystals their beautiful shapes.

ACTIVITY 4-1   SOLIDS, LIQUIDS AND GASES

Prepare a table with three columns. Label one column "Gas," one "Liquid," and the third "Solid." In the table, list the substances you see every day under the proper category: gas, liquid, or solid.

Elements and Compounds

Minerals have a definite composition, or at least a specific range of composition. Some minerals are simple elements. **Elements** are the basic substances that are the building blocks of matter. Gold, as all other elements, has just one kind of atom. The number of protons in the nucleus, or center of the atom, determines which element it is. There are 92 natural elements. Copper and sulfur are also minerals that are elements. Figure 4-3 on page 86 is a photograph of copper as it comes from the ground. This is called native copper. Graphite and diamond are different crystal structures of the same element, carbon.

A second group of minerals with a fixed composition are the chemical compounds. **Compounds** are substances that

Figure 4-3 Native copper is a mineral that is also an element.

are made up of more than one kind of atom (element) combined chemically into larger units called molecules. Each molecule of a specific compound has the same number and kinds of atoms. For example, quartz (SiO_2) is a very common mineral compound in which each molecule has one atom of silicon (Si) and two atoms of oxygen (O). Every molecule in each compound is exactly the same. That is why compounds have a fixed composition. (See Figure 4-4.)

The third group of minerals is made up of mineral families that have a variable composition. For example, the olivine family is a mixture of two chemical compounds. Olivines usually contain one compound of magnesium, silicon, and oxygen and another compound of iron, silicon, and oxygen. These compounds mix readily to form a single mineral family in which the properties of the mineral (color, hardness, density, etc.) are relatively consistent. These two compounds are not only similar in composition, but they are also similar in appearance and other physical properties. Figure 4-5 shows how eight elements combine to form the common minerals and mineral families.

The feldspars, the most common family of minerals, always contain the elements oxygen, silicon, and aluminum along with varying amounts of potassium, sodium, and cal-

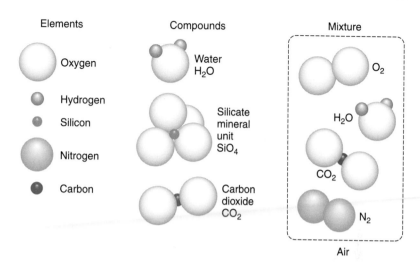

Figure 4-4 Elements, compounds, and mixtures are different forms of matter.

cium. (See Figure 4-6 on page 88.) All feldspar samples have some common properties. Feldspars are light colored, a little harder than glass, and have a density of about 2.7 g/cm³. Outside the laboratory, it can be very difficult to tell the different members of the feldspar family apart. For this reason, some references list feldspar as a specific mineral rather than a family of minerals.

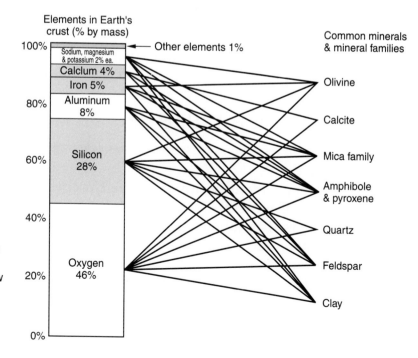

Figure 4-5 These eight elements and eight minerals make up the majority of surface rocks. Follow the lines to learn which elements are in each mineral listed.

Figure 4-6 Feldspar is the most common mineral in Earth's crust.

Minerals are defined as natural, inorganic, crystalline solids that have a specific range of composition and consistent physical properties. A surprising result of this is that ice is a mineral. It is a natural solid substance that forms crystals. Frost and snow are good examples of crystalline ice. As the solid form of water, each molecule of ice (H_2O) is composed of two atoms of hydrogen (H) and one atom of oxygen (O). But, unlike other minerals, ice has a low melting temperature, and therefore, it is not a mineral of which rocks are formed.

 ## WHAT ARE THE PROPERTIES OF MINERALS?

People identify things by their properties. In the identification of minerals, some properties are more useful than are others. The following section will concentrate on those properties that are most important in identifying minerals. The "Properties of Common Minerals" chart in the *Earth Science Reference Tables* lists some of the most helpful characteristics used to distinguish one mineral from another.

Rocks and minerals weather when they are exposed to the conditions at Earth's surface. When identifying a mineral, it is important to use a fresh, unweathered sample. It is the fresh surface that best shows the properties of the mineral. Abrasion caused by rocks colliding crushes the minerals into a powder that is much more difficult to identify than the original sample. Furthermore, as rocks are broken down, minerals are often mixed in the process.

When the minerals are broken or powdered, a greater surface is exposed to air and water. Chemical reactions that are very slow on a fresh surface occur far more rapidly when the mineral is powdered. These chemical changes degrade the mineral by transforming it into a weathering product, just as iron turns to rust. Many minerals react with moisture to form clay minerals. If you try to identify a mineral in a weathered sample, you may see the properties of a weathering product rather than the properties of the original mineral.

Mineral identification tests can be separated into two categories: those that leave the sample unchanged and those that change the sample. Observing the color of a mineral, or testing to see if it is attracted by a magnet does not affect the mineral. You can get an idea of how a mineral has been broken by looking at surfaces where it broke apart. So far you have not done anything that makes the mineral more difficult to identify by the next person who sees it. Other tests are destructive. If you actually break the mineral into smaller pieces to observe its properties, you have performed a destructive procedure. While breaking the mineral may help you to identify the mineral sample, this destructive procedure also makes it more difficult for the next person using the mineral sample. Unfortunately, some useful tests are destructive. For example, to observe a mineral reacting with acid, the mineral must take part in a chemical reaction. That reaction degrades a part of the mineral into a weathering product.

When working with classroom samples, you should not perform destructive tests unless permitted by your teacher. Rough handling or destructive procedures can change a beautiful and valuable mineral sample into useless fragments.

Color

Color is perhaps the first thing you notice about a mineral. For some minerals, color is very useful in their identification. For example, the brassy golden color of pyrite is very distinctive. No other common mineral has this color. Sulfur is one of the few yellow minerals. Almandite garnet is often identified by its red color. However, many minerals can be the same color. A black mineral might be magnetite, biotite, amphibole, pyroxene, or a number of other less common minerals.

Many minerals are colorless or white, including pure samples of quartz, halite, gypsum and calcite. These light-colored minerals present another difficulty of using color to identify minerals. The color of these minerals can be changed by impurities. Impurities are small amounts of other substances found in the mineral. Smoky quartz is gray to black. Rose quartz is pink. Other impurities can make quartz orange, purple, or green. Agate is a banded form of quartz in which white layers often alternate with brown or other colors. As with many other light-colored minerals, identifying quartz by its color alone is very difficult. Dark-colored minerals are less likely to show variations in color from sample to sample because impurities usually do not cause them to change in color.

Streak

A different way of looking at color is **streak**, the color of the powdered form of a mineral. The method used to test for streak involves rubbing a corner of a mineral sample across an unglazed, white porcelain tile, called a streak plate. (Glazing is the glassy covering on plates, cups, and other porcelain kitchenware.) (See Figure 4-7.) An unglazed surface is used because mineral particles do not rub off on a surface that is too smooth. When performing the streak test, you are looking for the color of the powdered form of the mineral left on the streak plate rather than the color of the solid surface of the mineral sample.

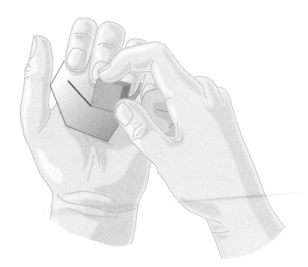

Figire 4-7 The streak test is used on a mineral with a metallic luster to find the color of the mineral's powdered form. A corner of the sample is rubbed across an unglazed porcelain streak plate.

For some minerals, the color of the streak is very different from the color of a fresh surface. For example, galena, a very dense ore of lead, has a shiny gray, bright metallic surface. When galena is drawn across a streak plate, the powder that is left behind is dark gray to black. Samples of hematite, an ore of iron, can vary in color depending upon how and where they formed. Sometimes hematite is red or brown and sometimes a dark metallic silver. However, the streak color of hematite is always reddish-brown. Pyrite is a metallic, brassy, yellow mineral that has a green to black streak. Most nonmetallic minerals leave a streak that is the same color as the sample itself. The streak test is therefore most useful with minerals that have a metallic luster. These are the minerals most likely to show one color on the surface of the sample and a different color streak.

A few minerals are too hard for the streak test. Samples of topaz, corundum, and diamond will scratch the streak plate rather than leaving behind a powder. (Obviously, this would not be good for the streak plate.)

Examining streak is a destructive test procedure. Each time you rub a mineral sample on a streak plate a little bit of the sample is lost and some of the powder produced in this procedure remains on the mineral sample. If the streak test is done carefully, the damage to the sample is small and acceptable.

Luster

Luster can be one of the more difficult observations. However, luster is a very useful property. Shine is a part of luster, but luster is more than how shiny a mineral appears. Luster also includes how light penetrates a fresh surface. **Luster** is defined as the way light is reflected and/or absorbed by the surface of a mineral. Is all of the light reflected? Does some of the light penetrate and some reflect? Is most of the light absorbed? The answers to these questions describe the characteristics of a mineral that determine its luster.

Luster is divided into two categories: metallic and nonmetallic. Minerals with a metallic luster reflect light only from their outer surface. These minerals may have the hard look of a polished metal surface. Silver, copper, gold, galena, and pyrite have metallic luster. Light does not penetrate their surfaces, and you cannot see anything below the surface. Luster is sometimes described as what a mineral looks like it is made from. Minerals with a metallic luster look as if they are made from a metal. Please note that luster is independent of other properties of metals. For example, most metals are also relatively dense. But a mineral can have a metallic luster and have a density lower than most metals. Minerals that have metallic luster can be hard or soft. Like color, the only way to observe luster is with your eyes.

Minerals with nonmetallic luster can be sorted into several groups. Some minerals, such as quartz and feldspar, have a glassy luster. You have probably observed that while glass is shiny, light can penetrate glass. If this were not true, windows would not let light into a room. Few people would see a piece of glass and think that it is a metal. They just do not look the same. Metal and glass are shiny, but the glass surface does not have the hard look of a metallic surface. A porcelain dinner plate has a shiny finish, but it does not appear to be made of metal. Porcelain has a glassy luster that gives it a softer look. While glass and porcelain are shiny, they do not have a metallic luster.

Luster can also be dull. Clay has a dull, or earthy, luster because it is not shiny at all. Dull luster is obviously nonmetallic. You would never mistake clay for a metal surface.

Figure 4-8 Pyrite with quartz—the quartz at the center of the sample has a glassy luster. The pyrite surrounding the quartz has a metallic luster.

There are a few other terms used to describe nonmetallic luster. Talc has a pearly luster. Garnet has a waxy luster. These terms are used to bring to mind substances that reflect and transmit light in the same way as the mineral sample. Figure 4-8 illustrates the difference between metallic and glassy luster.

ACTIVITY 4-2   LUSTER OF COMMON OBJECTS

Make a list of objects in and around the classroom that can be described by the different categories of luster. For each object, record why it fits into that category of luster.

Aluminum foil is a good example. Notice how one side is more reflective than the other is. But also notice how the side that is dull still does not allow light to penetrate below its surface.

 **Hardness**

The **hardness** of a mineral is its resistance to being scratched. The sharp corner of any mineral will scratch a substance that is softer. However, a mineral can be scratched only by a substance that has a greater hardness. This is the standard method used to test hardness.

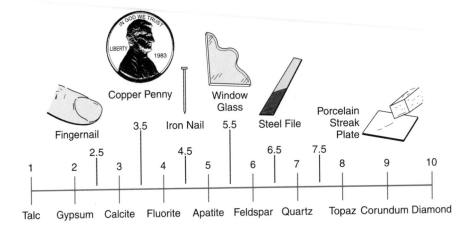

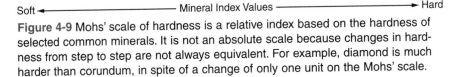

Soft ◄─────────── Mineral Index Values ───────────► Hard

Figure 4-9 Mohs' scale of hardness is a relative index based on the hardness of selected common minerals. It is not an absolute scale because changes in hardness from step to step are not always equivalent. For example, diamond is much harder than corundum, in spite of a change of only one unit on the Mohs' scale.

The mineralogist Friederich Mohs developed a special scale of hardness used to identify minerals. This scale is known as **Mohs' scale** of hardness. (See Figure 4-9.) The scale lists 10 relatively common minerals ordered from soft to hard. The softest mineral on Mohs' scale is talc. Assigned a hardness of 1, talc is so soft that any other mineral on the Mohs' scale can scratch it. A streak plate has a hardness of about 7.5 on the scale. The hardest mineral on Mohs' scale is diamond, which was assigned a hardness of 10. Diamond is the hardest known natural substance. A diamond can scratch every other mineral, but no other mineral can scratch a diamond.

Feldspar and quartz are among the most common minerals. Feldspar has a Mohs' hardness of 6. The hardness of quartz is 7. This means that both minerals are relatively hard to scratch. However, the corner of a fresh sample of quartz when drawn across a smooth surface of feldspar leaves a scratch. On the other hand, feldspar will not scratch quartz even when a fresh edge is used because quartz is harder than feldspar. If two mineral samples have the same hardness, for example two pieces of the same mineral, each will be able to scratch the other.

Consider an example of how the hardness test works. Suppose you found a mineral that can scratch samples of talc, gypsum, and calcite. But the sample will not scratch fluorite or other minerals harder than fluorite. The hardness of the sample must be between 3 and 4 on Mohs' scale. If the sample scratched fluorite and fluorite scratched the mineral sample, the sample's hardness is 4. Perhaps the mineral you are trying to identify is fluorite.

Unfortunately, testing for hardness is a destructive test. After an object has been scratched many times, new scratches can be difficult to observe. One way to make the test less destructive is to observe whether a fresh edge of the mineral will scratch a transparent glass plate. Glass is inexpensive and easy to replace. Most glass has a hardness of 5.5 on Mohs' scale. Therefore, this is a good way to distinguish harder minerals from softer minerals with relatively little damage to the mineral sample.

Please note that hardness is not a resistance to breakage. Steel is a tough material because it does not break easily. Many of our most durable products are made from steel. Steel is not shattered by hard impacts. Diamonds are much harder than steel, but a blow that might dent or bend steel could shatter the diamond into small shards. That is why hardness is determined by the scratch test, not by any kind of collision.

 **Crystal Shape**

If you have visited a mineral display in a museum, you have seen what some consider to be among the most beautiful substances in the natural world. The regular, geometric shape of a crystal is the result of the ordered alignment of atoms and molecules within the crystal. In fact, crystals provide visible evidence that matter is composed of atoms. The variety of colors and shapes of mineral crystals has led some people to call crystals the "flowers of the physical world." One of the reasons that large crystals are valued is because they are rare. Most of the rocks that we see in the lithosphere do not have crystals large enough to be obvious. The crystals may be too small to see without magnification. Even if crystals are visible they are probably not the perfectly shaped crystals you see in museums.

Mineral crystals form in several different environments. Cubic crystals of halite (the mineral in rock salt) are left behind when a salty lake or lagoon evaporates. Other crystals form when water at Earth's surface or circulating underground, deposits the minerals from solution. If the process is slow enough, layer upon layer of atoms and molecules build up to form visible crystals. Some crystals grow when molten rock cools slowly. Slow cooling allows a mineral to form large networks of ordered atoms and molecules. When lava or magma (hot liquid rock) comes to Earth's surface, dissolved gases expand and leave holes in the rock. Holes that are connected may allow water, especially heated water, to circulate through the rock. This hot water can dissolve a mineral from the large mass of rock and deposit it in the holes as crystals. Natural quartz crystals sold in some science, gift, and rock shops were formed in this way. If a large mass of magma that contains dissolved water cools slowly enough, the whole rock can be made of intergrown crystals several centimeters or more in length. Water is not required for crystal growth, but it helps form large crystals.

Minerals can be identified by their characteristic crystal shapes. Quartz and calcite are colorless, white, or light-colored minerals with a glassy luster. Both minerals are very common. Unless you look beyond these similarities, you might get quartz and calcite confused. Quartz forms six-sided, hexagonal crystals. In cross section, they are shaped like a wooden pencil. Calcite crystals are usually four-sided with very different angles between the crystal faces.

Cleavage and Fracture

When minerals are broken, they tend to break in characteristic patterns. The term **cleavage** refers to the tendency of some minerals to break along smooth, flat planes. Cleavage surfaces can be recognized because they reflect light like a flat sheet of glass.

The number of cleavage directions and the angles between them are the most important features of cleavage. Minerals in the mica family, such as biotite and muscovite, show perfect cleavage in only one direction. Large crystals of mica are sometimes called books of mica because they can be split again and again into very thin sheets. The feldspars have two cleavage directions that meet at nearly a right angle. The feldspar minerals break into pieces that have a rectangular cross section. The ends of the shards are not cleavage surfaces, so they are not as smooth and do not reflect light like the two cleavage planes.

Some minerals show three cleavage directions. When they break into pieces, the particles have shiny surfaces all around. Halite, the primary mineral in rock salt, and galena, a silvery metallic mineral, split into little cubes. These minerals show three cleavage planes that meet at right angles. Calcite is the primary mineral in limestone. When calcite crystals are broken, three cleavage directions are observed. But calcite's cleavage planes do not meet at right angles. Calcite breaks into rhombohedrons that look like rectangular solids that have been pushed to one side. (See Figure 4-10.)

Why do minerals show cleavage? Mineral crystals are composed of atoms arranged in neat rows. The crystals tend to break parallel to the rows of atoms. Minerals also break along surfaces where the bonds, or attachments, of atoms are relatively weak. Many minerals such as halite, calcite, and mica, cleave parallel to their crystal faces.

Not all minerals break along definite cleavage directions. Minerals that break along curved surfaces or surfaces that are not parallel are said to show **fracture**. Natural quartz crystals have six flat sides. But when quartz crystals are broken they fall apart along curved surfaces that are not parallel to the flat sides. This is known as conchoidal fracture.

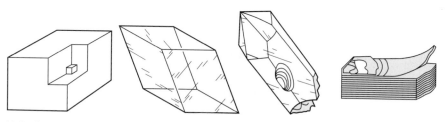

Figure 4-10 These four common minerals are easy to identify because of their crystal shape and cleavage.

Halite forms cubic crystals that cleave parallel to the crystal faces.

Calcite crystals are rhombohedral and cleave parallel to the crystal faces.

Quartz crystals are hexagonal but fracture along curved surfaces.

Minerals in the mica family split into thin, flexible sheets.

Garnet breaks smoothly, and the shards have shiny surfaces. But the surfaces are not flat and parallel. Therefore, garnet shows fracture rather than cleavage.

Some substances break into fragments or a powder that shows neither cleavage nor fracture. For example, clay disintegrates into a fine powder. Clay is among the minerals that show neither cleavage nor fracture.

To test a mineral for cleavage or fracture you must break it. Crystals are rare and beautiful. Breaking these samples destroys them. It may be better to observe the natural breakage surfaces and the angles at which they meet rather than actually testing a mineral in a destructive procedure.

ACTIVITY 4-3   BREAKAGE OF HOUSEHOLD SUBSTANCES

For this activity you will need a simple magnifying lens made of glass or plastic.

Use the magnifier to observe a variety of granular substances in your home such as sugar, salt, baking soda, and soap flakes. Check with your parents to determine which substances are safe to handle and to help you avoid making a mess.

Make a list of the substances you observed, and tell why you think each shows cleavage, fracture, or neither.

 **Density**

In Chapter 1, density was defined as the concentration of matter. Objects that are dense are heavier than less dense objects of the same size. If you pick up a sample of magnetite or pyrite, it will weigh about twice as much as a same size piece of most other minerals. Gold is denser than magnetite or pyrite. A popular method used to separate gold from other minerals in a stream is called panning. Figure 4-11 shows a man panning for gold. This method is used in places where particles of gold have been washed into streams. A pan with slanted sides, such

Figure 4-11 Because of its high density, gold settles to the bottom of the pan as the lighter minerals are washed away.

as a pie pan, can be used to scoop up sand and gravel from a streambed. When the gravel is agitated in a pan and washed with water, the lighter particles rise to the top and they fall over the edge of the pan. At the same time, the densest particles settle into the crease in the pan. If the person using the process is very lucky he or she might find tiny flakes of "color," the prospector's name for small particles of gold.

| ACTIVITY 4-4 | SEPARATING MINERALS BY PANNING |

You can use panning to separate minerals by density. You will need a metal or plastic pan with sloped sides and a crease where the sides meet the bottom. Place a mixture of different kinds of sand-size particles in the pan and swirl them with lots of water. (You could use beach sand.) In your report, describe how the mineral particles that rise to the top appear different from the particles that collect at the bottom in the crease of the pan. Can you determine any differences among the particles other than density?

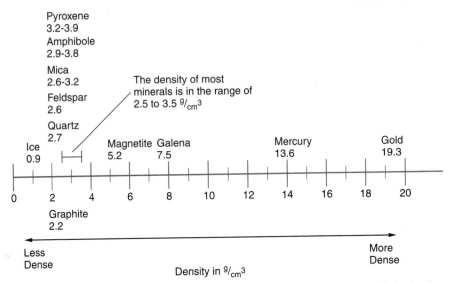

Figure 4-12 The density of many common minerals is between 2.5 and 3.5 g/cm³.

In Chapter 1, you also learned that the density of an object could be calculated by dividing its mass by its volume. When mass is measured in grams and volume is measured in cubic centimeters, the unit for density is grams per cubic centimeter (g/cm^3). For example, the density of water is $1 \ g/cm^3$. The density of the most common minerals is in the range of 2.5–3.5 g/cm^3. Many minerals have about the same density. Identifying a mineral by its density is helpful if the mineral is unusually dense. Figure 4-12 gives the density of many common minerals.

Magnetite and galena are about twice as dense as the more common minerals. As mentioned previously, gold is the densest mineral substance you are likely to encounter. The density of gold is 19.3 times as dense as water and roughly six or seven times as dense as the most common minerals.

 Special Properties

Some minerals have special properties that are relatively rare. These properties can be useful in identifying these minerals. For example, graphite and talc are unusual because they feel greasy or slippery. Minerals in the mica family (muscovite and biotite) cleave into thin, flexible sheets. Magnetite is an ore of iron that is attracted to a magnet. Sulfur has a distinctive odor and melts at a low temperature.

It is a good rule not to taste or eat anything in a science lab. This is especially true of laboratory chemicals. (You should ask permission from your teacher before you taste any mineral substances.) Halite (the principle mineral in rock salt) can be identified by its salty taste.

Acids can be dangerous substances. If you are allowed to use them, handle them with great care because a strong acid can burn your skin and make holes in clothing. When a drop of acid is placed on calcite or a rock that contains calcite, a chemical reaction occurs that gives off bubbles of gas.

A fresh surface of plagioclase feldspar may have small striations that look like parallel scratches on its surface. You are unlikely to find uranium in an Earth science lab because it is radioactive, which means that it gives off invisible rays and particles called ionizing radiation. (Ionizing means that the radiation is a high-energy form that can damage atoms.) This radiation can be detected with special instruments, and it is a strong indication of the presence minerals that contain uranium or several related elements.

WHAT ARE THE MOST COMMON MINERALS?

A variety of minerals have been mentioned in this chapter that illustrate the properties used to identify minerals. Of the thousands of known minerals, just five make up about 85 percent of Earth's crust. These are the minerals that you are most likely to see. (See Figure 4-13 on page 102.)

The two most abundant elements in Earth's crust are oxygen and silicon. The group of minerals that contains both oxygen and silicon is known as the **silicate** minerals. As you might expect, silicates are the most abundant minerals in

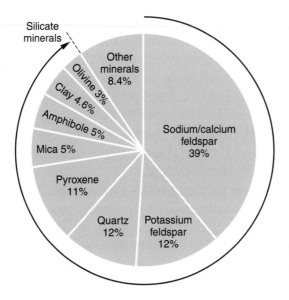

Figure 4-13 Relative abundance of minerals in Earth's crust.

Earth's crust. Of the following minerals, only calcite is not a silicate.

 **Feldspars**

The feldspar family of silicate minerals makes up more than half of Earth's crust. In fact, the name feldspar comes from the Swedish words for field and mineral. The name refers to the fact that Swedish farmers had to move large quantities of rock rich in feldspar from their fields before they could plant crops. The feldspar family of minerals is divided into groups according to composition. Plagioclase feldspars contain, in addition to silicon and oxygen, sodium and calcium in variable proportions. Potassium feldspar contains potassium instead of sodium or calcium along with silicon and oxygen. Distinguishing between these groups can be difficult. In this book, they will be described by their common properties. The feldspar minerals are generally white to pink in color and cleave in two directions at nearly right angles. Look again at Figure 4-6 on page 88. Feldspars have a glassy luster. With a hardness of 6 on Mohs' scale, the feldspar minerals will scratch glass but will not scratch quartz.

 **Quartz**

Quartz, another silicate, is second only to the feldspar minerals in abundance. If pure, quartz contains only silicon and oxygen; it has the chemical formula SiO_2. It is usually colorless or light in color and sometimes transparent. Quartz is often translucent, which means light can pass through but you cannot see objects on the other side as you could through a window. (See Figure 4-14.) Impurities can make quartz almost any color. The semiprecious gems amethyst, agate, and onyx are examples of quartz that is colored by small amounts of impurities. Like the feldspars, quartz has a glassy luster. With a Mohs' hardness of 7, quartz can scratch most other minerals. Although most samples of quartz are not crystals, when quartz does form crystals, they are hexagonal, or six-sided, and can be pointed on one or both ends.

 **Micas**

Minerals in this silicate family are easy to identify because they are the only common minerals that can be split into thin, flexible sheets. We say that mica has perfect cleavage in one direction. Like quartz and feldspar, the mica minerals have a glassy luster but they are relatively soft, with a hardness of only 2.5 on Mohs' hardness scale. Muscovite mica is rich in potassium and aluminum giving it a light color. Bi-

Figure 4-14 Quartz sometimes occurs in large, hexagonal crystals.

otite mica is dark in color due to the presence of iron and magnesium. Rocks that are rich in mica can have a reflective sheen on fresh surfaces.

 ## Amphiboles and Pyroxenes

The silicates amphibole and pyroxene are the most common dark-colored minerals. Both have a Mohs' hardness value between 5 and 6. They break into stubby splinters. Amphibole minerals, such as hornblende, can be distinguished from the pyroxene minerals, such as augite, by the angle at which their two cleavage surfaces meet. In the amphibole minerals, cleavage surfaces meet at 60° and 120° angles. The angle between cleavage surfaces in the pyroxene family is perpendicular, or 90°.

 ## Calcite

Although calcite is not as common throughout Earth's crust as are the minerals above, it is relatively common at the surface. Calcite is also the only mineral in this section that is not a silicate. By chemical composition, calcite is calcium carbonate ($CaCO_3$). This means that calcite contains carbon and oxygen rather than the silicon and oxygen combination that defines the silicate minerals. Calcite is the most common mineral in the carbonate group. In very pure form, calcite may look transparent and colorless, but like quartz it can have a variety of colors due to impurities. A soft mineral, calcite has a hardness of only 3. In very pure samples, breakage along cleavage directions can result in a rhombohedral shape that looks like a rectangular solid that has been pushed toward one side. Limestone and marble are common rocks that are made primarily of calcite.

 ## Olivine

Olivine is extremely common within Earth. However, it is not usually visible at the surface because it quickly weathers into clay. Olivine can often be observed in unweathered igneous rocks and it is readily identified by its olive-green color, glassy luster, and granular texture.

Clay

Clay is a family of soft, earthy minerals that usually forms as a decomposition product of a wide range of other minerals including feldspar and mica. It forms when these minerals combine chemically with water in the weathering process. Clay is a major component of soil and the primary mineral in shale. Clay power is composed of very fine particles that become sticky and flexible when a small amount of water is added. It often has an earthy odor.

Using a Flowchart to Identify Minerals

The simplified flowchart in Figure 4-15 on page 106 can help you identify seven of the most common minerals. In this chart, you start on the left and proceed through the chart to the right. For example, if you unknowingly had a sample of calcite, you would start by noticing that your sample is light in color. So at the first branch you would take the top choice. Your sample is too soft to scratch glass so you move to the top of the next division. Your sample cleaves in three directions, which takes us to the lower portion of the final division. Finally you know that your sample bubbles when tested with acid. Calcite is the only common mineral that has this combination of properties.

Using the Reference Tables

The most useful properties in mineral identification are incorporated in the "Properties of Common Minerals," Figure 4-16 on page 107, also found in the *Earth Science Reference Tables*. You should be able to use this chart to identify fresh samples of any of the 21 minerals listed in the "Mineral Name" column. Among the thousands of known minerals, these are the minerals you are most likely to observe both in the natural environment and in the science lab.

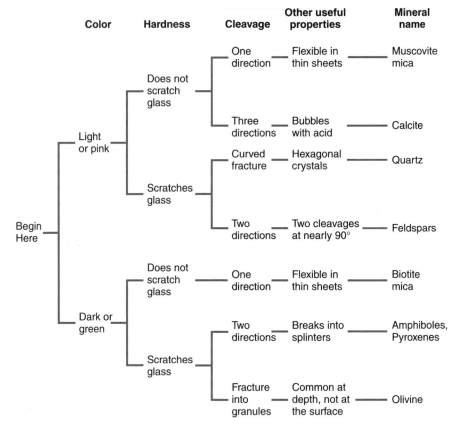

Figure 4-15 This flowchart can help you identify some of the most common minerals. Start at the left side and progress through the properties of the sample.

To use this chart, start on the left side by identifying the luster of the sample in question. Then work your way to the right checking each property listed (hardness, cleavage or fracture, color, etc.) This chart contains more than information to help you identify mineral specimens. It also includes information about the chemical composition and the principal uses of each mineral.

ACTIVITY 4-5   MINERAL IDENTIFICATION

Suggested materials: set of minerals, small magnifier, streak plate, small glass plate, magnet

Your teacher will provide you with a kit containing several minerals. Use the "Properties of Common Minerals" chart to identify

	HARD-NESS	CLEAVAGE	FRACTURE	COMMON COLORS	DISTINGUISHING CHARACTERISTICS	USE(S)	MINERAL NAME	COMPOSITION*
Metallic Luster	1–2	✔		silver to gray	black streak, greasy feel	pencil lead, lubricants	Graphite	C
	2.5	✔		metallic silver	very dense (7.6 g/cm^3), gray-black streak 	ore of lead	Galena	PbS
	5.5–6.5		✔	black to silver	attracted by magnet, black streak	ore of iron	Magnetite	Fe_3O_4
	6.5		✔	brassy yellow	green-black streak, cubic crystals 	ore of sulfur	Pyrite	FeS_2
Either	1–6.5		✔	metallic silver or earthy red	red-brown streak	ore of iron	Hematite	Fe_2O_3
Nonmetallic Luster	1	✔		white to green	greasy feel	talcum powder, soapstone	Talc	$Mg_3Si_4O_{10}(OH)_2$
	2		✔	yellow to amber	easily melted, may smell	vulcanize rubber, sulfuric acid	Sulfur	S
	2	✔		white to pink or gray	easily scratched by fingernail	plaster of Paris and drywall	Gypsum (Selenite)	$CaSO_4 \cdot 2H_2O$
	2–2.5	✔		colorless to yellow	flexible in thin sheets 	electrical insulator	Muscovite Mica	$KAl_3Si_3O_{10}(OH)_2$
	2.5	✔		colorless to white	cubic cleavage, salty taste 	food additive, melts ice	Halite	NaCl
	2.5–3	✔		black to dark brown	flexible in thin sheets 	electrical insulator	Biotite Mica	$K(Mg,Fe)_3$ $AlSi_3O_{10}(OH)_2$
	3	✔		colorless or variable	bubbles with acid 	cement, polarizing prisms	Calcite	$CaCO_3$
	3.5	✔		colorless or variable	bubbles with acid when powdered	source of magnesium	Dolomite	$CaMg(CO_3)_2$
	4	✔		colorless or variable	cleaves in 4 directions	hydrofluoric acid	Fluorite	CaF_2
	5–6	✔		black to dark green	cleaves in 2 directions at 90° 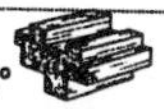	mineral collections	Pyroxene (commonly Augite)	$(Ca,Na)(Mg,Fe,Al)$ $(Si,Al)_2O_6$
	5.5	✔		black to dark green	cleaves at 56° and 124° 	mineral collections	Amphiboles (commonly Hornblende)	$CaNa(Mg,Fe)_4(Al,Fe,Ti)_3$ $Si_6O_{22}(O,OH)_2$
	6	✔		white to pink	cleaves in 2 directions at 90°	ceramics and glass	Potassium Feldspar (Orthoclase)	$KAlSi_3O_8$
	6	✔		white to gray	cleaves in 2 directions, striations visable	ceramics and glass	Plagioclase Feldspar (Na-Ca Feldspar)	$(Na,Ca)AlSi_3O_8$
	6.5		✔	green to gray or brown	commonly light green and granular	furnace bricks and jewelry	Olivine	$(Fe,Mg)_2SiO_4$
	7		✔	colorless or variable	glassy luster, may form hexagonal crystals 	glass, jewelry, and electronics	Quartz	SiO_2
	7		✔	dark red to green	glassy luster, often seen as red grains in NYS metamorphic rocks	jewelry and abrasives	Garnet (commonly Almandine)	$Fe_3Al_2Si_3O_{12}$

*Chemical Symbols:

Al = aluminum	Cl = chlorine	H = hydrogen	Na = sodium	S = sulfur
C = carbon	F = fluorine	K = potassium	O = oxygen	Si = silicon
Ca = calcium	Fe = iron	Mg = magnesium	Pb = lead	Ti = titanium

✔ = dominant form of breakage

Figure 4-16 Properties of common minerals.

each of the samples. Record all the information you used to help you identify each mineral. But record only what you actually observed. For example, you probably have only a glass plate to determine the hardness of the minerals. Therefore, you will not be able to record the specific Mohs' hardness number as listed on the chart. However, you can record whether the mineral could scratch glass.

Please handle the minerals with care and avoid damaging the samples when you perform destructive procedures. For example, rather than breaking the sample to observe fracture or cleavage, look at the surface of the sample to see how it broke when the mineral sample was prepared for the kit.

TERMS TO KNOW

compound	**element**	**hardness**	**mineral**	**silicate**
cleavage	**fracture**	**luster**	**Mohs' scale**	**streak**

CHAPTER REVIEW QUESTIONS

1. Based in information in the *Earth Science Reference Tables* (Figure 4-16) which mineral is an iron ore that has a characteristic reddish brown streak?

   (1) magnetite   (3) hematite
   (2) pyrite   (4) olivine

2. Halite has three cleavage directions at 90° to each other. Which model best represents the shape of a broken sample of halite?

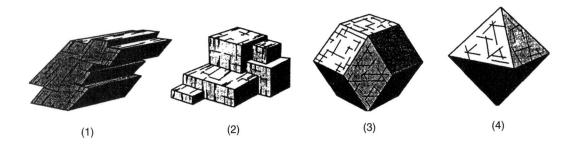

      (1)   (2)   (3)   (4)

Base your answers to questions 3–5 on Mohs' hardness scale and on the chart below showing the approximate hardness of some common objects.

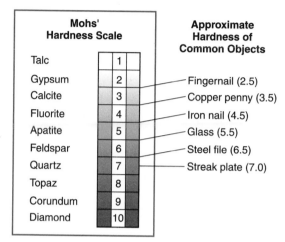

MINERAL HARDNESS

3. Which statement is best supported by this scale?

    (1) A fingernail will scratch calcite, but not quartz.
    (2) A fingernail will scratch quartz, but not calcite.
    (3) A piece of glass can be scratched by quartz, but not by calcite.
    (4) A piece of glass can be scratched by calcite, but not by quartz.

4. Mohs' scale would be most useful for

    (1) identifying a mineral sample.
    (2) finding the mass of a mineral sample.
    (3) finding the density of a mineral sample.
    (4) counting the number of cleavage surfaces of a mineral sample.

5. The hardness of these minerals is most closely related to the

    (1) mineral's color.
    (2) mineral's abundance in nature.
    (3) amount of iron the mineral contains.
    (4) bonding of the mineral's atoms.

6. What mineral leaves a green-black powder when rubbed against an un-glazed porcelain plate?

    (1) galena   (3) hematite
    (2) graphite   (4) pyrite

7. Which statement about the minerals plagioclase feldspar, gypsum, biotite mica, and talc can best be inferred from the "Properties of Common Minerals" chart in the *Earth Science Reference Tables* (Figure 4-16)?

   (1) They have the same physical and chemical properties.
   (2) They have different chemical properties but similar physical properties.
   (3) They have different physical and chemical properties, but they have identical uses.
   (4) The physical and chemical properties of these minerals determine how they are used.

8. Which mineral is white or colorless, has a hardness of 2.5, and splits with cubic cleavage?

   (1) calcite
   (2) halite
   (3) pyrite
   (4) biotite

9. What property do nearly all rocks have in common?

   (1) They show cleavage.
   (2) They contain minerals.
   (3) They were organically formed.
   (4) They formed on Earth's surface.

10. Looking at a mineral sample on a table without touching or moving it, what property could you observe?

    (1) hardness
    (2) atomic structure
    (3) density
    (4) luster

11. Which of the following is not a true mineral?

    (1) biotite mica
    (2) crude oil
    (3) potassium feldspar
    (4) sulfur

12. Which of the following is a mineral that, if pure, contains only one kind of atom?

    (1) quartz
    (2) calcite
    (3) pyrite
    (4) sulfur

13. Which of the following substances is a mineral?

    (1) sandstone
    (2) steel
    (3) quartz
    (4) glass

14. What kind of rock is composed mostly of material that is not a mineral?

    (1) granite   (3) sandstone

    (2) organic limestone   (4) molten magma

15. Which of the following minerals are you most likely to eat every day?

    (1) calcite   (3) galena

    (2) quartz   (4) halite

Open-Ended Questions

16. Describe how to test the hardness of a mineral sample.

17. What is the difference between cleavage and fracture?

18. Magnetite is a dark-colored mineral that shows metallic luster. Biotite is a dark-colored mineral that has a glassy luster. Use these two minerals to explain how a glassy luster differs from a metallic luster.

19. State three ways in which quartz and plagioclase feldspar are similar.

Base your answers to questions 20 and 21 on the data table below, which shows the volume and mass of three different samples, A, B, and C, of the mineral pyrite.

Pyrite

Sample	Volume (cm³)	Mass (g)
A	2.5	12.5
B	6.0	30.0
C	20.0	100.0

20. On a piece of graph paper, plot the data (volume and mass) for the three samples of pyrite and connect the points with a line.

21. According to your graph, what is the mass of a sample of pyrite that has a volume of 10.0 cm³?

Chapter 5

The Formation of Rocks

 ## WHAT IS CLASSIFICATION?

In Chapter 1 you learned that science is a way of making, organizing, and using observations. One of the important skills of a scientist is the ability to classify. In the process of **classification**, objects, ideas, or information is organized according to their properties. Things that are similar in some characteristic are grouped together. Things that do not have this characteristic are put into a different group. In this process, it is important to specify what properties are being used to group things. For example, in the last chapter you read about the classification of minerals according to those with metallic luster and minerals with nonmetallic luster.

ACTIVITY 5-1	CLASSIFICATION

Your teacher will give you about a dozen objects to classify. Divide the objects into groups based on their observable properties. Start by listing all the objects. Each time you divide the objects into groups, state a single property that allows you to clearly decide to which group an object belongs. As you divide each group into smaller groups, limit the number of subgroups to two or three.

Continue the process until each object is alone in a group. Create a flowchart that will allow someone else to separate and/or identify all the objects.

 # WHAT ARE ROCKS?

A **rock** is a substance that is or was a natural part of the solid Earth, or lithosphere. Rocks come in a great many varieties. Some rocks are unusual enough for geologists and people interested in geology to collect them. Rocks can also be colorful or attractive. Landscapers often make use of rocks in planning homes or parks.

Most rocks are composed of a variety of minerals. In some rocks, you can recognize the minerals as variations in color or other mineral properties within the rock. Granite is a good example. If you look carefully at a sample of granite with a hand lens, you will probably observe some parts of the rock that are transparent. This is probably the mineral quartz. Other parts are white or pink with angular cleavage. These are properties of plagioclase feldspar and potassium feldspar. Dark mineral grains that occur in thin sheets are biotite mica. Dark minerals that occur in stubby crystals are probably hornblende, the most common mineral in the amphibole family. Other minerals can occur in granite, but they are not as common as these four.

Other rocks are composed of a single mineral. Some varieties of sandstone, particularly if they are very light in color, can be nearly 100 percent quartz. Very pure limestone is nearly all calcite. Rock salt can be nearly pure halite.

ACTIVITY 5-2 MAKING A ROCK COLLECTION

As you read through this chapter, collect rocks from around your community. If you wish, you can add rocks from your travels or rocks that you have obtained elsewhere. You may recall that when you worked with minerals you tried to use fresh, unweathered samples. It is important to follow the same rule in collecting rocks. Weathered rocks tend to crumble, and they are also harder to identify than fresh samples.

This rock collection will have two purposes. First, it should help you discover the variety of rocks around you. The second purpose will be to use the information in this chapter to classify your rocks.

Your samples should be divided into the three categories you will learn about in this chapter: igneous, sedimentary, and meta-morphic rocks. You may collect as many rocks as you wish, but the final submission will consist of six small rocks, each of which shows a properties of its category. Therefore, for each rock submitted classify it as igneous, metamorphic, or sedimentary and tell why each sample fits into its group. Your samples may be placed in a half-dozen egg carton.

You are encouraged to check your progress on this assignment with your teacher and to carefully observe the rock samples you use in class to clarify the properties used to classify rocks. Your teacher will collect your samples after the class has completed the chapter.

Classification of Rocks

In the previous chapter you learned that minerals are classified on the basis of their observable properties such as luster, hardness, color, and cleavage. Geologists have found it more useful to separate rocks into three groups based on how they formed. The way rocks form is called their **origin**.

When any material is heated enough, it will melt. This is true even of rocks. If the melted material cools enough, it will solidify. The process of solidification is sometimes called crystallization. You may recall that minerals were defined as crystalline solids. The formation of a true solid involves atoms and molecules moving into fixed and ordered positions. While the networks of atoms and molecules may not be large enough to form visible crystals, it is the movement of atoms and molecules into these fixed and ordered positions that defines the formation of a solid. Therefore solidification is the formation of a crystalline solid.

Igneous rocks form from hot, molten (liquid) rock material that originated deep within Earth. Only igneous rocks have this origin. Hot, liquid rock is called **magma**. (At Earth's surface magma is known as **lava**.) In Chapter 2 you learned that Earth's temperature increases as you go deeper within the planet. In some places within Earth, it is hot enough to melt rock. When this molten rock rises to or near Earth's surface where it is cooler, the liquid rock material changes to solid rock. Igneous rocks are especially common around volcanoes and in places where large bodies of rock that have melted and then solidified underground have been pushed to the surface.

In Chapter 7 you will learn that most of Earth's interior is in the solid state. If temperatures underground are hot enough to melt rock, why is the interior not mostly liquid? The reason that Earth is mostly in the solid state is the increase of pressure with depth. While the increasing temperature tends to melt the rock, the increase in pressure prevents melting.

Rocks weather and break down when they are exposed at Earth's surface. The weathered rock material is usually washed away by rainfall and carried into streams. Eventually the weathered material accumulates as layers of sediment somewhere, usually in a large body of water such as an ocean. As more layers of weathered rock are deposited on top of the older layers, the lower layers are compressed by the weight of sediments accumulating above them. Furthermore, water circulating through them carries rock material into the spaces between the particles filling them with natural cement. Compression and cementing of weathered rock fragments or the shells of once living creatures is the origin of most **sedimentary rocks**. The second group of rocks is called sedimentary rocks.

If sedimentary or igneous rocks are buried so deeply that heat and pressure cause new minerals to form, the result is **metamorphic rocks**. Metamorphic rocks are the only group that forms directly from other rocks (igneous or sedimentary). But if most metamorphic rocks form within Earth, how can they be found at the surface? The answer to this question involves two important Earth-changing processes. The first is uplift. Earth contains a great deal of heat energy. As heat escapes from the interior of the planet, it sometimes pushes up rocks to form mountains. The second process is weathering and erosion. Weathering and erosion wear down the mountains exposing rocks at the surface that actually formed at depths of 10 km or more underground. Wherever you find a large mass of metamorphic rock at the surface, you are probably looking at the core of an ancient mountain range.

Not all rocks fit easily into one of these three categories. Some volcanoes throw great quantities of ash into the air. The ash falls, settles in layers, and hardens. The settling part of ash layers' origin is similar to the processes that form sedimentary rock. However, because the material came from a volcano, volcanic ash is classified as an igneous rock. Another example is the gradual change from igneous or sedimentary rock to metamorphic rock. It may not be clear at what point the rock should no longer be classified as its original parent rock and when it should be called a metamorphic rock. In spite of these occasional difficulties, the classification of rocks by their origin has generally served geologists and Earth science students well. Figure 5-1 illustrates the way to classify rocks as igneous, sedimentary, or metamorphic by their appearance. Rocks can be classified by their appearance because the way they look offers clues to their formation.

HOW ARE IGNEOUS ROCKS CLASSIFIED?

Igneous rocks are classified by their color and texture. The colors in rocks come from the minerals that make up the rocks. Minerals rich in aluminum (chemical symbol, Al) are commonly light-colored, sometimes pink. These minerals are called **felsic** because the feldspars are the most common light-colored minerals. the word "felsic" comes from *fel*dspar and *si*licon. Minerals rich in magnesium (chemical symbol, Mg) and iron (chemical symbol, Fe) such as olivine and py-

Because **igneous rocks** have formed from molten magma or lava, they are composed of intergrown crystals. Rapid cooling, however, can make the crystals too small to be visible. Igneous rocks are usually quite hard and dense, and layering is rare. Gas bubbles may give igneous rocks a frothy texture.

Most **sedimentary rocks** are composed of rounded fragments cemented in layers. In fine-grained rocks, the individual grains may be too small to be readily visible. Rocks made by chemical precipitation are composed of intergrown crystals, although these crystals are relatively soft. A rock that contains fossils is almost certainly a sedimentary rock.

  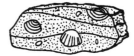

Metamorphic rocks, like igneous rocks, are usually composed of intergrown crystals. But, like sedimentary rocks, they often show layering, banding, or foliation. The layers may be bent, or distorted.

Figure 5-1 Rocks can be classified by their appearance.

roxene families are called **mafic** (MAY-fic) The word "mafic" comes from a combination of *ma*gnesium and *fer*ric, which is used to describe iron. Mafic minerals are often dark colored.

The next characteristic in classifying igneous rocks is their texture. **Texture** describes the size and shape of the grains and how they are arranged. Texture answers the following questions. Is the rock composed of different kinds of grains? How large are these grains, and what shape do they have? Do they show any kind of organization?

In igneous rocks, the size of the crystals is a result of how quickly the rock solidified. If the magma cooled slowly, the atoms and molecules had enough time to form crystals that are visible without magnification. Granite is a good example of a rock that cooled slowly. Granite is a popular building stone because it resists wear and weathering and because it is attractive. Granite has a speckled appearance. The different colors in granite come from the different minerals of which it is composed. Crystals from $\frac{1}{4}$- to 1-cm long are common in granite. If the granite is pink, it probably contains a large amount of potassium feldspar, which can be pink or white.

Most granite forms in large masses within Earth. The movement of magma to a new position within Earth's crust is called **intrusion**. Intrusion occurs totally underground, or inside Earth. Sometimes a large quantity of hot magma rises to a place near the surface where it slowly cools to form solid rock.

In other cases, granite originates from a mass of rock buried deeply enough to melt. As the mass cools and crystallizes, it slowly forms granite. Because coarse-grained igneous rocks such as granite form deep underground, they are classified by origin as intrusive or **plutonic** rocks. (The term plutonic comes from the name of Pluto, the Roman god of the underworld.)

Basalt is also a common igneous rock, especially under the oceans. The ways that basalt differs from granite can help you understand how igneous rocks are classified. Basalt usually forms from magma that rises to or very near the surface. **Extrusion** is the movement of magma onto Earth's surface. At the surface, lava cools quickly, and the resulting crystals are too small to be visible without magnification. Fine-grained igneous rocks such as basalt are therefore called extrusive or **volcanic** rocks. Basalt is rich in mafic minerals that give it a dark color, generally dark gray to black. See Figure 5-2.

Figure 5-3 is a chart from the *Earth Science Reference Tables* that can help you understand and classify igneous rocks

Figure 5-2 Granite and basalt. Granite is a light-colored, coarse-grained igneous rock. Basalt contains much smaller crystals and is relatively dark in color.

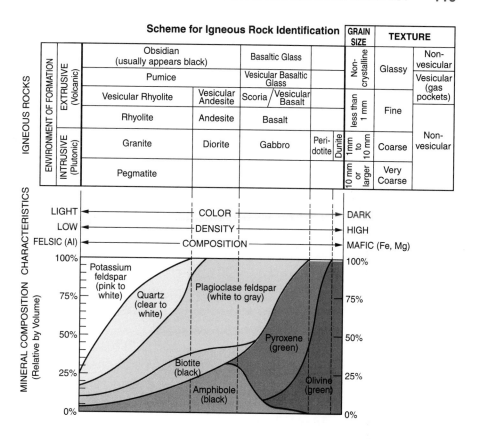

Figure 5-3 You can use this chart to identify the most common igneous rocks and to estimate their mineral composition.

primarily by color and texture. The rocks with the smallest crystals (fine-grained, extrusive rocks) are at the top of this chart. In these rocks, the grains of the different minerals are too small to be readily visible without magnification. Below them are the coarse igneous rocks in which it is easy to see the different minerals.

Variations in mineral composition occur between the light-colored (felsic) rocks, such as granite and rhyolite on the left side of the chart, and the mafic rocks, such as basalt and gabbro on the right side. Along with this variation in composition (felsic to mafic) comes a change not only in color (light to dark) but also a change in density. You may not be able to feel that a mafic rock is heavier than a felsic rock of the same size, but the difference is measurable. The difference in density will become important when we consider the interior of our planet in Chapter 7.

Among the terms that you will find on the chart is vesicular. A **vesicular** texture refers to the gas pockets, called vesicles, that are common in extrusive igneous rocks. When magma rises, the decrease in pressure causes trapped gases to form bubbles. This is similar to what happens when you open a bottle of carbonated soda. Bubbles in soda are trapped carbon dioxide. But the gas bubbles that form in lava are mostly water vapor, which escapes into the atmosphere. Scoria has large vesicles, and may look like cinders from a fire. Pumice has smaller gas bubbles, and it can be so light it may float on water. Pumice is sometimes sold as an abrasive used to scrape the gratings of barbecue grills. If you forget what vesicular means, look at the top, right portion of the chart where you will find the word "vesicular" just above the two words "gas pockets" in parentheses.

At the top of the chart is a texture called glassy. Sometimes lava cools so quickly it forms a rock that looks like a shiny, dark, glass material. This is obsidian. If obsidian (also known as volcanic glass) contains crystals, they are too small to be seen even under a microscope. The properties of the other igneous rocks listed in the Scheme for Igneous Rock Identification can be determined from the rock's position on the chart. For example, pegmatite appears at the bottom left of the chart. Like other igneous rocks on the left, pegmatite is relatively light in color. Its position at the bottom means that pegmatite is composed of very large crystals.

The bottom section of the Scheme for Igneous Rock Identification is called Mineral Composition. This section shows the minerals that are common in igneous rocks. For example, granite usually contains potassium feldspar, quartz, plagioclase, biotite, and amphibole. If you imagine a vertical line running directly below the word granite and into this section, you will see that quartz and potassium feldspar make up about 66 percent of the volume of granite. The percent of each mineral is indicated by the scale that appears on each side of the Mineral Composition section of the chart. The composition of basalt is under the word "Basalt" near the other side of the chart. Basalt is mostly plagioclase and pyroxene. The mineral composition of igneous rocks is variable.

Figure 5-4 Scoria is an igneous rock that contains large air pockets. As the magma rises to the surface the decrease in pressure causes the gases, such as water vapor and carbon dioxide, to expand, forming pockets. The gases escape into the atmosphere.

The various compositions of each rock are enclosed by the dotted lines.

This chart is a good example of how much information is available in the *Earth Science Reference Tables*. The good news is that these charts should be available to you whenever they can help you on labs or tests. However, you will need to understand how to use the charts and what the words on the chart mean.

The most common igneous rocks have characteristics that will help you identify them as igneous. Igneous rocks are made of mineral crystals such as those you can observe in granite, although the crystals may be too small to see without magnification. Most igneous rocks do not show layering. While lava flows may occur in pulses that turn into thick layers of igneous rock, small samples seldom show layering. Some fine-grained igneous rocks contain rounded holes made by the escape of gases trapped in the magma. If you observe these textures, you are probably looking at an igneous rock. (See Figures 5-4 and 5-5.)

Figure 5-5 Four textures common in igneous rocks.

Glassy
(Obsidian)

Fine grained
(Basalt)

Coarse grained
(Granite)

Very coarse
(Pegmatite)

Common Igneous Rocks

Of all the igneous rocks named on the Scheme for Igneous Rock Identification in the *Earth Science Reference Tables*, or Figure 5-3 on page 119, you are most likely to encounter just seven in your Earth science course. These seven igneous rocks are very easy to tell apart, and they show the range of properties of igneous rock.

Granite is a coarse-grained, felsic (light-colored) igneous rock. Its overall color is likely to be light gray or pink. Because of slow cooling the mineral crystals are large enough to be visible without magnification.

Rhyolite is the fine-grained equivalent of granite. Rhyolite is light colored and felsic in composition. Rapid cooling of the magma has resulted in very small mineral grains that are unlikely to be readily visible.

Gabbro, like granite, is composed of large crystals because of slow cooling of the magma. (It is coarse grained.) Unlike granite, gabbro is mafic in composition, which means that it is composed primarily of the dark minerals rich in iron and magnesium.

Basalt has a mineral composition similar to gabbro's, so it is also relatively dark in color. However, basalt cooled so quickly that, as in rhyolite, the individual mineral grains might be too small to see without magnification.

The next two igneous rocks share an unusual feature. Scoria and pumice are both full of air pockets. This is an indication that they probably formed from magma rich in dissolved gases, such as water vapor, and ejected from a volcano during a violent eruption. The pockets in pumice are small enough that individual pockets are not obvious. Scoria has larger pockets and looks like cinders.

Volcanic glass is also called obsidian. The term glass describes its smooth texture, which results from rapid cooling of lava that had little dissolved water or gases. It is usually black due to the even distribution of dark minerals, even through the mineral composition of volcanic glass is most often felsic. See Figure 5-6.

Figure 5-6 The glassy texture of obsidian (volcanic glass) indicates that it cooled very quickly.

 # WHAT IS THE BOWEN REACTION SERIES?

Figure 5-7 illustrates the Bowen reaction series. American geologist N. L. Bowen devised this chart. This diagram shows that as magma of mixed composition cools, different minerals crystallize at different temperatures. Those minerals at

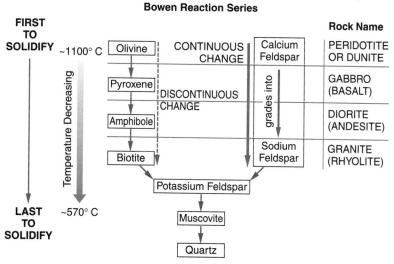

Figure 5-7 The Bowen reaction series illustrates the sequence of crystallization of different minerals in magma. As the magma cools, minerals near the top of the chart solidify first, followed by those below them.

the top of the chart, such as olivine, crystallize at a temperature of about 1100°C. As the magma cools, solidification continues. The minerals near the bottom of the chart, such as quartz, remain liquid until the magma reaches a temperature of about 570°C. The minerals along the left side of the chart crystallize one after the other. This is called a discontinuous change. The single box on the right side of the chart shows a gradual change in the crystallization of feldspar from calcium-rich plagioclase to the more sodium-rich plagioclase. This is called a continuous change because as the temperature drops, there is a gradual increase in the proportion of sodium feldspar.

The work of Dr. Bowen helped geologists solve a difficult problem in geology. Nearly all magma comes from deep within Earth. At those depths, the rock is rich in mafic minerals, such as olivine and pyroxene. Geologists could not explain how the solidification of this mafic magma could form granite, which contains relatively little of the mafic minerals.

The solution to this problem involved understanding that magma can rise toward the surface as it cools slowly. When magma cools to about 1100°C, crystallization begins. The first crystals to form are the mafic minerals near the top of the chart, leaving the magma depleted of those minerals. If the magma is moving upward toward the surface and cooling as it rises, it will become more felsic because the mafic minerals crystallized first and were left behind. From this interpretation scientists inferred that the deeper within Earth an igneous rock forms, the more likely it is to be rich in mafic minerals. Similarly, light-colored, granitic rocks originate from magmas that have risen slowly. As the magmas moved upward, they lost their mafic minerals along the way, and the magma therefore became enriched in felsic minerals. This is the origin of felsic magma.

This separation of minerals can happen only if the magma rises slowly, allowing partial solidification to occur. If the magma emerges quickly, the whole range of minerals is carried to the surface. The composition of the resulting igneous rock will be a good indication of the composition of the original magma deep within Earth.

ACTIVITY 5-3 IDENTIFICATION OF IGNEOUS ROCKS

Obtain a set of igneous rocks from your teacher. Please handle them carefully and let your teacher know if any samples are badly damaged.

Use the information you learned in this chapter and the appropriate chart in the *Earth Science Reference Tables* to identify each of the igneous rocks in your set. List the name of each rock, such as granite or scoria, along with the characteristics you observed that allowed you to identify it.

 # WHAT ARE SEDIMENTARY ROCKS?

Within Earth's crust, igneous rock is the most common rock type. However, most of the surface of our planet is covered with a relatively thin layer of sedimentary rocks. Unlike igneous rocks, it is difficult to give a precise definition of sedimentary rocks. Most sedimentary rocks are made of the weathered remains of other rocks that have been eroded and later deposited as sediment in layers. Over time, the sediments are compressed by the weight of the layers above them. In addition, the layers may be cemented by mineral material left by water circulating through the sediments. The cementing material is usually silica (fine-grained quartz), clay, or calcite. All sedimentary rocks are formed at or near Earth's surface. Although this description applies only to the clastic, or fragmental, group of sedimentary rocks, these are the most common rocks of sedimentary origin. **Fossils** are any remains or impressions of prehistoric life. If fossils are present in a rock, the rock is almost certainly a sedimentary rock. The processes that create igneous and metamorphic rocks usually destroy any fossil remains.

You can recognize sedimentary rocks because they are usually composed of particles, often rounded particles, compressed and cemented into layers. Shale, the most common rock on Earth's surface, is made of particles of sediment too small to be visible without magnification. Shale breaks easily into thin layers.

 ## Clastic (Fragmental) Rocks

Clastic and **fragmental** are terms applied to the group of sedimentary rocks that are composed of the weathered remains of other rocks. These are the most common sedimentary rocks. Clastic rocks are formed by the processes of deposition, compression, and cementing of sediments. Although some sediments are deposited by wind, glaciers, or even as rock falls, most are the result of deposition in water. Seas or parts of the ocean once covered large parts of the continents. Streams and rivers carried sediments from the surrounding land into these bodies of water. The particles of sediment settled to the bottom of the water, forming fine-grained sedimentary rocks. Where deposition is rapid or currents are fast, the particles of sediment that are deposited are larger. Clastic, or fragmental, rocks are classified by the size of the sedimentary particles from which they are formed.

Information about the range of sizes of the various particles in sedimentary rocks is found in the sedimentary rock chart in the Earth Science Reference Tables and also in Figure 5-8. For example, according to this chart, sand can be defined as particles of sediment that range between 0.006 cm and 0.2 cm in size.

Conglomerate is the coarsest grained clastic rock. It is dominated by particles that are readily visible: about 0.2 cm or larger. Conglomerate sometimes looks like artificial cement with rounded pebbles embedded in it. Silica (very fine quartz), clay, and calcite (the mineral in limestone) are common cements that hold the larger particles together. There is no upper limit to the size of the particles in conglomerate, but cemented pebbles are the most common texture of conglomerate. If the particles are angular (a sign that they have not been transported very far before deposition) the term breccia (BRETCH-ee-a) is used instead of conglomerate.

Scheme for Sedimentary Rock Identification

INORGANIC LAND-DERIVED SEDIMENTARY ROCKS					
TEXTURE	GRAIN SIZE	COMPOSITION	COMMENTS	ROCK NAME	MAP SYMBOL
Clastic (fragmental)	Pebbles, cobbles, and/or boulders embedded in sand, silt, and/or clay	Mostly quartz, feldspar, and clay minerals; may contain fragments of other rocks and minerals	Rounded fragments	Conglomerate	
			Angular fragments	Breccia	
	Sand (0.2 to 0.006 cm)		Fine to coarse	Sandstone	
	Silt (0.006 to 0.0004 cm)		Very fine grain	Siltstone	
	Clay (less than 0.0004 cm)		Compact; may split easily	Shale	
CHEMICALLY AND/OR ORGANICALLY FORMED SEDIMENTARY ROCKS					
TEXTURE	GRAIN SIZE	COMPOSITION	COMMENTS	ROCK NAME	MAP SYMBOL
Crystalline	Varied	Halite	Crystals from chemical precipitates and evaporites	Rock Salt	
	Varied	Gypsum		Rock Gypsum	
	Varied	Dolomite		Dolostone	
Bioclastic	Microscopic to coarse	Calcite	Cemented shell fragments or precipitates of biologic origin	Limestone	
	Varied	Carbon	From plant remains	Coal	

Figure 5-8 This table can help you identify the ten most common sedimentary rocks.

Although sandstone is defined by a precise limit of particle sizes (0.2–0.006 cm), you can identify it by its gritty feel, like sandpaper. Shale feels smooth because the clay particles of which it is composed are so tiny they are invisible without strong magnification. (See Figure 5-9 on page 128.) Rocks made of particles larger than those in smooth shale but smaller than those in gritty sandstone are classified as siltstone.

Unlike igneous rocks, clastic sedimentary rocks are not classified by their mineral composition. Any clastic sedimentary rock can contain quartz, feldspar, or clay, all of which are the remains of the weathering of other rocks. Nor does color help to tell them apart. The mineral content of the rocks influences their color. Pure quartz is usually light in color while clay generally makes the rocks gray or black. Iron staining is common in sedimentary rocks, giving many of these rocks a red to brown color.

Figure 5-9 Shale is a clastic sedimentary rock made of clay-size particles. It usually breaks into thin layers.

Chemical Precipitates

The next group of sedimentary rocks listed in the Scheme for Sedimentary Rock Identification is not as common as clastic rocks. This group forms as the water evaporates, leaving dissolved solids behind. When evaporation occurs, the compounds left behind become too concentrated to remain in solution. Therefore, the solids deposit as mineral crystals. This process is called chemical **precipitation**. Precipitation forms rocks know as the **crystalline sedimentary rocks**. Thick layers of underground rock salt are mined in Western New York State to be used as food additives and to melt ice on roads. These deposits as well as similar salt layers found worldwide identify places where large quantities of salt water have evaporated.

Rock gypsum and dolostone form in a similar process. However the minerals gypsum and dolomite form from salt water with a different composition of dissolved mineral. Unlike clastic sedimentary rocks, which are classified by grain size, the chemical precipitates are classified by their chemical or mineral composition.

In general, rocks composed of crystals are not sedimentary. However, the chemical precipitates are the only sedimentary rocks made of intergrown crystals. Sedimentary rocks of chemical origin are formed of crystals that are relatively soft and often white in color. Furthermore, they are found among other layers of sedimentary rock. For these reasons, the chemical precipitates are seldom mistaken for igneous or metamorphic rocks, which are also composed of intergrown crystals. Sedimentary precipitates, such as rock salt, are usually composed of a single mineral.

 Organic Rocks

Bioclastic rocks are formed from material made from or by living organisms. When you find a seam (layer) of coal, you are probably looking at the remains of an ancient swamp environment where plants grew, died, accumulated layer upon layer, and were compressed and turned to stone. The green color of living plants is due to chlorophyll. But chlorophyll quickly breaks down when plants die. The carbon content of the plant remains, which gives coal its black color. Coal is mined as a fuel. In addition, it is used in making a variety of plastics and medicines. Fossil remains of extinct plants are especially common in coal in which plant impressions can be preserved in great detail. These are sometimes known as the organic group of sedimentary rocks.

A second bioclastic sedimentary rock is limestone. Limestone is usually formed by the accumulation and cementation of the hard parts of animals, such as the external skeletons of coral colonies and seashells. This organic material can be transformed into the mineral calcite, the primary mineral in limestone. Layers of limestone indicate the long-term presence of an active biological community in shallow seawater. This kind of active biological environment cannot be found in deep water because these ecosystems need sunlight. Sunlight cannot penetrate to the bottom in a deep-water environment. Figure 5-10 on page 130 illustrates some common sedimentary rocks.

Fine-grained and layered (shale)   Layers of gritty particles (sandstone)   Cemented pebbles (conglomerate)   Thick layers with fossils (limestone)

Figure 5-10 Four common types of sedimentary rocks.

ACTIVITY 5-4   **IDENTIFICATION OF SEDIMENTARY ROCKS**

Obtain a set of sedimentary rocks from your teacher. Please handle them carefully and let your teacher know if any samples are badly damaged.

Use the information you learned in this chapter and the appropriate chart in the *Earth Science Reference Tables* to identify each of the sedimentary rocks in your set. List the name of each rock, such as shale or rock salt, along with the characteristics you observed that allowed you to identify it.

 # HOW DO METAMORPHIC ROCKS FORM?

Have you ever baked cookies? You may know that the cookie dough you put into the oven has very different properties from the baked cookies that come out. In a similar way, rocks subjected to conditions of heat and pressure within Earth are changed to metamorphic rocks. In fact, the term metamorphism means changed in form. Metamorphic rocks are the only kind of rocks that begins as other rocks. Heat and pressure cause changes, which transform rocks from one rock type to another. This usually happens either deep underground where both the temperature and the pressure are high, or close to an intrusion of hot magma at or near the surface. It is important to remember, however, that if the heating melts the rock, cooling and solidification will form an igneous rock. Metamorphic rocks do not form from magma.

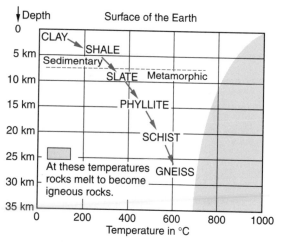

Figure 5-11 This diagram shows the transformation of clay or shale into different kinds of metamorphic rock as the rock is buried progressively deeper within Earth. The scales of depth and temperature show the conditions in which four of the most common metamorphic rocks originate.

Conditions necessary to make various kinds of metamorphic rocks, or to make magma, which will cool to become igneous rock.

The metamorphic process causes visible changes. Minerals that are stable at the surface undergo chemical changes when they are subjected to intense heat and pressure. Figure 5-11 shows a progression of rock types that occur when clay or shale is subjected to increasing heat and pressure by being buried deeper and deeper within Earth where both temperature and pressure increase.

You learned in the last section that shale is formed by the compaction of clay-sized particles under the weight of overlying layers. With deeper burial, chemical changes begin that transform shale through a series of metamorphic rocks. The clay minerals begin to change to mica, as the rock becomes harder and denser, forming the metamorphic rock slate.

At this point, a new feature of the rock starts to develop. In most slate, you can see the original bedding planes of the sedimentary rock. The parent rock, shale, usually breaks apart along these sedimentary bedding planes. But the growth of mica crystals is likely to be in a different direction than that of the original layers. Mica crystals grow in response to the forces on the rock. Even though these mineral crystals may be too small to be visible, they do affect the way that slate breaks apart. Breakage in a direction that crosses the original bedding planes signals that mineral changes happened. This alignment of mineral crystals is called **folia-**

Figure 5-12 The way schist breaks in layers is a result of the parallel growth of mica crystals. This parallel alignment of mineral crystals is called foliation.

Figure 5-13 Gneiss is a high-grade metamorphic rock that often shows banding. Banding is a separation of light and dark minerals into layers.

tion. Foliation is a feature of texture common to many metamorphic rocks.

Continued burial produces a rock called phyllite (FILL-ite). The growth of mica crystals gives phyllite a sheen like silk and may destroy the original sedimentary layering. Although the mica crystals are still too small to be visible without magnification, the shiny appearance of the rock and the even more pronounced breakage along the foliation direction indicates that mineral changes continued. Schist, the next rock to form, has mica crystals that can be seem without magnification. Continued growth of mica crystals in a single direction enhances the foliated appearance. The layering in schist is sometimes wavy. (See Figure 5-12.)

The final metamorphic product is gneiss (NICE). Not only can you see evidence of parallel crystal growth (foliation) in gneiss, but also the minerals may have separated into light- and dark-colored layers, parallel to the foliation. This is a property called **banding**. The light-colored bands are mostly quartz and feldspar. Feldspar is a new mineral that is not evident in schist. The dark bands are mostly biotite, amphibole, and pyroxene. Figure 5-13 shows banding in gneiss. Some samples of gneiss do not show banding.

Gneiss may also look like granite with abundant feldspar, but without banding. But even these samples will show a parallel alignment of mineral crystals.

The change from clay to mica and then to feldspar is not an isolated progression. Other minerals, many of them unique to metamorphic rocks, form and disappear along the way. Red garnet is a good example. Garnet can often be seen as little red pods in schist or gneiss.

Changes during the formation of a metamorphic rock destroy original structures in a rock, such as sedimentary layering and fossils. Gradually, these features are eliminated by foliation and crystal growth as the rock is subjected to more heat, more pressure, and more time.

The series of metamorphic rocks explained above presents the most common examples of the foliated metamorphic rocks. But some kinds of metamorphic rocks do not show foliation. When limestone is subjected to intense heat and pressure, calcite crystals grow and the rock changes from limestone to marble. The growth direction of calcite crystals is not affected by the force of the overlying rock or by movements of Earth's crust. This is why foliation does not occur in marble. (See Figure 5-14.) Sometimes marble shows a swirled layering, but this is probably due to differences in composition of the original limestone layers. When sandstone changes to quartzite, and conglomerate to metaconglomerate, these metamorphic products do not show parallel crystal alignment.

Figure 5-14 Marble is a metamorphic rock that does not show foliation.

Figure 5-15 Unlike sedimentary conglomerate, metaconglomerate breaks through the pebbles.

Therefore marble, quartzite, and metaconglomerate are *nonfoliated* metamorphic rocks. (See Figure 5-15.)

 ## Origins of Metamorphic Rocks

Metamorphic rocks can be separated into two groups by their origin. (See Figure 5-16.) Sometimes large-scale movements of Earth's crust cause a vast region of rock to sink into the Earth. When this occurs, a large mass of rock experiences increased heat and pressure. This process is called **regional metamorphism.** As the rock is drawn deeper into Earth, chemical changes in the minerals, crystal growth, and compaction cause the original parent rock to be metamorphosed.

If these metamorphic rocks form deep within Earth, why do we find them at the surface? In Chapter 7, you will learn that large-scale movements of Earth's crust are related to heat flow from deep within the planet. The same forces that push rock to the depths where metamorphism occurs can also push metamorphic rocks upward along with the rocks covering them to form mountain ranges. After uplift occurs, weathering and erosion wear down the mountains to expose the regional metamorphic rocks. This process may take millions of years.

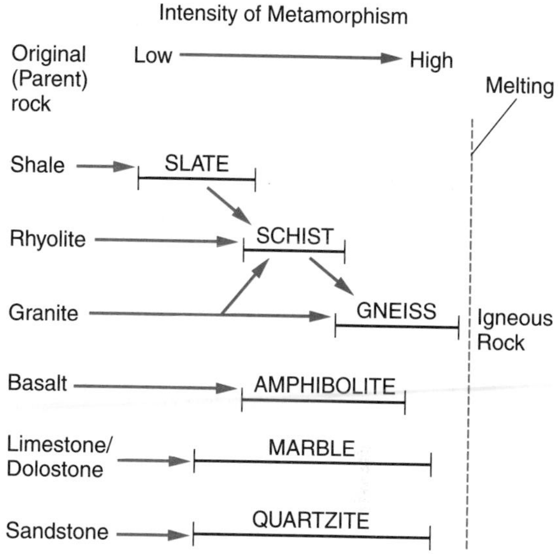

Figure 5-16 Origins of metamorphic rocks. This diagram shows the most common varieties of metamorphic rocks as well as the parent rock (sedimentary or igneous) from which they formed. The "I-beam" below the rock name indicates the range of metamorphic change in the formation of each.

The next group of metamorphic rocks occurs over a smaller area. An intrusion of hot, molten magma will change the rock with which it comes in contact. This process is called **contact metamorphism.** In this environment, rocks are not exposed to the intense pressure that is found deeper within Earth. Therefore, rocks that have undergone contact metamorphism do not show foliation. The farther you go from the heat source (intrusion), the less the parent rock has changed. In fact, it is common to find metamorphic rock grading into the original sedimentary or igneous rock within a few meters of the heat source. This change can sometimes be observed as a decrease in crystal size as you move from the intensely baked rock next to the intrusion into rock that has been altered less by the heat. Hornfels is a name often applied to contact metamorphic rock of various mineral compositions.

The *Earth Science Reference Tables* contains the Scheme for Metamorphic Rock Identification. (See Figure 5-17 on page 136.) The four rock types at the right in the top half of the chart are the four foliated metamorphic rocks. They are listed in order of increasing metamorphic changes and increasing grain size. These four rocks show the progressive metamorphism of shale that was explained earlier in this section. Note the shaded bars in the "composition" column that indicate the mineral makeup of these rocks. These bars show that minerals

Scheme for Metamorphic Rock Identification

TEXTURE		GRAIN SIZE	COMPOSITION	TYPE OF METAMORPHISM	COMMENTS	ROCK NAME	MAP SYMBOL
FOLIATED	MINERAL ALIGNMENT	Fine	MICA QUARTZ FELDSPAR AMPHIBOLE GARNET PYROXENE	Regional	Low-grade metamorphism of shale	Slate	
		Fine to medium		(Heat and pressure increase with depth)	Foliation surfaces shiny from microscopic mica crystals	Phyllite	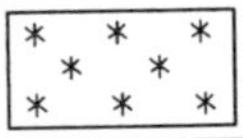
					Platy mica crystals visible from metamorphism of clay or feldspars	Schist	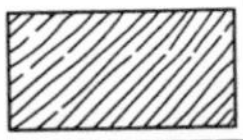
	BAND-ING	Medium to coarse			High-grade metamorphism; some mica changed to feldspar; segregated by mineral type into bands	Gneiss	
NONFOLIATED		Fine	Variable	Contact (Heat)	Various rocks changed by heat from nearby magma/lava	Hornfels	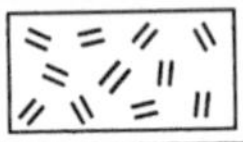
		Fine to coarse	Quartz	Regional or Contact	Metamorphism of quartz sandstone	Quartzite	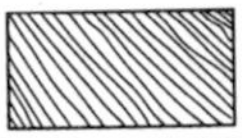
			Calcite and/or dolomite		Metamorphism of limestone or dolostone	Marble	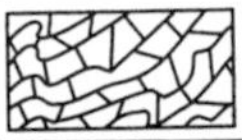
		Coarse	Various minerals in particles and matrix		Pebbles may be distorted or stretched	Metaconglomerate	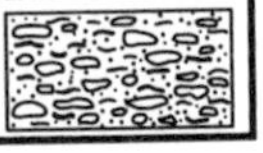

Figure 5-17 This chart can help you identify metamorphic rocks by their observable properties and how they formed. It also identifies the most important minerals in the foliated rocks.

in the mica family are found in all four foliated metamorphic rock types. Quartz, feldspar, amphibole, and garnet are not common in slate, but are common in the three foliated rock types below slate. Of the six minerals shown here, pyroxene is the best indicator of extreme conditions of heat and pressure.

Unlike the rocks in the top half of this chart, the four rocks named at the right in the bottom half of the chart do not show a progressive change. Each has a different mineral and chemical composition. These are the four most common metamorphic rocks that do not show foliation.

ACTIVITY 5-5   IDENTIFICATION OF METAMORPHIC ROCKS

Obtain a set of metamorphic rocks from your teacher. Please handle them carefully and let your teacher know if any samples are badly damaged.

Use the information you learned in this chapter and the appropriate chart in the *Earth Science Reference Tables* to identify each of the metamorphic rocks in your set. List the name of each rock, such as schist or marble, along with the characteristics you observed that helped you to identify it.

WHAT IS THE ROCK CYCLE?

There is a popular saying, "Nothing is unchanging except change itself." This saying reminds us that whether we are aware of it, everything around us is changing. Many geological changes occur so slowly that they are difficult to observe. But nearly everywhere, rocks are slowly changing as they adjust to the conditions and environment in which they are found. These changes are shown in a model of Earth environments and materials called the rock cycle.

Planet Earth receives only a very small amount matter from outer space in the form of meteorites. At the same time, a small amount of Earth's atmosphere escapes. Therefore, in terms of mass, the planet is nearly a closed system. (A closed system is one that has no exchange with the environment outside itself.) Rocks change from one form to another. As you will learn later in this course, the exchange of energy is very different. Earth receives light and heat from the sun, and Earth radiates energy into space. These are the primary exchanges that drive weather systems.

In Figure 5-18 on page 138 within the rectangles are the three major categories of rocks: sedimentary, metamorphic, and igneous. Sediments and magma are shown in rounded figures because although they are important substances within the rock cycle, they are not actually kinds of rock. The lines and arrows show how materials within the rock cycle can change. The terms printed along these lines tell you what changes are represented and the order in which they occur.

For example, magma can change to a different substance in only one way: it forms igneous rock by the process of solidifica-

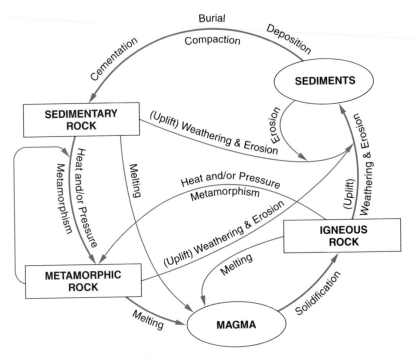

Figure 5-18 The rock cycle.

tion, or crystallization. However, igneous rock can change to another substance by any of three paths. If it is heated and melts, it can go back to molten magma. If the igneous rock is subjected to intense heating and possibly pressure, but remains below its melting temperature, the process of metamorphism will transform the igneous rock into a metamorphic rock. In the third possible path, the igneous rock is exposed to the atmosphere, probably by being pushed up to the surface, where air, water, and weather break it down and carry it away as sediments. The term "uplift" appears in parenthesis because uplift to expose the rock at the surface is a likely event, but it is not essential. In the case of a lava flow onto the surface, no uplift would be needed to expose the rock to weathering and erosion.

The rock cycle diagram illustrates that nearly all rocks are made from the remains of other rocks. In the rock cycle, as you follow the arrows that show changes, notice that each begins and ends with Earth materials. None of the arrows comes in from outside the diagram. None of them take Earth materials out of the system. It is a closed system.

There is one group of sedimentary rocks that does not fit into this rock cycle. The organic sedimentary rocks, such as limestone and coal, are formed from the remains and plants and animals not other rocks.

Some rocks show a complex origin. You may recall that conglomerate is made from pebbles that are held together by a cementing material, such as silica (very fine quartz), clay, or calcite. If the conglomerate contains pebbles of gneiss, granite, and sandstone, each component of the conglomerate shows a different process of rock formation found in the rock cycle.

TERMS TO KNOW

banding
bioclastic
classification
clastic
contact metamorphism
crystalline sedimentary
  rocks
extrusion
felsic
foliation

fossil
fragmental
igneous rock
intrusion
lava
mafic
magma
metamorphic
origin

plutonic
precipitation
regional metamorphic
  rock
rock
sedimentary rock
texture
vesicular
volcanic

CHAPTER REVIEW QUESTIONS

1. What feature in igneous rocks is the best indication of how rapidly the magma crystallized?

    (1) color
    (2) density

    (3) grain size
    (4) mineral composition

2. Which igneous rock is likely to contain about 50% pyroxene and about 40% plagioclase feldspar?

    (1) granite
    (2) peridotite

    (3) diorite
    (4) basalt

3. What type of igneous rock is likely to contain both amphibole and potassium feldspar?

    (1) diorite
    (2) granite

    (3) gabbro
    (4) andesite

4. If rock fragments approximately 0.01 cm in diameter were cemented into a fragmental sedimentary rock, what kind of rock would it be?

    (1) sandstone
    (2) conglomerate

    (3) limestone
    (4) granite

5. Which kind of rock is most likely to contain a trilobite fossil?

    (1) sedimentary
    (2) igneous

    (3) metamorphic
    (4) plutonic

6. Which sequence of events would make a sedimentary rock?

    (1) uplift, weathering, and erosion
    (2) melting, solidification, and uplift
    (3) deposition, burial, and cementation
    (4) crystal growth, compaction, and melting

7. Which clastic sedimentary rock is made of the smallest grains?

    (1) conglomerate
    (2) sandstone

    (3) siltsone
    (4) shale

8. Base your answer on the pictures of four rocks. Each circle shows a magnified view of a rock. Which rock is metamorphic and shows evidence of foliation?

Rock 1	Rock 2	Rock 3	Rock 4
Bands of coarse intergrown crystals of various sizes	Particles of 0.01-cm to 1.0-cm size cemented together	Intergrown crystals less than 0.1 cm in size	Intergrown crystals mostly 2.0 cm in size

    (1) 1
    (2) 2

    (3) 3
    (4) 4

9. Which sedimentary rock is most likely to change to slate during regional metamorphism?

(1) breccia   (3) dolostone
(2) conglomerate   (4) shale

10. Which sequence of change in rock type occurs as shale is subjected to increasing heat and pressure?

(1) shale → schist → phyllite → slate → gneiss
(2) shale → slate → phyllite → schist → gneiss
(3) shale → gneiss → phyllite → slate → schist
(4) shale → gneiss → phyllite → schist → slate

11. What kind of rock results from the contact metamorphism of limestone?

(1) schist   (3) granite
(2) dolostone   (4) marble

12. Which two rocks are composed primarily of a mineral that bubbles with acid?

(1) limestone and marble   (3) sandstone and quartzite
(2) granite and dolostone   (4) slate and conglomerate

13. The diagram below shows four rock samples.

Sample A   Sample B   Sample C   Sample D

Which sample best shows the physical properties normally associated with regional metamorphism?

(1) A   (3) C
(2) B   (4) D

14. Which of the following changes would *not* occur during metamorphism?

(1) formation of new minerals   (3) crystal growth
(2) increasing density   (4) melting

15. Which of these locations is a good place to find rocks changed by contact metamorphism?

(1) at the core of a mountain range   (3) near an intrusion of magma
(2) in a sandy desert   (4) at the center of a lava flow

Open-Ended Questions

16. State one observable characteristic, other than crystal size, that makes granite different from basalt.

17. State two processes that must occur to change sediments into a clastic sedimentary rock.

Base your answers to questions 18 and 19 on the rock cycle diagram below

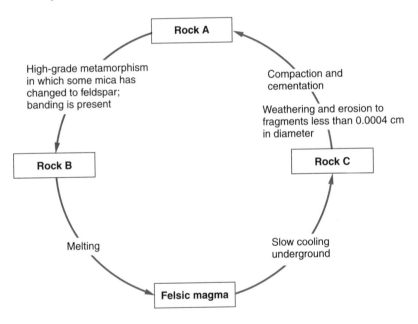

18. State the specific names of rocks A, B, and C in the diagram. Do *not* write the terms "sedimentary," "igneous," or "metamorphic."

19. State one condition or process that would cause the high-grade metamorphism of rock A.

20. What mineral is common in schist and gneiss, but not common in slate and phyllite?

Chapter 6

Managing Natural Resources

WHAT IS A NATURAL RESOURCE?

Every aspect of your life depends on the materials obtained from planet Earth. The food you eat, the products you buy, and the clothes you wear are all the result of technology and Earth's natural resources. A **natural resource** is any material from the environment that is used to support people's lives and lifestyles. Consider a visit to the supermarket. One of your first actions is to take a shopping cart. The steel frame of the cart is made from iron and other metals obtained from mines. The nonmetal parts of the cart are manufactured from plastics, which come from petroleum obtained from oil wells. The energy to open the doors as well as to heat and cool the building is largely from electricity generated by the power plants. Those power plants burn coal, oil, natural gas, or they extract energy from nuclear fuels, which also come from the ground. As you approach the fresh fruits and vegetables, consider the importance of soil, water, and air in their growth.

The word *resource* comes from an old French word that means "to arise in a new form." It is the responsibility of all people to make the best use of our natural resources while conserving them for future generations and protecting the environment. At every step, there are important decisions and compromises to be made. How much will be used now and how much will be conserved? How can these resources be extracted while maintaining a clean and beautiful environment? Not everyone will agree with the decisions that are made. But people can all make better decisions if they understand the issues involved in using and conserving resources. Earth's natural resources can be divided into two groups; nonrenewable and renewable.

WHAT ARE NONRENEWABLE RESOURCES?

Nonrenewable resources exist in a fixed amount, or if they are formed in nature, the rate of formation is so slow that the use of these resources will decrease their availability. Rocks, minerals, and fossil fuels are the primary nonrenewable resources. In nearly every case, the use of these resources is increasing while the natural supply (reserves) are decreasing. This situation forces us to look for lower quality reserves to meet our growing demands.

ACTIVITY 6-1   ESTABLISHING A LOCAL NATIONAL PARK

Imagine that the National Park Service has chosen your community as the location for a new national park. It is your task to choose a local feature that might be of interest to the public either for preservation or for recreational development. Identify the advantages and disadvantages that would be involved in this project. Take into account the points of view of different parts of the community including landowners, small business owners, environmentalists, local government personnel, and students.

Metal Ores

Many of the minerals and rocks you studied in Chapters 4 and 5 are resources that are important to the economy. Rocks that are used to supply natural resources are called **ores**. The resources must be extracted, or separated, from their ores. Metallic elements that are economically important are obtained from minerals. Some elements, such as gold, silver, and copper, occur as native metals, that is, uncombined with other elements.

COPPER   The element copper is used extensively in electrical wiring, plumbing, coins, and a wide variety of other applications. While copper is not as good a conductor of electricity and heat as gold is, it is much cheaper than gold because it is more plentiful. Copper can be drawn into wire (ductility) and pressed into thin sheets (malleability). Copper is mixed with other metals such as zinc and tin to make alloys (mixtures of metals). These alloys, brass and bronze, are relatively hard, resist being deformed, and resist weathering. Like most other geological resources, copper is extracted from rocks that have unusually high concentrations of copper. In some places like northern Michigan, small amounts of native copper are found. But most copper is obtained from the mineral chalcopyrite, which was deposited by hot water circulating deep underground. In the early years of mining, some locations yielded ores containing as much as 10 percent copper. However, as the most concentrated ores were depleted, new technologies were developed along with large-scale mining operations. Now, refiners can profitably extract copper from ores that contain less than 1 percent copper.

The earliest copper mines were tunnels that followed rich veins of copper ore into the rocks. These mines were dark and dangerous places in which to work. Most copper today is produced from open pit mines where giant machines scoop up tons of copper ore with each bite. The ore is transported to refineries where it concentrated and purified. A variety of mechanical, chemical, and electrical process extract the copper

Figure 6-1 Gold is a mineral element valued for its physical and chemical properties. These gold nuggets were found in Canada's Yukon region.

as well as smaller amounts of other valuable substances, such as sulfur, gold, and silver, that can increase the profitability of the process.

GOLD Gold is sometimes found in sand and gravel deposits in streams and along shorelines. (See Figure 6-1.) Due to its density, gold that is transported by running water settles in places where the stream water slows down. The pieces of gold get caught in cracks in bedrock. (Bedrock is the solid rock that can be found everywhere under the soil and sediments.) Gold is also recovered from solid rock where it has been brought in by hot water. Because of its mechanical and electrical properties, gold is valuable. Gold is also an attractive substance that does not corrode. Therefore, gold is often used in jewelry. Of the estimated 100,000 tons of gold that have ever been produced, nearly all of it is still in circulation.

IRON The refining of iron ore began thousands of years ago. The process most likely occurred by accident when a rock with a high concentration of iron was placed in a fire pit. Today, iron is the principal metal of construction and technology. It is used in making frames for buildings, for automobiles, and even in eating utensils. Iron is found worldwide, and is second only to aluminum in abundance in Earth's crust. Iron ore is easy to find and is inexpensive to extract. Combining iron with small amounts of other metals makes

Figure 6-2 Banded iron ore is not renewable because it formed at a time when Earth's atmosphere contained little or no oxygen.

steel. Steel is harder and more resistant to breakage and corrosion than is iron. The most important iron ore deposits are the banded iron formations that settled out of ancient seas early in Earth's history. (See Figure 6-2.) At that time, the atmosphere contained very little oxygen. While there are great quantities of iron ores, the fact that almost no new ores are being formed makes iron a nonrenewable resource.

 Using Mineral Resources

One of the important jobs of geologists is to locate ore deposits and estimate the amounts of metal that can still be extracted. Aluminum and iron are so abundant in Earth's crust that geologists are not concerned about running out of them. However, the best reserves of copper and gold in the United States have already been used up. For this reason, the use of imported gold and copper is increasing. These rare metals are often recycled. About 25 percent of the copper produced in the United States is the product of recycling. The amount of a mineral resource that is recycled depends on its cost. The more it is worth, the more is recycled. Because of gold's high price, people do not toss away a gold object the way they dis-

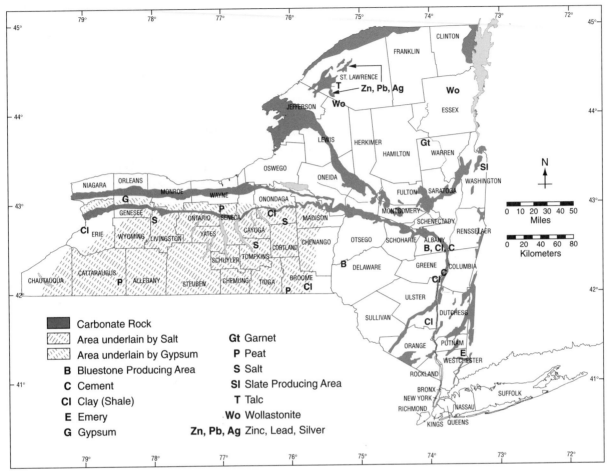

Figure 6-3 This map shows the locations of most important mineral resources found in New York State.

card aluminum soda cans and iron objects. Figure 6-3 shows locations of the most important mineral resources in New York State.

ACTIVITY 6-2   ADOPT A RESOURCE

Select a specific nonrenewable resource. Then ask permission from your teacher to prepare a report on the resource. Your report should include where the resource is found, how it is refined or changed for use, and what the resource is used for.

Fossil Fuels

Coal, petroleum, and natural gas are our principal sources of energy for the production of electricity, heating, and transportation. The energy stored in these substances today represents sunlight absorbed by prehistoric plants. Because of their organic origin, coal, petroleum, and natural gas are not minerals, although they are found in sedimentary rock. They are certainly important natural resources found in Earth, and they are vital to our economy.

PETROLEUM When Iraq invaded Kuwait in 1990, the response from North America and Europe was swift and massive. It was almost as if our own country had been invaded. The key to this response is the importance of Middle Eastern oil to our economy. While the United States is the world's largest user of petroleum, it has only about 3 percent of the world's oil reserves. A serious interruption of the flow of Middle Eastern oil would have had a devastating effect on our economy.

Petroleum is thought to be the remains of microscopic organisms that lived in the oceans millions of years ago. Sinking to the bottom, the remains escaped complete decay in the cold, dark environment of the ocean bottom. The deposition of more layers of sediments trapped the organic remains. Over time, the organic remains changed to a complex low-density liquid and natural gas. These substances moved upward until an overlying layer of fine-grained rock, such as shale, trapped them. If an oil company has enough skill and luck to drill into a region where oil and gas are present, it can pump the oil and gas out of the ground. Petroleum products are so valuable on world markets that petroleum is sometimes called "black gold."

COAL Coal is formed from the remains of plants that once grew in great coastal swamps. A lack of dissolved oxygen and the presence of hydrogen sulfide in the water prevented complete decay of the plant remains. Burial and compression changed the decaying plant material to coal. While the United States has large deposits of coal, it is used mostly in large

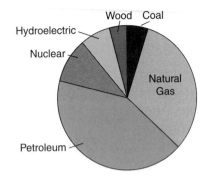
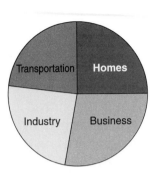

Figure 6-4 The fossil fuels, coal, oil, and natural gas supply most of the energy used in New York State.

power plants to generate electricity. In these power plants, large quantities coal are handled efficiently and air pollution controls are used to reduce air pollution. The direct use of coal for home heating and cooking have decreased because oil, gas, and electricity are easier to use and they also cause less air pollution.

The fossil fuels are considered nonrenewable resources because we are using them more quickly than natural processes can replace them. Some scientists estimate that in a few hundred years Earth could run out of fossil fuels that took 500 million years to form. The two pie graphs in Figure 6-4 illustrate the major sources and uses of energy in New York State.

 Construction Materials

In terms of total mass, more Earth resources are used in construction than in any other application. The material used generally depends on the needs of the builder and the materials available in the area. Sand, gravel, and crushed stone are spread as a base under paved roads as a road base. These materials are used because they do not compress under heavy traffic, and they allow water to drain through them. A proper base layer helps the roads last longer than roads built directly on the ground.

Concrete is made from a mixture of limestone (or dolostone) and clay that has been baked to drive off the water and carbon dioxide content of these materials. When water is added to the concrete, it "sets" to form a resistant building material. Although it is more expensive than brick, concrete, or steel, cut stone is sometimes used for the outer facing of buildings.

WHAT ARE RENEWABLE RESOURCES?

Renewable resources are those resources that can be replaced by natural processes at a rate that is approximately equal to the rate at which they are used. For example, sunlight is our planet's principal source of energy. Solar heat collectors make use of this energy. But, as fast as it is used, more solar energy arrives. Although renewable resources may never run out, our use of these resources can be limited by how quickly they are generated. If your community uses a nearby river for its municipal water supply, the amount of water used cannot be greater than the total amount of water flowing in the river.

 **Wood**

Before people made extensive use of fossil fuels, wood was generally the fuel of choice. In most areas, wood could be cut locally. People burned wood to heat homes, prepare hot water, and cook food. Wood is a renewable resource. Once the trees are cut, new trees can grow back in as little as 10 years, or perhaps as long as a century. As long as usage and replacement are in balance, the supply would be unlimited.

 Soil

For now, soil will be considered as the loose material at Earth's surface that is capable of sustaining plant growth. Later in the book you will move to a definition that is related to how soil is formed. Biological productivity is an essential feature of soil as a resource. If you look at a food chain, it begins with the organisms called producers. Plants, as producers, are the only organisms capable of living in an environment without needing other organisms for food. The producers use the substances they obtain from soil, along with air, water, and sunlight to create living tissue. All animals (including humans) depend on producers to transform the resources of the physical environment into food.

In some places, the best soil has been carried away by running water or by dust storms. This is often the result of careless agricultural practices. A farmer may plow the land up and down the slopes of hills, which allows runoff to gain speed and carry away large quantities of soil. To avoid this, farmers can plow the land along the contours of the hills, making each ridge of soil function as a dam to hold back the water and its load of sediments. Another careless procedure is leaving land without any plant cover. Without plants and their roots to slow runoff and hold soil in place, soil can be quickly washed or blown away.

The processes of weathering, infiltration of groundwater, and biological activity can restore soil, but the process could easily take hundreds of years. Although some people consider soil to be a renewable resource, the best practice is to protect soil through conservation.

 **Water**

Water is another essential natural resource. All living things need water or they depend upon other organisms that require water. Water is used by plants during photosynthesis. People also use it for drinking, transportation, waste disposal, industrial development, and as an ecological habitat for our food sources. Salt water in the oceans and in several landlocked bodies of water, such as the Great Salt Lake in Utah, makes up 97 percent of Earth's water. However, this water is unfit for many of needs such as drinking and agriculture. Glaciers and the polar ice caps hold another 2 percent

Figure 6-5 This canal diverts a major part of the Colorado River into the Imperial Valley of California. Normally, none of the river's water reaches the Pacific Ocean.

of Earth's water. Of the remaining 1 percent, most is in the ground where it is not as easy to tap as are sources at the surface. Most of our water needs are fulfilled by the 0.008 percent of the total that is found as surface water in streams and lakes.

The Colorado River flows from the Rocky Mountains southwest thought the desert toward the Pacific Ocean. Five states depend upon Colorado River water. Through a series of governmental agreements the states of Colorado, Utah, Arizona, Nevada, and California as well as Mexico have tapped into the river. Their residential, agricultural, and industrial use of the water consumes the total volume of water flowing in the river. Unless there is a flood, none of the water from the Colorado River reaches the ocean. (See Figure 6-5.) This area is the fastest growing region in the country. How will this area find more water to support its growth?

Water is usually considered a renewable resource. As water is drawn from surface sources, such as streams and lakes, more water flows in to replace it. There is an assumption here that once water is used, the water cycle replenishes it. It is not always that simple.

In the 1920s, the federal government encouraged people to move into the Great Plains region of the United States by offering them loans and inexpensive land. Rich soil and good precipitation made this region seem a natural location for agricultural development. In the following decade, there was

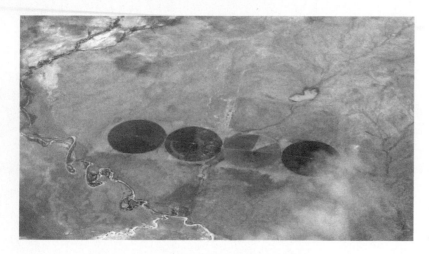

Figure 6-6 These circles are irrigated fields in western Texas. Water is supplied by a well at the center of each circle.

a series of years of reduced rainfall. Meanwhile the new farm owners had plowed the land and it was exposed to wind erosion. The dust storms of the 1930s blew away much of the most productive soil. Some farmers drilled wells to bring water to the surface. That part of the United States is over a natural groundwater reservoir called the Ogallala aquifer. While farming continues today, it is often depends on irrigation water drawn from the aquifer. (See Figure 6-6.) Each year the water level in the aquifer goes down because more water is withdrawn than nature can restore.

 ## HOW CAN WE CONSERVE RESOURCES?

Conservation is the careful use, protection, and restoration of our natural resources. To practice conservation, people need to estimate their resources and project their needs. Only then can they decide what needs to be done. For example, although the use of aluminum is increasing, Earth's crust contains so much aluminum that there seems little danger of running short. However, the future for gasoline and other petroleum products is not as bright. Some people estimate that petroleum will start to run short in just a few decades. There is a popular saying "Necessity is the mother of invention."

Necessity to replace petroleum fuels may be getting close. But it is not clear what "inventions" will carry us through this crisis. Conservation generally involves one or more of the three "R's" of ecology: reduce, recycle, and replace.

 ### Reduction

If you could see a soda can that was made 50 years ago, you might be surprised at how heavy it would be. Changes in the manufacturing process and the shape of soda cans have allowed beverage companies to use much less metal, generally aluminum, than ever before. People who carpool and/or drive small cars not only save money, but they also contribute to extending our limited and petroleum reserves.

 ### Recycling

A significant fraction of the copper we use has been recovered from old buildings, automobiles, and electrical parts. If you grew tired of a piece of gold jewelry, you would probably not put it into the trash. More likely, you would sell it for its value in gold. While aluminum is very abundant on our planet, it is costly to extract and purify from its ore. This has contributed to programs to recycle aluminum cans. (Recycling programs also help the environment by making it less likely that people will discard used soda cans along highways.) Each time we use recycled metals we delay the inevitable time when these resources will run out.

 ### Replacement

Replacement is often the long-term solution to dwindling resources. In the past, most people heated their homes and cooked their food by burning wood. Wood is a renewable resource. However, as the human population grew and people moved into cities, nearby wood sources could not keep up with the growing need for fuel. Coal and then petroleum fuels replaced wood as the fuel of choice. In many applications, plastics are replacing wood and leather. Necessity will certainly continue to bring about new solutions to our problems.

ACTIVITY 6-3	WATER USE IN THE HOME

Measure or estimate the amount of water that your family uses at home on a typical day. To do this you will need to identify all the household devices and activities that use water. You must also find ways to determine how much water these uses require. (You may find useful information in reference sources, on the Internet, on your water bill, or from your water meter.) After you have determined your family's daily water usage, suggest the best ways to reduce your water consumption.

WHAT ARE THE EFFECTS OF ENVIRONMENTAL POLLUTION?

We can define **pollution** as a sufficient quantity of any material or form of energy in the environment that harms humans or the plants and animals upon which they depend. Pollution can be classified according to the part of the environment has been affected: ground, water, and/or air.

Ground Pollution

The city of Niagara Falls, New York, became an industrial area because of the availability of inexpensive hydroelectric power. Chemical plants were built in the area to take advantage of this local resource. Unfortunately, the local government did not monitor or regulate the industry and its disposal of chemical waste materials. Drums of toxic chemicals were buried when a local waterway called Love Canal was filled in. Eventually the land was given to the city of

Niagara Falls. The city built low-cost housing and an elementary school on the land. It was only when toxic liquids began to seep out of the ground that the residents realized there was a problem. When the toxic chemicals were identified and the health dangers became apparent, the residents were shocked and angry. They did not want their families, especially the children, to be exposed to these chemical waste products. As the tragedy became news, the value of homes in the area decreased dramatically. People could not sell their homes and they could not afford to buy homes in another area. The homeowners felt let down by their local government, which should have informed them about the chemical danger at Love Canal.

Another example of ground pollution can be found in agricultural areas where insecticides have been sprayed to protect crops from insect damage. Over the years, residues of these chemicals may remain in the soil and make the soil unfit for other crops. Furthermore, wind-blown dust can be a hazard to farm workers and it can carry the chemicals to nearby residential areas. Children and senior citizens are especially vulnerable to diseases brought on by exposure to dangerous substances.

 Water Pollution

Near the middle of the twentieth century, the growth of the economy and demand for electrical power led General Electric (GE) to seek a new substance to cool electrical transformers. It is much easier to send electrical power over long distances if the voltage is stepped up to levels that would be dangerous in homes. GE needed to build transformers to change the high-voltage current sent over transmission lines into ordinary household current. Transformers can overheat if they are not placed in a liquid that will carry away excess heat. Water is a good absorber of heat, but it can evaporate or boil away. GE used liquid chemicals called polychlorinated biphenyls (PCBs). PCBs are good absorbers of heat and are less likely to boil away than is water. Because PCBs are chemically stable, the company felt that there was little danger if these substances leaked into the environment. Two

Figure 6-7 The Hudson River has been contaminated by PCBs that washed into the river north of Albany.

manufacturing plants along the Hudson River north of Albany, New York, were using the chemicals. (See Figure 6-7.) Neither GE nor the state was aware that PCB pollution would become a serious environmental issue.

After PCBs had been used for several decades, it was discovered that they could lead to serious health problems if people consumed them. Scientists also found that as fish in the river absorbed the PCBs by eating plants, or by eating other fish, the chemicals built up in their bodies. It became apparent that PCBs were not as safe as they at first appeared to be. Meanwhile, dangerous amounts of PCBs had washed into the Hudson River at the two manufacturing plants. The government warned people about the danger of eating fish from the river and looked for a long-term solution. GE and environmental groups have clashed on what to do with the contamination. Environmentalists (and the courts) have held that GE should remove the PCBs from the river's ecosystem by dredging the most contaminated mud and disposing of it on land. GE wants to leave the sediments where they are. The company contends that the PCBs will eventually wash out of the river.

 Air Pollution

Air pollution has been a concern in most urban areas. Exhaust gases from automobiles along with the airborne discharge from homes, businesses, and industries can lead to elevated levels of ozone and oxides of nitrogen. Serious air pollution events have occurred in many cities. The response to rising air pollution is the use of cleaner fuels and pollution control devices on motor vehicles and electrical power plants that burn fossil fuels.

Some air pollution issues have been dealt with quickly while other forms of pollution are more difficult to address. It was discovered recently that chlorofluorocarbon gases (CFCs) given off by some kinds of spray cans and air conditioners were causing Earth's protective ozone layer to weaken. (Ozone is needed in the upper atmosphere because it helps protect Earth from harmful shortwave solar energy.) Government and businesses have found other chemicals to use in air conditioning systems and spray cans that do not break down ozone in the upper atmosphere.

Our planet faces more difficult issues. Through the use of fossil fuels humans have increased the carbon dioxide content of Earth's atmosphere by about 25 percent. Most of that increase has occurred in the last 50 years. The concern is that carbon dioxide absorbs heat energy that might otherwise escape into space. The increase in carbon dioxide concentration in the air is likely to cause **global warming**, an increase the average temperature on Earth. While some locations might benefit from a warmer climate, productive farmland could be changed into desert. Another consequence could be the melting of polar ice caps and a rise in sea level, which would drown coastal cities. Unfortunately, our society depends on fossil fuels for heating, electrical power, and transportation. At this time, there is no clear alternative to the use of fossil fuels.

TERMS TO KNOW

conservation	**nonrenewable resources**	**pollution**
global warming	**ore**	**renewable resource**
natural resource		

CHAPTER REVIEW QUESTIONS

1. Which of the following is a natural resource?

   (1) plastic
   (2) gasoline

   (3) water
   (4) steel

2. Which group of substances comes from mineral ores?

  (1) plastics   (3) metals
  (2) fossil fuels   (4) soil

3. Which metal is most common in Earth's crust?

  (1) iron   (3) aluminum
  (2) copper   (4) magnesium

4. Which of the following is a fossil fuel?

  (1) wood   (3) uranium
  (2) coal   (4) solar energy

5. Which is the best way to prevent soil erosion on a farm?

  (1) Plow fields along level contour lines.
  (2) Leave plowed fields barren every second year.
  (3) Use irrigation water throughout the year.
  (4) Cut down any forested areas near the farm.

6. Freshwater is usually considered a renewable resource. In what way is water like a nonrenewable resource?

  (1) Earth has a limited supply of water.
  (2) Water circulates through the water cycle.
  (3) Most of Earth's freshwater is in glacial ice.
  (4) Water can exist as a solid, liquid, or gas.

7. Which of the following is usually considered a renewable resource?

  (1) coal   (3) petroleum
  (2) wind   (4) aluminum

8. What is Earth's most important renewable energy source?

  (1) coal   (3) sunlight
  (2) natural gas   (4) ozone

9. Conservation includes which practice?

  (1) Making estimates of the availability of natural resources
  (2) Giving tax benefits to those who use natural resources

(3) Importing ever greater amounts of crude oil

(4) Eliminating recycling programs

10. What does this quote mean? "We do not inherit the environment from our parents. We borrow it from our children."

(1) We are running out of some resources.

(2) The cost of natural resources always increases.

(3) Conservation is important to future generations.

(4) Many natural resources are renewable.

11. Where is pollution of the environment usually most serious?

(1) the tops of high mountains

(2) cities and industrial areas

(3) state and national parks

(4) the deepest parts of the oceans

12. What local food in the New York City area is most likely to be contaminated with PCBs?

(1) meat

(2) vegetables

(3) fruits

(4) fish

13. Why did manufacturers stop using chlorofluorocarbon gases (CFCs) in air conditioners and spray cans?

(1) CFCs contribute to global warming.

(2) CFCs weaken the ozone layer.

(3) CFCs are toxic to plants and animals.

(4) CFCs cause ground pollution.

14. Which gas in the atmosphere tends to absorb heat radiation?

(1) oxygen

(2) nitrogen

(3) argon

(4) carbon dioxide

15. What gas is a pollution problem when we have too much at ground level, but causes other problems when there isn't enough of it high in the atmosphere?

(1) carbon dioxide

(2) oxygen

(3) ozone

(4) nitrogen

Open-Ended Questions

16. If gold is a better conductor than copper, why do we use more copper than gold in electrical devices?

17. The fossil fuels are considered geological resources, but not mineral resources. Why?

18. Why is wood considered to be a renewable resource?

19. What is the danger of using water from an aquifer faster than it can be replaced?

20. How might adding carbon dioxide to the atmosphere cause flooding in some areas?

Chapter 7

Earthquakes and Earth's Interior

 ## WHAT CAUSES EARTHQUAKES?

Have you ever felt an earthquake? An **earthquake**, or seismic event, is a sudden movement of Earth's crust that releases energy. The mild motion of a small seismic event can be exciting. But what if you felt that your life was in danger? Large earthquakes can be very frightening. Early in the morning on Friday, March 27, 1964, one of the largest seismic events on record struck coastal Alaska. For the residents of Anchorage, it started as a gentle rolling motion. Then, for the next 5 minutes the ground shook and heaved. One side of the street in the main shopping district moved downhill 3 m (about 10 ft), causing large cracks to open. Windows broke and facing materials fell from buildings onto the sidewalks. A large housing development slid down toward the bay; some homes fell into the water. The nearby port of Valdez was badly damaged by the shaking, and completely destroyed by a series of gigantic waves that left a large ship sitting on a city street. In this chapter, you will learn why earthquakes occur, how they are measured, and what they tell us about Earth's interior.

ACTIVITY 7-1	ADOPT AN EARTHQUAKE

Choose an important, historic earthquake that you would like to report about. Before you begin your report, let your teacher know what earthquake you chose. Your teacher may require that each student select a different seismic event. After your choice is approved, use library resources or the Internet to gather information about that event. Your report must include a bibliography of your sources of information. Your teacher will specify what information will be needed for each entry in the bibliography.

 Earth's Internal Energy

You may recall from physical science that energy always flows away from the hottest places and toward cooler locations. Our planet is like a gigantic heat engine. Lava erupting from volcanoes or deep gold mines with walls that are hot enough to cause burns, make us aware of energy stored in our planet. In fact, at Earth's center the temperature is greater than 6000°C. This is the energy that causes earthquakes as well as slow deformation of the crust.

There are two principal sources of Earth's internal heat energy. Some of the energy is left over from the formation of the planet about 4.5 billion years ago. To that heat is added the energy of radioactive decay. Some large atoms, such as uranium, are unstable. They spontaneously break down into more stable substances, for example, lead.

Heat energy can travel in three ways. Some heat travels through Earth and to the surface by conduction. **Conduction** occurs when heated molecules pass their vibrational energy to nearby molecules by direct contact. Most metals are good conductors of heat energy. This is one of the ways that heat can travel through the Earth toward the surface. **Radiation** is the movement of electromagnetic energy through space. The sun radiates light and other forms of electromagnetic energy into space. Earth absorbs some of that energy and changes it to heat. Earth radiates heat energy into the atomsphere. The third form of heat flow is convection. In **convection**, heated matter circulates from one place to another due to differences in temperature and density. On Earth's surface the wind circulates energy by convection. Within Earth, heated material rises toward the surface in some places, and in other places, cool material sinks into Earth. Horizontal currents complete the circulation by connecting the areas of vertical heat flow. This pattern of circulation is known as a **convection cell**. Although convection motion within Earth occurs at only a few centimeters per year, over time heat flow by convection is capable of causing catastrophic changes such as earthquakes.

Stress and Strain

Students often consider Earth's mantle and crust to be solid. However, forces applied to solid rock at elevated temperatures and over long periods of time can bend, or deform, rocks. This slow bending builds up forces on rocks near the surface. This kind of force is called **stress**. Even over long periods of time, surface rocks do not deform as readily as the rocks inside Earth that are under intense heat and pressure. Surface rocks respond to stress by elastic bending called strain. Therefore, the stress on rocks near the surface builds until the rocks break, and an earthquake occurs. What you feel as an earthquake is the sudden movement of the ground, which releases stored energy. (See Figure 7-1.)

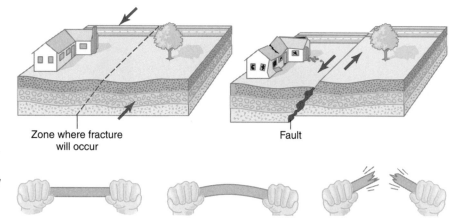

Figure 7-1 When stress exceeds the strength of Earth's crust, rocks suddenly break along a zone of weakness and they move. Sudden shifting along a fault releases energy that radiates as an earthquake.

Zone where fracture will occur

Fault

Earthquakes do not occur randomly. Cracks in solid rock are very common. Cracks in Earth's crust along which movement occurs are known as **faults**. (See Figure 7-2.) Faults are usually zones of weakness. Therefore, the places most likely to break are those that have broken in the past. Perhaps the most active fault in the United States is the San Andreas Fault, which extends from the Gulf of California past Los Angeles and through the San Francisco area. (See Figure 7-3.) The land on the western side of the San Andreas Fault is moving northward at an average rate of several centimeters per year. Due to the resistance of Earth's crust, this motion is not uniform. Stress builds until it is greater than the friction that holds the rocks in place. When this occurs, the Pacific side jolts northward as much as several meters in a single earthquake. Large earthquakes occur at various places along the San Andreas Fault roughly once every 10 to 20 years.

The underground place where the rock begins to separate is known as the **focus** (plural, foci) of the earthquake. Directly above the focus, at the Earth's surface, is the **epicenter**. (See Figure 7-4.) An earthquake is felt most strongly at or near the epicenter. The intensity of the shaking is also

Figure 7-2 Earthquakes occur when the ground breaks and moves along geologic faults such this small fault in Utah.

Figure 7-3 The San Andreas Fault follows this valley south of San Francisco.

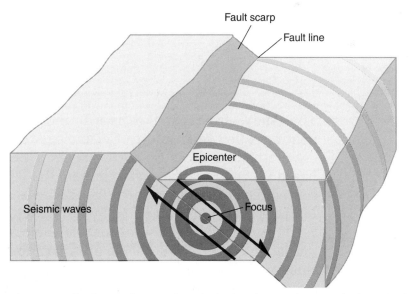

Figure 7-4 The focus of an earthquake is the place underground where the break begins. Directly above the focus is the epicenter where seismic energy first reaches the surface. Energy from the rupture travels outward in all directions as energy waves.

influenced by the nature of the ground. Loose sediments usually experience more violent movement than nearby solid bedrock.

HOW ARE EARTHQUAKES MEASURED?

Some seismic events are stronger and do more damage than others. The most violent earthquake is modern times was probably a 1960 event centered in the Pacific Ocean off the coast of Chile. But this earthquake did not have the human impact of two earthquakes that occurred in China, one in 1556 and the other in 1976. Each Chinese earthquake had its epicenter near a major city. Many buildings in these cities were made of building blocks without a wooden or steel frame to hold the bricks together. This kind of construction works well for vertical forces. But it does not do well when there is horizontal motion. During these earthquakes, buildings were destroyed and hundreds of thousands of people lost their lives.

Earthquake Scales

The study of earthquakes is a branch of geology and engineering called **seismology**. Seismologists use two very different methods to measure earthquakes. The **Mercalli scale** is based on reports of people who felt the earthquake and observed the damage it caused. The 12 levels of intensity on the Mercalli scale are designated by Roman numerals. (See Table 7-1.) Intensity I is felt by only a few people who are at rest, but awake and alert. Most people are unaware of an earthquake this small. Everyone feels intensity VI, but there is little damage. An earthquake of intensity XII causes great damage to all buildings, and the shaking is so violent that large objects are thrown into the air. The major advantage of the Mercalli intensity scale is that no instruments are required and information can be gathered after the earthquake. Furthermore, everyone who felt the event or observed the damage becomes a virtual measuring instrument.

TABLE 7-1. An Abbreviated Mercalli Scale

Intensity	Effects
I–II	Almost unnoticeable.
III–IV	Vibrations are noticeable; unstable objects are disturbed.
V–VI	Dishes can rattle; books may be knocked off shelves; damage is slight.
VII–VIII	Shaking is obvious, often prompting people to run outside; damage is moderate to heavy.
IX–X	Buildings are knocked off foundations; cracks form in ground; landslides may occur.
XI–XII	Wide cracks appear in ground; waves seen on ground surface; damage is severe.

The **Richter scale** of earthquake magnitude is based on measurements made with an instrument called a **seismograph**. (See Figure 7-5 on page 170.) Within a mechanical seismograph is a weight suspended on a spring. As the ground shakes, the weight tends to remain still. A pen attached to the weight traces a line on paper attached to a rotating drum. The larger the earthquake the more the line appears to move from its rest position. Richter magnitude is based on the size of the waves recorded on the drum. The advantage of the Richter scale is its precision based on true measurements. There is no human bias or need to adjust for places that have different kinds of building construction. One disadvantage is that measurements are taken only in those places that have an instrument. In addition, the Richter scale was developed for earthquakes in California, and is based on measurements made with a seismograph that no longer represents the latest in technology.

In recent years, seismologists have developed a magnitude scale called **seismic moment**. This scale is based on the

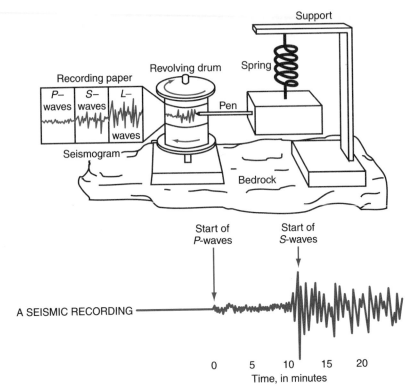

Figure 7-5 Mechanical seismographs record the shaking of the ground with a large mass suspended on a spring. As the ground shakes, a pen makes a record on a rotating drum. The recording that is produced is called a seismogram.

total energy released by the earthquake. For small and moderate events, the seismic moment and Richter scales give very similar numbers. Both scales are based on the use of instruments, although the seismic moment scale is usually based on modern electronic instruments. These electronic instruments are better than mechanical seismographs at measuring low-frequency vibrations and large movement of the ground. Low-frequency motions are an important part of the largest seismic events.

Both Richter and seismic moment scales are **logarithmic**. In a logarithmic scale an increase of one unit on the scale translates to a 10-fold increase in the quantity measured. Therefore an increase of one unit on the Richter of seismic moment scales means that there was 10 times as much ground movement. It also means that an earthquake one magnitude number smaller caused only $\frac{1}{10}$ the movement of the ground. Both scales are open-ended scales. In theory, there is no limit to how large or small the earthquake can be. However, it appears that there is a limit to how much energy the crust can absorb before it breaks. For this reason, scientists do not expect any earthquakes larger than magnitude 10. At the other end of the scale, small earthquakes get lost in vibrations caused by passing vehicles and people walking nearby.

 ## HOW DO EARTHQUAKES RADIATE ENERGY?

Have you ever tossed a stone into a still pond and observed the waves it makes? Starting from the point at which the stone hit the water, water waves carry energy away in all directions. If the stone falling into the water represents an earthquake, it would be the arrival of those waves that are felt and measured as the earthquake.

Seismic waves are vibrations of the ground generated by the break that started the earthquake. The energy that is released is carried away in all directions by these vibrations. Three kinds of seismic waves carry earthquake energy. **Primary waves**, or *P*-waves, travel the fastest. (They are called primary because they are the first waves to arrive at any location away from the epicenter.) *P*-waves cause the ground to vibrate forward and back in the direction of travel. This vibration direction makes them a form of longitudinal waves. *P*-waves are like sound waves because they alternately squeeze and stretch the material through which they pass. This is the kind of motion that you might observe if a speeding car ran into a line of parked cars and the energy of the collision traveled along the line of cars. *P*-waves can travel through any material—solid, liquid, or gas.

Secondary waves, also known as *S*-waves, are the next to arrive. They travel at roughly half of the speed of *P*-waves. As the *S*-waves pass, the ground vibrates side to side, the way that waves are passed along a rope. Like light, *S*-waves are *transverse* waves, which cause vibration perpendicular to

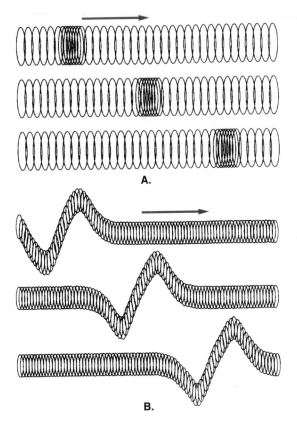

Figure 7-6 Seismic waves are classified by the direction of vibration. As *P*-waves carry energy from the focus, they cause the ground to vibrate forward and back parallel to the direction of motion as shown in A. However *S*-waves cause the ground to move perpendicular to the direction in which they move as shown in B.

the direction in which they are moving. *S*-waves can travel through a rigid medium (solids) but they cannot pass through a fluid, which includes liquids or gases. Collectively, *P*- and *S*-waves are known as body waves because they pass *through* Earth rather than along its surface.

When *P*- and *S*-waves reach the surface, they are transformed into surface waves, which are slower and have a circular motion. Surface waves usually do the most damage because they include both vertical and horizontal motion. (See Figure 7-6.)

| ACTIVITY 7-2 | MODELING SEISMIC WAVES |

Make a line of at least 5 to 10 students across the front of the classroom. Students should stand with their hands on the shoulders of the person in front of them. Model *P*-, *S*-, and surface wave motions passing along the line. Seismic waves may also be demonstrated with a long spring toy. The spring can be stretched and released quickly to show how *P*-waves travel. It can be moved side-to-side to show how *S*-waves travel.

 Adding and Subtracting Time

Working with time readings may take some practice. Most of the numbers you work with have a base of 10. However, there are 60 seconds in a minute and 60 minutes in an hour. Furthermore, the hour is never greater than 12. For these reasons, doing calculations with time is not the same as most of the calculations you do.

Time can be written in the form of hours:minutes:seconds. For example, the time 06:15:20 means 6 hours, 15 minutes, and 20 seconds. (In this chapter it will not be important whether the time is A.M. or P.M.) The time 12:45 can be expressed to the nearest second as 12:45:00.

To add 2 hours:10 minutes:15 seconds to 9:45, the addition can be set up as follows:

$$
\begin{array}{r}
02{:}10{:}15 \\
+\ 09{:}45{:}00 \\
\hline
11{:}55{:}15
\end{array}
$$

But suppose that the numbers are a little larger:

$$
\begin{array}{r}
08{:}50{:}15 \\
+\ 09{:}45{:}00 \\
\hline
17{:}95{:}15
\end{array}
$$

You may have noticed that there can be no time written as 17:95:15. The hour can be no larger than 12 and the minutes' and seconds' columns cannot exceed 60. It is therefore necessary to add a minute and/or an hour to the next column on the left. Instead of 95 seconds, it becomes 1 hour and 35 minutes. Adding an hour changes the hour number from 17 to

18. You are not done yet. Most clocks do not read 18 hours. So 12 hours is subtracted (a full cycle of the clock), making it 6 hours:

$$
\begin{array}{r}
08{:}50{:}15 \\
+\ \underline{09{:}45{:}00} \\
06{:}35{:}15
\end{array}
$$

This subtraction problem is quite simple because none of the numbers exceed 60 seconds, 60 minutes, or 12 hours:

$$
\begin{array}{r}
11{:}34{:}50 \\
-\ \underline{06{:}27{:}10} \\
07{:}07{:}40
\end{array}
$$

In subtraction problems you may also need to borrow from the column to the left. Consider the next calculation.

$$
\begin{array}{r}
11{:}14{:}20 \\
-\ \underline{06{:}27{:}50}
\end{array}
$$

If you subtract a larger number from a smaller number, you get a negative number. However, negative numbers are not used to express time. So you need to change the top number by borrowing from the units to the left while keeping the same value. When you borrow 1 minute and add 1 minute to the right column, 20 seconds becomes 80 seconds. Having added a minute to the seconds column, you need to subtract 1 minute from the minutes column. So 14 minutes becomes 13 minutes.

But now you have a problem in the minutes' column because 27 minutes is greater than 13 minutes. It is now necessary to borrow 1 hour (60 minutes) from the hours' column. The 13 minutes becomes 73 minutes, and 11 hours becomes 10 hours. This problem now can be written as follows:

$$
\begin{array}{r}
10{:}73{:}80 \\
-\ \underline{06{:}27{:}50} \\
04{:}46{:}30
\end{array}
$$

✓ *Practice Problem 1*

Add 4 hours:15 minutes:35 seconds to 10 hours:50 minutes:35 seconds.

✓ *Practice Problem 2*

Subtract 5 hours:30 minutes:45 seconds from 10 hours:25 minutes:40 seconds.

To find earthquake origin times you will need to subtract one time from another. Using this system of borrowing from the column to the left should allow you to solve any problem of this type.

 # HOW ARE EARTHQUAKES LOCATED?

Seismologists (scientists who study earthquakes) know that an earthquake has occurred when they feel the ground shaking or they see their instruments detect energy waves. Earthquakes can happen anywhere. Some of them occur in populated areas. But more often they happen in the oceans or in places where few people live. Furthermore, the small wave could have come from a small, nearby earthquake or from a large earthquake located thousands of kilometers away. In this section, you will learn how seismologists can locate an earthquake with recordings from as few as three seismic stations.

The first step is to determine how far away an earthquake is. The fact that *P*-waves travel much faster than *S*-waves enables seismologists to determine the distance to an earthquake's epicenter. When an earthquake occurs, energy waves move away in all directions. Both *P*- and *S*-waves start out from the epicenter at the same time. A person at the epicenter would feel both waves immediately. There would be no separation between them. But the farther you are from the epicenter, the longer the delay between the arrival of *P*-waves and the arrival of *S*-waves. **Travel time** is the time inferred

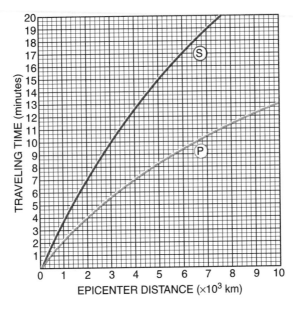

Figure 7-7 Travel time graph for *P*- and *S*-waves.

between the breaking of the rocks that causes an earthquake and when the event is detected at a given location.

Figure 7-7 is a graph from the *Earth Science Reference Tables* that shows how long it takes *P*- and *S*-waves to travel to places as far as 10,000 km away. Note that the scale of distance from the epicenter is on the bottom axis and travel time is on the vertical axis. Along the bottom, the distance from one dark line to the next is 1000 km (10^3 km). Each 1000-km division is split into five subdivisions. Therefore each thin line represents 1000 km ÷ 5 = 200 km. On the vertical scale, each thick line represents 1 minute and there are three divisions. What is the value of each thin line? Remember there are 60 seconds in each minute. Therefore each vertical division is one-third of a minute, since 1 minute (60 seconds) ÷ 3 = 20 seconds.

For example, this graph tells you that if an earthquake epicenter is 4000 km away, the *P*-wave will take 7 minutes to get to you. That is, at your location you will feel the first vibrations 7 minutes after the rupture occurs at the epicenter. The graph shows that the *S*-waves need about 12 minutes: 40 seconds to travel the same distance. Therefore, if the time separation between the arrival time of the *P*- and *S*-waves is 5 minutes:40 seconds, the epicenter must be 4000 km away.

An observer closer to the location of the epicenter would notice a smaller delay between the arrival of the *P*- and *S*-waves. A more distant recording station would register a greater separation between them.

Imagine that you did observe a 5-minute:40 second time delay. What does this tell you about the location of the epicenter? It tells you that the epicenter of the earthquake is 4000 km from your location. Therefore, on a map you could draw a circle centered on your location with a radius of 4000 km and know that the epicenter is somewhere on that circle. However, you do not know where on the circle the epicenter is, but you do know it has to be that distance away from your station.

The text and photographs that follow will demonstrate a procedure that can be used to locate the epicenter of an earthquake. Notice that Figure 7-8 shows a seismogram recorded at Denver, Colorado, on which there are two divisions in each minute. Therefore each division is 30 seconds. On this seismogram you can see that the *P*-wave arrived at 8 hours:16 minutes:0 seconds, and the *S*-wave arrived at 8 hours:19 minutes: 30 seconds. The time separation between *P*- and *S*-waves is therefore 3 minutes:30 seconds.

There is only one distance at which this time separation can occur. To find it, place the clean edge of a sheet of paper along the vertical distance scale. Make one mark at 0 min-

Figure 7-8 By reading the seismogram you can see that *P*-waves arrive 3 minutes:30 seconds before the *S*-waves at this recording location.

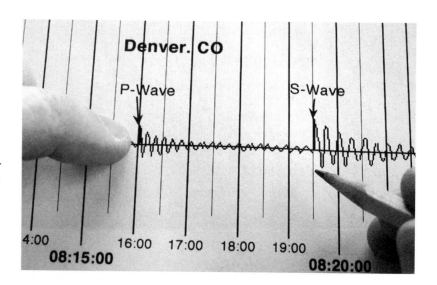

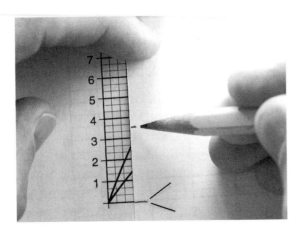

Figure 7-9 The space between the two marks on the edge of the sheet of paper represents the time difference between the arrival of the *P*- and *S*-waves.

utes and another at 3 minutes:30 seconds as shown here. (See Figure 7-9.) Perform this step carefully to avoid problems further along in the procedure.

Next, as you keep the edge of the edge of the paper vertical, move the paper to the position at which each marks rests on a graph line. As shown in Figure 7-10, the 0 mark will be on the *P*-wave line and the mark for 3 minutes:30 seconds will be on the *S*-wave line. Notice that this time separation occurs only if the epicenter is 2200 km away.

The next objective is to draw a circle around the Denver seismic station with a radius equivalent to 2200 km. Figure 7-11 shows a drawing compass stretched along the map scale

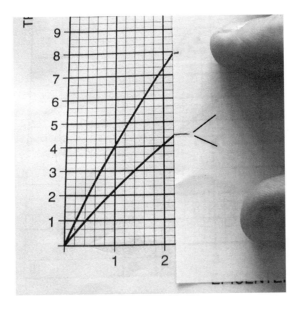

Figure 7-10 The time separation marks are moved to the right until they line up with *P*- and *S*- waves graph lines. The time separation indicated on the edge of the paper will occur at a specific distance, which can be read below on the distance scale.

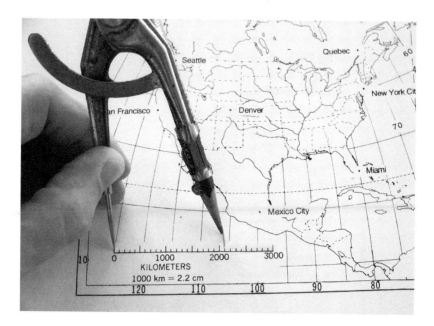

Figure 7-11 Use the map scale of distance to stretch the compass to the correct epicenter distance.

to 2200 km. (See Figure 7-11.) Do this with care because it is another step where errors are common.

Place the point of the compass on the map location of the Denver seismic station. Be sure the compass opening has not moved since it was set at 2200 km. Draw a circle around Denver at a distance equivalent to 2200 km. (See Figure 7-12.)

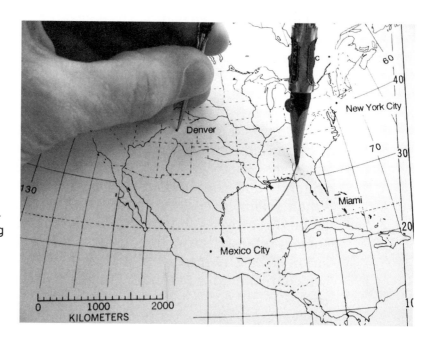

Figure 7-12 Place the point of the compass at the recording station and draw a circle at the epicenter distance. The epicenter must be somewhere on this circle.

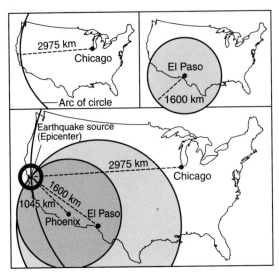

Figure 7-13 When the distance from the epicenter of an earthquake is determined for three recording stations, the intersection of the three circles on the map will mark the epicenter.

Now you have drawn a circle drawn on a map that shows how far the epicenter is from Denver. The epicenter can be anywhere on that circle. You still need two more seismic stations to locate the epicenter. Following the same steps with another seismogram of this earthquake recorded at Miami would give you a second circle. Most likely those two circles will intersect at two points. The epicenter must be at (or near) one of those points. Using the same procedures with a seismogram from a third location will enable you to draw another circle. These three circles should intersect at one point on the map. That point is the epicenter. (See Figure 7-13.)

If you have the data, you can draw more than three circles. That is a good way to check your work. Is it really that neat and simple? The theory is, but in practice the circles may not meet exactly. More often the three circles make a small triangle, and the epicenter is located inside that triangle. When seismologists work with real seismic recordings and make these circles to locate actual seismic events, the circles seldom intersect at a single point. Differences in the composition of Earth's interior and a wide range of other variables can contribute to errors that prevent a perfect meeting of the circles.

Finding the Origin Time

The **origin time** is the time at which the fault shifted, producing the earthquake. Knowing when the *P*- or *S*-waves arrived at your location and how far away the epicenter is, you can calculate the origin time. Suppose that a friend arrived at your house at 10:30 and told you that the trip took him 20 minutes. You could determine that your friend left home at 10:10. In the same way, if you know the time that a seismic wave arrives and you know how long it traveled, you can tell when the earthquake began at the epicenter.

When a *P*- or *S*-wave arrives, how can you tell how long it was traveling? The key to this is knowing how far away the epicenter is and using the travel time graph. You have already learned that the distance to an epicenter can be determined from the time delay between the arrival of the *P*- and *S*-waves. If you know the distance to the epicenter, you can use the travel time graph to discover how long it took for the *P*- and *S*-waves to travel from the epicenter to your location: the travel time.

Let us continue with an example in the "How are Earthquakes Located?" section. In the first example in that section, the separation between *P*- and *S*-waves determined that the epicenter was 4000 km from your location. You also need to know when the *P*- and *S*-waves arrived. So let us work with a *P*-wave arrival time of 02:10:44 and an *S*-wave arrival time of 02:14:04. By reading the graph you can tell that at a distance of 4000 km a *P*-wave takes 7 minutes of travel time:

> 02:10:44 (*P*-wave arrival time)
> − 00:07:00 (travel time)
> 02:03: 44 (origin time)

You can check this time by using the *S*-wave data. To travel 4000 km an *S*-wave takes 12 minutes:40 seconds:

> 02:16:24 (*S*-wave arrival time)
> − 00:12:40 (travel time)

Borrowing 1 minute and adding it to the seconds' column makes this

$$02:15:84 \ (S\text{-wave arrival time})$$
$$-\ 00:12:40 \ (\text{travel time})$$
$$02:03:\ 44 \ (\text{origin time})$$

This is a single earthquake so you should expect the two origin times to agree. If they are not at least close in time, you probably need to go back and look for an error in your work.

 ## WHAT IS INSIDE EARTH?

For a long time people have wondered about what is inside Earth. In Chapter 2 you learned that Earth has a radius of more than 6000 km. The world's deepest mine shaft in South Africa reaches to about 5 km below the surface. The Russians have drilled to a depth of 12 km, but even this is much less than 1 percent of the distance to Earth's center.

If a doctor needs to see inside your body without cutting you open, he or she can use such tools as X-rays and CAT scans. However, these techniques are not able to penetrate thousands of kilometers of solid rock. Scientists have found another way to examine Earth. Earthquake waves do travel through Earth's interior, and they can be used to obtain information about its structure. From observing which earthquake waves travel through the planet, where they are detected, and the speed at which they travel, scientists make inferences about Earth's structure. Based on direct observations and analysis of seismic waves, scientists have divided the solid planet into four layers: crust, mantle, outer core, and inner core, as shown in the *Earth Science Reference Tables*.

 ### The Crust

The crust is the outermost layer of Earth. In all of the mining and drilling humans have done, this is the only layer that has been observed directly. In most places a thin layer of sedimentary rocks covers metamorphic and igneous rocks. The deeper you dig, the greater the chance of finding igneous rocks. Within the continental crust, the overall composition is close to the composition of granite. Under the sediments and sedimentary rocks of the oceans, the crust is more dense. This is verified by observing that most volcanoes that erupt in the ocean basins bring lava to the surface that hardens to form basalt. Geologists therefore infer that the crust under the oceans is mostly basalt or has a similar composition. Furthermore, from very deep mines, scientists know that the temperature within the crust increases with depth at a rate of about 1°C per 100 m. Based on this rapid rate of temperature change, it seems logical that Earth's interior is very hot.

The bottom of the crust is inferred from observations of earthquake waves. Seismic waves that travel through just the crust are slower than are those that dip below the crust and enter the mantle. This change in seismic wave speed occurs at depths as shallow as 5 km under the oceans to as deep as 60 km under some mountain ranges on land. The Croatian geophysicist Andrija Mohorovicic first noticed this boundary, and it is now named in his honor. However, most people shorten the name of the boundary to the **Moho**.

 ## The Mantle

The mantle extends from the bottom of the crust to a depth of about 2900 km. In fact, it contains more than half of Earth's volume. However, the inability to penetrate below the crust with mines or drills means that investigations of the mantle must be conducted by other means.

From studies of volcanic eruptions, scientists believe that magma that originates below the crust is rich in dense, mafic minerals such as olivine and pyroxene. These are the minerals on the right side of the Scheme for Igneous Rock Identification in the *Earth Science Reference Tables*. Some magma originates in the crust beneath mountain areas where Earth's crust is thick and felsic in composition.

A particularly violent eruption of material from deep within Earth is responsible for most natural diamonds. These gas-charged intrusions from deep within the mantle are called kimberlites. They are named for Kimberley, a city in South Africa, where diamonds have been mined for many years. Diamonds can form only under conditions of extreme heat and pressure, such as the conditions within the mantle. The fact that diamonds are found in these volcanic rocks indicates that the magma came to the surface from deep within the mantle.

Most of the meteorites that fall to Earth are composed of dense, mafic minerals such as pyroxene and olivine. These are called stony meteorites. Scientists believe that these meteorites are the material from which Earth and the other planets formed billions of years ago, or they are the remains of a planet that was torn apart by a collision with another object. If this is the case, you would expect meteorites to have a composition similar to the planets, including Earth.

The diameter of planet Earth has been known for centuries. From this value, it is easy to calculate Earth's volume. The mass of Earth has been determined based on its gravitational attraction. Knowing Earth's mass and volume, scientists have calculated that Earth's overall density is about 5.5 g/cm^3. That is about twice as dense as most rocks in the crust. Therefore, scientists expect the mantle, which includes most of Earth's volume, to be composed of minerals that are more dense than those in the crust are.

Both *P*- and *S*-waves travel through the mantle. Because *S*-waves will not travel through a liquid, the mantle is known to be in the solid state. Furthermore, earthquake waves travel faster in the mantle than they do in Earth's crust. This indicates that the rock in the mantle is more brittle than crustal rocks. Again, olivine and pyroxene fit the observations.

All these observations provide strong evidence that Earth's mantle is composed primarily of iron- and magnesium-rich, mafic silicate minerals. Therefore, geologists infer that the mantle is composed mostly of olivine and pyroxene in the solid state.

 The Core

Starting at 2900 km and extending to Earth's center is the next layer. Seismologists noticed that *P*-waves can penetrate the whole planet, but *S*-waves do not travel through Earth's core. Therefore the core must be liquid.

Starting from the atmosphere, each of Earth's layers is more dense than the layer above it: atmosphere, hydrosphere, crust, mantle. You might therefore expect the core to be even more dense than the mantle.

Meteorites come in two basic varieties. You read earlier that the more abundant stony meteorites are thought to be composed of the material found in the mantle. The second group of meteorites are mostly iron. Scientists therefore believe Earth's core is also mostly iron with some nickel.

The behavior of *P*-waves deep within the core leads us to infer that the inner part of the core is a solid. (You may recall that *S*-waves cannot get there because a liquid outer core surrounds the solid inner core.) It appears that extreme pressure at the very center of the core makes it a solid. The solid state of the inner core may also result from different elements mixed with iron. Geologists therefore infer that a liquid outer core surrounds a solid inner core, both composed primarily of iron.

The conditions of heat and pressure deep within Earth prevent scientists from any direct observations. But they have been able to duplicate these conditions in laboratories. Very small amounts of the substances geologists believe to be present in the mantle and core can be squeezed under extreme pressure between flat faces of two diamonds. The instrument is called a diamond anvil. Heat is applied with a laser beam. Then X-rays are used to probe the crystal structure of the substance. These studies confirm, for example, that pressure at Earth's center can cause the iron-rich inner core to solidify.

Earthquake Shadow Zones

When a large earthquake occurs, seismic waves travel through Earth's layers to recording stations worldwide. The behavior of these waves provides important information about Earth's structure. For example, scientists know that *S*-waves will not travel through the outer core because the outer core is a liquid.

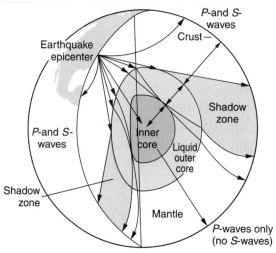

Figure 7-14 Earth's shadow zones. As *P*- and *S*-waves travel through Earth, the liquid outer core blocks *S*-waves. Refraction (bending) of seismic waves at the mantle-core boundary causes a shadow zone where neither *P*- nor *S*-waves are received.

The zones where direct *P*-waves are recorded are more complicated. These *P*-waves arrive at locations as far as half way around Earth. These locations also receive *S*-waves. *P*-waves are also received in a region directly opposite the epicenter. Those are the *P*-waves that travel through the liquid outer core (and the solid inner core). But there is a shadow zone that does not receive any direct *P*- or *S*-waves. Figure 7-14 shows that beyond the reach of *S*-waves is a zone that circles Earth where no direct *P*- or *S*-waves are received. This shadow zone circles Earth like a donut with a hole in the middle. While you might expect this region to receive at least direct *P*-waves, the bending of seismic waves prevents even them from reaching the shadow zone.

The bending of energy waves is known as **refraction**. It can be observed whenever energy waves enter a region in which they speed up or slow down. From Figure 7-14 you can see how seismic waves curve as they travel through the mantle and core. You can also see the *P*-waves bend when they move from the mantle into the core as well as when they move from the core back into the mantle. The result of this bending (refraction) of seismic waves is the shadow zone where no direct *P*- or *S*-waves are received.

TERMS TO KNOW

conduction	Mercalli scale	secondary waves
convection	Moho	seismic moment
convection cell	origin time	seismograph
earthquake	primary waves	seismologist
epicenter	radiation	seismology
fault	refraction	stress
focus	Richter scale	travel time
logarithmic		

CHAPTER REVIEW QUESTIONS

1. The diagram below shows the bedrock structure beneath a series of hills.

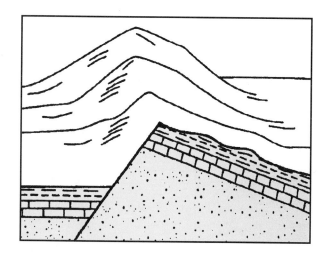

Which process was primarily responsible for forming the hills?

(1) folding   (3) deposition
(2) faulting   (4) vulcanism

2. Approximately how long does it take a *P*-wave to travel 6500 km from an epicenter?

(1) 6.5 minutes   (3) 10.0 minutes
(2) 8.0 minutes   (4) 18.5 minutes

Base your answers to questions 3–5 on the diagrams below.

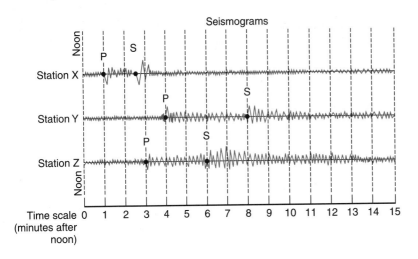

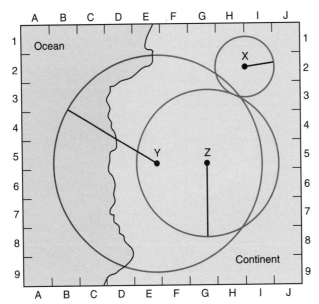

3. What is the average velocity of an *S*-wave in the first 4 minutes of travel?

(1) 1 km/minute

(2) 250 km/minute

(3) 500 km/minute

(4) 4 km/minute

4. Approximately how far from station Y is the epicenter?

(1) 1300 km

(2) 2600 km

(3) 3900 km

(4) 5200 km

5. Seismic station Z is 1700 km from the epicenter of an earthquake. Approximately how long did it take the *P*-wave to travel to station Z?

(1) 1 minute 50 seconds
(2) 2 minutes 50 seconds
(3) 3 minutes 30 seconds
(4) 6 minutes 30 seconds

6. On the map, which location is closest to the epicenter of the earthquake?

(1) E–5
(2) G–1
(3) H–3
(4) H–8

7.

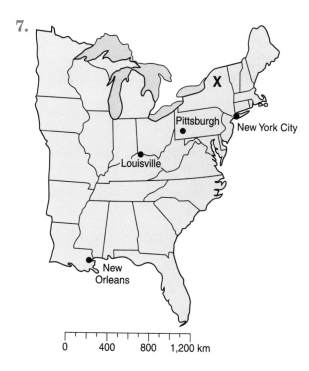

Letter X on the map above represents the epicenter of an earthquake. At which city is there a difference of approximately 3 minutes and 20 seconds between the arrival times of *P*-waves and *S*-waves?

(1) New Orleans
(2) Louisville
(3) Pittsburgh
(4) New York City

8. Which map correctly shows how the location of the epicenter was determined?

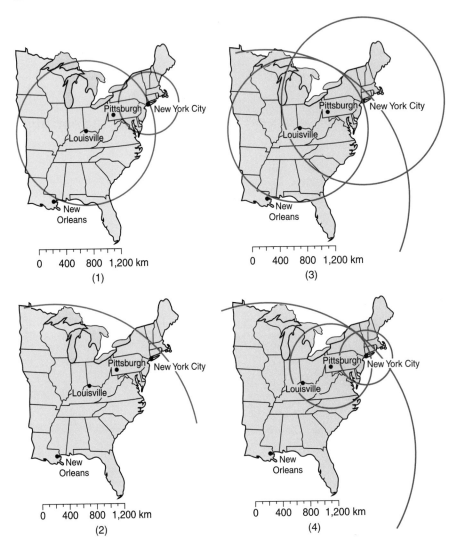

9. An earthquake's *P*-waves arrived at a seismograph station at 02 hours:40 minutes:00 seconds. The *S*-wave arrived at the station 2 minutes later. What was the approximate distance from the epicenter to the seismograph station?

(1) 1100 km

(2) 2400 km

(3) 3100 km

(4) 4000 km

10. An earthquake occurred at 09:15:30 at location X. At what time did the first *P*-waves arrive at station Y which is 2000 km from location X?

(1) 09:15:50   (3) 09:19:30

(2) 09:17:30   (4) 09:25:40

11. Which object is likely to have the same composition as Earth's mantle?

(1) granite   (3) an iron meteorite

(2) rhyolite   (4) a stony meteorite

12. An *S*-wave from an earthquake that travels toward Earth's center will

(1) be deflected by Earth's magnetic field.

(2) be totally reflected off the crust-mantle interface.

(3) be absorbed by the liquid outer core.

(4) reach the other side of the earth faster than *P*-waves.

13. Earth's outer core is best inferred to be

(1) liquid, with an average density of approximately 4 g/cm^3.

(2) liquid, with an average density of approximately 11 g/cm^3.

(3) solid, with an average density of approximately 4 g/cm^3.

(4) solid, with an average density of approximately 11 g/cm^3.

14. Compared to Earth's crust, its core is inferred to be

(1) less dense, cooler, and composed of more iron.

(2) less dense, hotter, and composed of less iron.

(3) more dense, hotter, and composed of more iron.

(4) more dense, cooler, and composed of less iron.

15. When a major earthquake occurred, seismic stations over most of Earth recorded *P*-waves, *S*-waves, or both. But some recording stations almost half way around Earth did not receive *P*-waves or *S*-waves. What was the reason for this circular shadow zone where seismic stations did not record any direct *P*- or *S*-waves?

(1) Seismic waves are bent as they move from one substance to another.

(2) No seismic waves can travel through Earth's liquid outer core.

(3) Liquids absorb *P*-waves but the can transmit *S*-waves.

(4) Liquids absorb *P*-waves, and solids absorb *S*-waves.

Open-Ended Questions

16. State one way in which seismic *P*-waves are different from seismic *S*-waves.

17. A seismic station in Massena, New York, recorded *P*- and *S*-waves from a single earthquake.

    If the first *P*-wave arrived at 1:30:00 and the first *S*-wave arrived at 1:34:30, determine the distance from Massena to the epicenter of the earthquake.

18. For the earthquake in question 17, state what additional information is needed to determine the exact location of the epicenter.

19. Over a period of several weeks a seismologist received both *P*-waves and *S*-waves from two different earthquakes with epicenters in the same location. How could the scientist tell which was the stronger earthquake?

20. How do most igneous rocks from the ocean basins differ from most igneous rocks in the continents?

Chapter 8

Plate Tectonics

 ## DO CONTINENTS MOVE?

In previous chapters, you learned about Earth's size and shape, the materials of which it is made, and how heat energy causes Earth's crust to move. You are learning that Earth is a dynamic planet: it is constantly changing. In this chapter, you will consider the patterns that scientists find when they look at movements of Earth's lithosphere and how these patterns relate to the features found on the surface.

This is a story of changing scientific ideas. A **paradigm** (PARA-dime) is a coherent set of principles and understandings. Until very recently scientists thought of the Earth's continents as stationary features attached to the interior by solid rock. However, as scientists found more and more evidence for moving continents, they had to change their ideas of what they mean by a planet "solid as rock." When the evidence for moving continents became so strong that scientists could no longer dismiss it, the geological community experienced a paradigm shift. They changed from thinking of Earth's surface features as fixed in position to proposing a planetary surface composed of moving segments.

Continental Drift

When explorers and mapmakers crossed the Atlantic Ocean in the 1600s, some of them noticed an interesting coincidence. The more complete their maps became, the more the Atlantic Ocean shorelines of the Americas and the Old World continents of Africa and Europe looked as if they would fit together like pieces of a jigsaw puzzle. Initially this was simply noted as a curiosity. Some of them may have wondered if the continents had ever been together. They thought of rock as solid and unmoving. The idea that whole continents could move was not taken seriously.

In 1912, Alfred Wegener (VEYG-en-er)—a German explorer, astronomer, and **meteorologist** (a meteorologist is a scientist who studies weather)—proposed that the continents we see are the broken fragments of a single landmass that he called Pangaea (pan-JEE-ah), which means all Earth. He hypothesized that over millions of years, the continents slowly moved to their present positions. Wegener supported his theory by pointing out that a particular fossil leaf (*Glossopteris*) is found in South America, Africa, Australia, Antarctica, and India. Bringing the continents together would put the areas where these fossil organisms are found next to each other as shown in Figure 8-1. The same placement would also line up distinctive rock types and mountain areas. Wegener also noted glacial striations (scratches on bedrock caused by the moving ice of a glacier) in places such as Australia and Africa that now have a climate too warm for glaciers. He suggested that these places that now have warm climates were once located near the South Pole. On the basis of this evidence, Wegener proposed his hypothesis that came to be known as continental drift.

The idea that continents could move was so radical that few geologists took it seriously. How could the landmasses move through the solid rock of the ocean basins? The causes suggested by Wegener could not provide the force needed to move continents. As another way to discredit Wegener's idea of continental drift, scientists pointed out that he was not a geologist. In 1930 when Alfred Wegener died on an expedition

Figure 8-1 If the continents were moved together by closing the Atlantic Ocean, mountain chains, similar rock types, and evidence of glaciation on several continents would come together.

to Greenland, very few geologists supported his ideas. His most enthusiastic supporters were in the Southern Hemisphere where the most convincing evidence for continental drift had been observed.

ACTIVITY 8-1   MATCHING SHORELINES

Obtain an outline map of the world. Cut out the continents and lay the pieces on your desk in their present locations. Then, slide them together to show how they might have fit to make a single land mass. Paste or tape them onto a sheet of paper in the position they might have been in before they separated.

Seafloor Spreading

The rocks of the ocean bottoms are very different from most rocks on land. When scientists look at the mineral composition of rocks from the continents they find that most have a composition similar to granite. That is, the continental rocks are generally felsic, light-colored rocks with a relatively low density. The low-density material of which continents are composed tends to rise above the ocean basins. Most oceanic rocks have a composition similar to basalt. They are mafic. You may recall from Chapter 5 that mafic rocks are darker in color and relatively dense. This is why the ocean basins tend to ride lower on the rocks of Earth's interior and are covered by the oceans.

At the time Wegener worked on his hypothesis, the oceans were vast expanses of unknown depth. After his death, the strongest evidence for his theory of moving continents came from the oceans. During World War II, there were important advances in technology and oceanic exploration. A moving ship could make a continuous record of the depth of the ocean by bouncing sound waves off the ocean bottom. After the war, scientists used this technology to map the world's ocean bottoms. Much of the ocean bottom is flat and featureless. But scientists discovered an underwater system of mountain ranges that circles Earth like the seams on a baseball. These are the **mid-ocean ridges**. Most of these 64,000-km-long features are under water, but they rise above sea level in Iceland and on several smaller islands. By looking at the landscape of Iceland, scientists can investigate processes that occur at the bottom of the ocean hidden from view.

In 1960, Princeton University geologist Harry Hess reconsidered Wegener's ideas. Hess had commanded a ship during World War II that measured ocean depths. He had seen the newly discovered shape of ocean bottoms including the mid-ocean ridges. Hess suggested that molten magma from the mantle rises to the surface at the mid-ocean ridges and erupts onto the ocean bottom. In many places, the ocean ridges are like two mountain chains separated by a valley. It is in the valley that the most active eruptions are taking place to form a strip of new lithosphere. Some of the magma solidifies as it comes into contact with the cold ocean water making pillow-shaped rocks. Wherever scientists find pillow lavas, they can be sure that magma moved into water and solidified quickly. Pillow lavas are common along the mid-ocean ridges but are uncommon elsewhere in the oceans. Hess suggested that the mid-ocean ridges are the places where new lithosphere is made and adds on to older material that moves

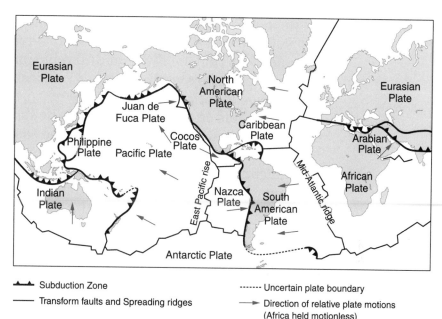

Figure 8-2 Earth's surface is composed of about a dozen lithospheric plates. These plates can move together, move apart, or slide past each other as shown by the arrows on this map.

Subduction Zone

Transform faults and Spreading ridges

Uncertain plate boundary

Direction of relative plate motions (Africa held motionless)

away from the ridges on both sides. Hess called this process **seafloor spreading**. (See Figure 8-2.)

MAGNETIC EVIDENCE New technology also enabled scientists to make continuous records of the magnetism of the ocean floor. In Chapter 3, you learned that Earth is a giant magnet. Scientists are not sure what causes Earth's magnetic field, but the likely cause is currents of molten iron in Earth's outer core. These currents tend to line up with Earth's spin axis and therefore make Earth's magnetic poles close to the geographic North and South Poles. When magma cools to make rock, iron forms the magnetic mineral magnetite. Within molten magma, iron is free to align itself with Earth's magnetic field and change its alignment as the magma flows or Earth's magnetic field changes. When magma rises to the surface and solidifies, the alignment of the iron in the rock can no longer change. The new rock holds a record of Earth's magnetism at the time it crystallized.

Earth's magnetic poles have reversed many times. The magnetic field weakens and then builds in the opposite direction. In other words, the north magnetic pole becomes the south magnetic pole, and then the south magnetic pole becomes the north magnetic pole. The time from one reversal to the next is not uniform, but most reversals occur after 100,000 to 1,000,000 years. A record of these reversals is preserved in igneous rocks that solidified at different times. In Chapter 18, you will learn how scientists use radioactive substances to find the age of rocks.

If the whole ocean floor formed at the same time, all the igneous rocks would be magnetized in the same direction. However, scientists see bands of polarity. **Polarity** is the direction of a magnetic field. The direction is determined with an instrument that measures Earth's magnetic field. Some of the bands are magnetized the way they would if they solidified with the present alignment of the magnetic poles. These areas have normal polarity. Other parts of the ocean floor show reversed polarity. These rocks crystallized when the magnetic north and south poles were opposite to the way they are now. These bands of normal and reversed polarity are the strongest near the ocean ridges. In addition, they show a symmetrical pattern. The striped pattern of normal and reversed polarity on one side of an ocean ridge is almost a mirror image of the pattern on the other side. (See Figure 8-3.) These magnetic bands provide evidence that new crust is forming as the plates move away from the ridges.

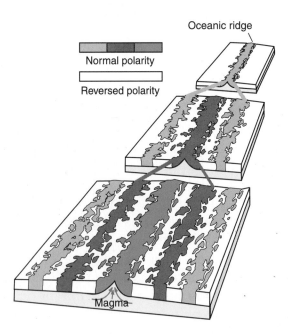

Figure 8-3 New lithosphere is created at the mid-ocean ridges. That portion of the lithosphere takes on Earth's magnetism and slowly moves away from the ridge. As the magnetic poles reverse, parallel bands of normal and reversed magnetic polarity are created at the mid-ocean ridge.

AGE OF ROCKS   Geologists have been able to determine the age of rocks recovered from the ocean floors. In the 1950s, the *Glomar Challenger* was able to drill into the basaltic rock of the ocean floor to recover the solid bedrock of the oceans. When laboratory work on these samples revealed their age, a pattern emerged. Geologists discovered that the rocks recovered close to the ridges were younger than those collected far from the mid-ocean ridges. As the distance from the mid-ocean ridges increases, so does the age of the ocean floor.

These observations support the idea that ocean floor is being created continuously at the mid-ocean ridges, and moves like a conveyor belt toward the continents. But how can new lithosphere be created if Earth is not getting larger? In Chapter 6, you read about recycling as a way to conserve resources. Earth recycles its lithosphere. Old lithosphere moves back into Earth's interior at locations known as **subduction zones**. As lithosphere is being created by magma moving to the surface at the mid-ocean ridges, the older lithosphere moves across the ocean bottom and away from the ridges. The rate of motion varies from ocean to ocean, but some move as fast as 10 cm/year. Typically, the plates move about as fast as your fingernails grow. Figure 8-4 illustrates that at the subduction zones sections of the lithosphere sink into Earth.

At ocean trenches, the more mafic parts of the lithosphere are absorbed into the mantle. The felsic continental litho-

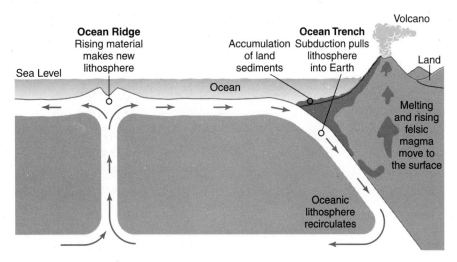

Figure 8-4 As new crust is created at the ocean ridges, the ocean floor moves like a slow conveyor belt toward the ocean trenches. The ocean trenches, also known as zones of subduction, are places where ocean bottom is drawn back into Earth. Continental rocks resist subduction.

sphere resists being drawn into Earth in the process of subduction. This lighter material forms intensely folded and faulted mountains near the subduction zones. Felsic rocks drawn into Earth often melt and come back to the surface in volcanic eruptions. Due to the water in these continental rocks changing to steam, volcanic eruptions of felsic rocks are often violent and explosive. Both the highest mountains and the deepest parts of the oceans occur near zones of subduction and both of them are related to the process of subduction.

Plate Tectonics

You may recall that one of the most important objections to Wegener's idea of continental drift was that there was no mechanism to explain how continents could move through the ocean basins. Seafloor spreading provided an explanation for the motion of Earth's lithosphere. When the discoveries of seafloor spreading were added to Wegener's ideas about continental drift, geologists recognized a larger theory of **plate tectonics**. (The word **tectonics** means large-scale motions of Earth's crust.) Tectonic forces are responsible for uplift and mountain building. (See Figure 8-5.) In 1965, Canadian geologist J. Tuzo Wilson proposed that the whole Earth is covered by about a dozen rigid sections called **lithospheric plates**. The plates include the crust as well as the rigid upper mantle.

Figure 8-5 The San Rafael Swell in Utah is the result of tectonic forces associated with the movement of Earth's plates pushing up these formerly flat-lying rock layers. Note the large truck for scale.

Plates are approximately 100 km in thickness, and ride over a portion of the mantle that is less rigid than the upper part of the mantle. Compared with Earth's 6000-km radius, the plates are relatively thin. The lithospheric plates are created at the ocean ridges and destroyed at the subduction zones. Some plates contain only ocean floor, but most of the major lithospheric plates have oceanic and continental lithosphere. Zones of geological activity, such as the mid-ocean ridges and subduction zones, form the boundaries of the plates. In addition to the major plates, a number of smaller plates exist where the larger plates do not meet precisely.

The motion of the plates is driven by Earth's internal heat. While conduction does allow some of Earth's energy to reach the surface, it is too slow to account for most of the heat flow from Earth's center. The major form of heat flow outward from Earth's interior is convection. The ocean ridges are places where convection brings heated matter to the surface. There the heat energy can radiate into space.

Convection occurs only in fluids. A **fluid** is any substance that can flow. Generally, scientists consider gases and liquids to be fluids, but not solids. Convection in air produces the winds. Ocean currents are the result of convection in water. But how can solid rock flow as a fluid? If you could plunge into the mantle with a rock hammer, the rocks at depth would seem just as solid as similar rocks at the surface. If you were to hit mantle rock with the hammer, the hammer would bounce off the rock and make a high-pitched clink. The hammer does not deform the rock because it creates only a short-term force. Although there are places where rock has melted, forming magma, nearly all the mantle is in the solid state.

You may recall S-waves do not pass through liquids. The passage of S-waves through Earth's mantle is an indication that the mantle is solid. Earth's mantle is a solid, but it can flow to carry energy by convection. How can this be? The key to resolving this contradiction is the extreme heat and pressure inside Earth. Under extreme pressure and high temperature, solid rock can flow slowly like a fluid. In addition, the rate of movement of the plates is only a few centimeters per year. That is about the rate at which your fingernails grow. This is slow enough to allow convection in a material that otherwise behaves like a solid.

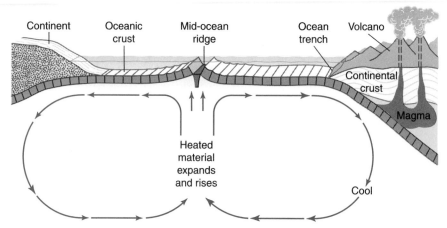

Figure 8-6 Slow-moving convection currents within Earth carry heat energy toward the surface. Material rises and creates new oceanic crust at the ocean ridges, and it moves back into Earth at ocean trenches. This circular pattern is called a convection cell. The continents are carried by this motion like passive rafts floating on water.

Convection is driven by density differences. Most materials expand when they are heated. Expansion increases volume and makes matter less dense. In places where a fluid is heated, it becomes less dense and tends to rise. Cooling usually causes matter to become more dense. This makes the substance heavier than an equal volume of the surrounding material. When fluids are cooled they usually sink. A heat source in one place, such as Earth's interior, and cooling in another place, such as Earth's surface, can set up a circular motion known as a *convection cell* as shown in Figure 8-6. In convection, the motion of the fluid occurs when heated material moves away from the source of heat. This is what is happening in Earth's asthenosphere.

 # WHAT IS EARTH'S INTERNAL STRUCTURE?

In Chapter 7, you read about the four layers within Earth as determined by seismic waves: crust, mantle, outer core, and inner core. Motion of the plates leads us to consider a different way to divide the outer layers of Earth into lithosphere, asthenosphere, and stiffer mantle. Note that Figure 8-7 is

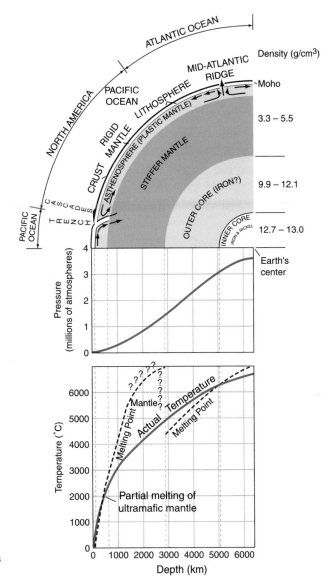

Figure 8-7 Inferred properties of Earth's interior

from the *Earth Science Reference Tables*. The following paragraphs explain this chart.

 Temperature

The word "inferred" in the caption reminds you that this diagram is based on laboratory simulations and other investigative work rather than on direct observations. An *inference*

is an interpretation or a conclusion based on indirect observations. This diagram has three parts: a temperature graph, a pressure graph, and a profile view inside Earth. All three parts of this figure refer to Earth's interior. The temperature graph shows how scientists think Earth's temperature changes with depth. It is easy to find an average temperature at Earth's surface because people can travel freely over the surface. Scientists have observed that the deeper Earth is penetrated with mines and wells the higher the temperature becomes. Beyond the depths of mines and wells, scientists must use laboratory instruments to simulate conditions inside Earth.

Observations of meteorites and the density of Earth support the inference that Earth's core is mostly iron. If scientists subject iron to the pressure estimated to exist at Earth's center, iron melts at just over 6000°C. That fact gives scientist enough information to begin to draw the dark line on this temperature graph. The dotted line on this graph shows the melting points of the substances geologists think occur in Earth's interior. At the surface, both lines are close together. Since the rocks at the surface are solid, the surface temperature is clearly below the melting point of these rocks. At a depth of about 100 km, is the region in which the temperature line (orange) is higher than the melting point (dashed line). Much of the magma that erupts from volcanoes comes from this depth. This is also where the lithosphere, the crust and the rigid part of the mantle that moves with the plates, ends. Below 100 km, there is partly melted rock that is close to the melting temperature of mafic silicates. The plastic nature of that layer allows rigid lithospheric plates to slowly move over Earth's surface. The **asthenosphere** is the plastic part of the mantle. Plastic refers to its state of matter. Have you ever played with Silly Putty? Silly Putty is a product that shatters like a solid when it is hit with a hammer. But leave a sphere of Silly Putty on a flat surface; it will slowly collapse into a puddle. A material is said to be **plastic** if it shows the properties of a solid under short-term stress, but flows like a liquid when stress is applied over a long period of time.

Below the asthenosphere, the melting point increases due to higher pressure, and the mantle loses its plastic nature.

This is because the temperature of this part of the mantle is well below the melting temperature. The question marks on the melting point line show that scientists are not sure about this part of the dotted line. Under these conditions of pressure, the melting temperature should be well above the solid temperature line. However, recent research has shown that some magma does seem to originate within the stiffer mantle. This might indicate that variations in composition occur here. Some parts of the stiffer mantle might melt at lower temperatures than other parts.

There is a sudden change in melting temperature at a depth of about 2900 km. Geophysicists, scientists who study the physical properties of Earth materials, believe that this is where the silicate composition of the mantle ends and the mostly iron outer core begins. Iron has a lower melting temperature than mafic silicate minerals. Therefore, the outer core is a molten liquid. Finally, there is a transition from the liquid part of the outer core to the solid phase of the inner core, even without a major change in composition. This indicates that the temperature at Earth's center is again below the melting point, and therefore the inner core is probably solid.

 Pressure

The graph at the center of the figure illustrates how the pressure within Earth changes with depth. This pressure is caused by the weight of the layers above. Therefore, as you might expect, pressure increases with depth. Note that the scale on the vertical axis is in millions of times normal atmospheric pressure.

The top of the diagram illustrates Earth's internal structure. By looking at the labels and surface features, you can tell that the diagram represents a region of earth from the middle of the Atlantic Ocean, across North America, and into the Pacific Ocean. The height of the Cascade Mountains is greatly exaggerated. It would be difficult to see the mountains if they were actually drawn to scale. Notice the arrows that show upwelling at the Mid-Atlantic Ridge and subduction at the trench. This is a part of the internal convection explained earlier.

The diagram also shows two ways to divide Earth's interior. From studies of earthquakes, scientists have learned much about Earth's crust, shown here by the dark line at the surface. (The thickness of this line has also been exaggerated.) The white, gray, and dotted sections are within Earth's mantle. Below the mantle are the outer and inner cores. A different way to show the interior is based on motions of the lithospheric plates. In the plate tectonic model, Earth's top layer is the lithosphere. The lithosphere includes the crust, shown in black, and the upper part of the mantle, shown in white. The asthenosphere and the stiffer, more solid part of the mantle are above the outer and inner cores.

 ## Hot Spots

A **hot spot** is a long-lived source of magma within the asthenosphere and below the moving lithospheric plates. If the plates were not moving, people would observe repeated eruptions of a single volcano. Instead, people see a line of volcanoes in which the oldest volcanoes are at one end and the youngest are at the other end. Scientists interpret this pattern as evidence of the motion of a plate over a stationary hot spot under the lithosphere.

The Hawaiian Islands (Figure 8-8) are an excellent example of volcanic eruptions at a hot spot. The oldest volcanoes in the Hawaiian chain are at the northwestern end of the chain. The volcanoes that produced the island of Kauai were active 3.8 to 5.6 million years ago. The Kauai volcanoes have not erupted since then. The major islands of Oahu, Molokai, and Maui were built by successive eruptions of volcanoes approximately 2.5, 1.5, and 1 million years ago, respectively. The youngest of the Hawaiian Islands is Hawaii, which gave its name to the whole chain of islands. Kilauea volcano on the island of Hawaii has been erupting continuously for nearly half a century.

If you did not know about plate motions, you would think that the source of lava was moving southeast. However, sci-

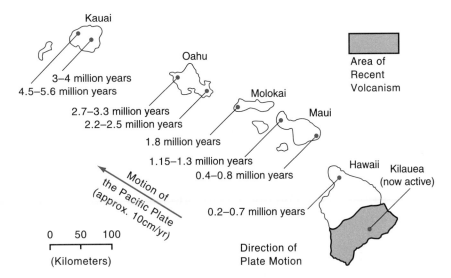

Figure 8-8 The progression of ages in the Hawaiian Islands shows that the Pacific Plate is moving over a deep hot spot below the lithospheric plate. Kilauea volcano on the largest island is currently active.

entists now realize that the northwestward motion of the Pacific Plate has transported the ocean floor over the hot spot. Successive eruptions created new islands while the older islands were carried northwest on the moving Pacific Plate.

ACTIVITY 8-2   **GRAPHING HAWAIIAN VOLCANOES**

Use the information in Figure 8-8 to draw a graph that shows ages of the Hawaiian Islands compared with their distance from Kilauea. The position of Kilauea is shown by a black dot on the island of Hawaii. Then, from your graph, determine the rate at which the Pacific Plate is moving over the stationary hot spot.

The line of Hawaiian volcanoes did not begin at Kauai. Oceanographers (scientists who study Earth's oceans) have analyzed basalt from the Emperor Chain of islands and seamounts that extends from the Hawaiian Islands northwestward nearly to Alaska. The age of these islands and seamounts steadily increases in that direction to a maximum age of approximately 60 million years. It seems clear that the

Hawaiian hot spot below the lithosphere has been supplying magma for at least the past 60 million years. During this time, the Pacific Plate has moved thousands of kilometers to the northwest. Even at a pace of a just few centimeters per year, over a period of millions of years the lithospheric plate can move great distances.

ACTIVITY 8-3   **ZONES OF CRUSTAL ACTIVITY**

You can find worldwide lists of active volcanoes and earthquakes, which provide latitude and longitude, on the Internet. Use this information to plot the position of at least 25 volcanoes or earthquake epicenters (choose one) on an outline map of the world. Compare the location of the features on your map with the features that others have plotted from different lists. Can you see similarities? Do the maps show any kind of pattern? Why do earthquakes and volcanoes often occur in the same places?

 Locating Plates and Plate Boundaries

A tectonic plate moves over Earth's surface as a single unit. The rigid nature of the plates generally transfers force applied anywhere in the plate to the edges of the plates. Although large earthquakes can and do occur within the plates, earthquake epicenters are more common where one plate meets another. When seismologists were able to record and locate earthquakes all over the globe, they noticed that there are distinct zones of crustal activity that stretch around the world. These zones of earthquake activity are also regions in which there are many volcanic eruptions and where tectonic mountain building is occurring. Compare Figure 8-9 with the map of plate boundaries in the *Earth Science Reference Tables*.

At the ocean ridges where new crust is forming, the plates are relatively thin. Deep earthquakes do not occur in this region because the upwelling material is relatively plastic, so it tends to flow and bend rather than break suddenly. Although there are many earthquakes at the mid-ocean ridges, their

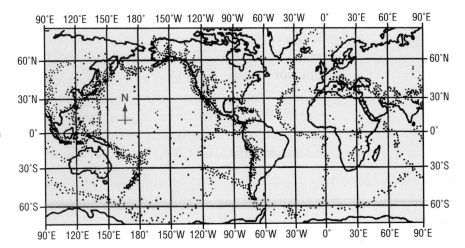

Figure 8-9 Dots on this map indicate the distribution of earthquake epicenters. Note that the greatest concentration of epicenters is at the boundaries of the lithospheric plates.

foci are shallow. However, where plates descend into the asthenosphere, the earthquake foci are deep. As they sink into Earth, plates remain rigid until they are heated enough to become plastic and pliable. Figure 8-10 shows the concentration of shallow seismic foci near a mid-ocean ridge and the persistence of deep foci where subduction occurs near an ocean trench. Seismologists can locate sinking plates by plotting deep focus earthquakes.

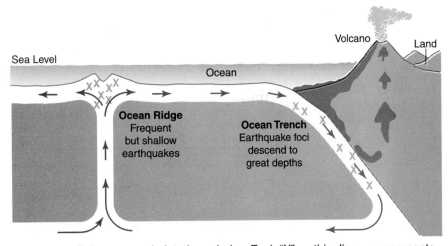

Figure 8-10 Epicenters and plate boundaries. Each "X" on this diagram represents an earthquake focus. At the mid-ocean ridges, heated lithosphere remains plastic until it rises to a position near the surface. Although earthquakes are numerous at the ridges, they are shallow. But a plate descending at an ocean trench remains cool long enough to support deep-focus earthquakes.

Types of Plate Boundaries

Figure 8-11 is a world map from the *Earth Science Reference Tables*. This map shows the boundaries of Earth's lithospheric plates. Plate boundaries are some of the most active zones of earthquakes and volcanic eruptions. These boundaries can be classified into four types based on the relative plate motions.

DIVERGENT BOUNDARIES   The double lines on the map indicate rift boundaries. These are places such as the Mid-Atlantic Ridge where heated material rises toward the surface. At the

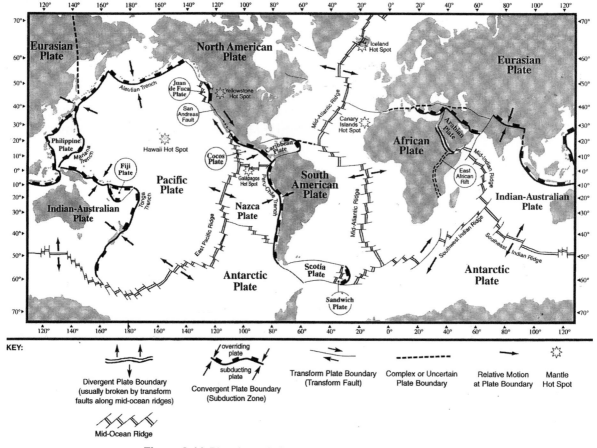

Figure 8-11 Plate boundaries are places where Earth's rigid lithospheric plates meet. At these boundaries, new lithosphere is created, destroyed, or the plate moves parallel to the boundary.

same time, the lithosphere is spreading away from the ridge allowing magma to come to the surface, cool, and make new lithosphere. At the mid-ocean ridges new ocean floor spreads away from the plate boundary. For this reason these areas are called **divergent plate boundaries**. Recall that this is also a region of many small and shallow earthquakes. Ocean ridges are where heat energy is escaping from Earth's interior most rapidly.

CONVERGENT BOUNDARIES The lines with black rectangles indicate subduction zones. These are the places where old, cool lithosphere sinks into Earth. Because the motion of the lithosphere is toward the plate boundary, zones of subduction are also known as **convergent plate boundaries**. It is common to find continental and oceanic lithosphere meeting near subduction zones. You may recall that the rock of the continents is granitic and therefore resists subduction.

A good example of a convergent plate boundary is the Peru-Chile Trench along the western coast of South America. The Nazca Plate is ocean floor and mafic in composition. It is therefore easily drawn into Earth. As it descends, the edge of the plate bends and shifts causing earthquakes. Descending plates are cooler than their surroundings so they stay rigid until the have absorbed heat. Seismologists therefore observe many earthquakes in these regions. Subduction zone earthquakes can be large because the brittle rock can absorb a great deal of energy before it breaks. The earthquakes can also have deep foci because the descending slab of lithosphere is cooler and less bendable than other rocks at the same depth. Observation of deep-focus earthquakes has given geologists an important tool that allows them to locate the plates as they move into Earth at subduction zones. Descending plates are the only place where earthquakes can originate from deep within Earth.

On the eastern side of the Peru-Chile Trench, most of the continental, felsic rocks are too light to be drawn down by subduction. They therefore bend, break, and pile up to make intensely folded and faulted mountains. In fact, the Andes Mountains along the western side of South America are second only to mountains of Asia in height. Some of the felsic rocks from the South American side of the subduction zone are drawn down.

You may recall from Chapter 5 that felsic minerals generally melt at a lower temperature than do mafic minerals. Therefore, as they absorb heat from their surroundings and from friction caused by plate movements, felsic rocks are the first to melt. The change to magma makes the rock fluid. Due to felsic magma's low density, it rises toward the surface. For this reason, subduction zones are regions of volcanic activity. Unlike the mafic volcanoes of Hawaii, which generate long-lasting streams of lava, felsic volcanoes tend to be more violent. Felsic magma is likely to contain water, which expands explosively when it reaches the surface. The 1980 eruption of Mount St. Helens in Washington is a good example of the eruption of less mafic lava.

The edges of converging plates may both be continental crust. Because continental crust resists subduction, this type of collision can produce a great mass of jumbled rock that builds the world's highest mountains. The Himalaya Mountains of Asia are the result of a collision between the north-moving Indian-Australian Plate and the giant plate that contains most of Europe and Asia. Measurements conducted in the Himalaya Mountains have shown uplift of several centimeters per year.

Volcanoes are not restricted to land areas. Volcanoes are common where the oceanic portion of one plate dives under another ocean floor segment. Partial melting of the descending plate may result in a curved line of volcanic islands known an **island arc**. The Aleutian Islands, which extend to the westward from Alaska, and the islands of Japan are island arcs created at subduction zones.

TRANSFORM BOUNDARIES   Some plates do not converge or diverge. A **transform boundary** occurs when two plates slip past each other without creating new lithosphere or destroying old lithosphere. If you could straddle a transform fault for a long enough time, you would see one foot heading in one direction parallel to the fault and your other foot going in the opposite direction. The San Andreas Fault in California is an excellent example of a transform boundary. In this area, the Pacific Plate is moving northwest with respect to the North American Plate. Motion along the fault is not continuous. At any place along the fault, the plates may be locked together by friction. When the force on the fault becomes great enough to overcome friction, the fault breaks suddenly and the plates move. This motion generates an earthquake.

DOES EARTH'S GEOGRAPHY CHANGE?

Scientists have used a wide variety of evidence to document how the continents have moved over millions of years. Ocean basins are known to be younger than the continents. This is because as new seafloor is created at the ocean ridges, old parts of the seafloor are drawn back into Earth's interior at the ocean trenches. The dense basaltic rock of the ocean floor is constantly recycling. However, the continents resist subduction due to their lower density. Large parts of the continents are composed of rock much older than the rocks found anywhere on the ocean floor.

Figure 8-12 on page 214 from the *Earth Science Reference Tables* shows the evolution of Earth's surface over a period of more than 300 million years. Notice that 362 million years ago North America was located along the equator. As time passed, North America moved north along with Africa and South America. In the past 200 million years, North America separated from Africa and Europe, opening the North Atlantic Ocean. Africa and South America split apart more recently, forming the South Atlantic Ocean.

Scientists cannot be sure how the plates will move in the future. But, if present trends continue, the Atlantic Ocean will become wider as separation continues at the Mid-Atlantic Ridge. As the American Plates continue to push into the Nazca and Pacific plates, the mountains near the western edges of Americas could grow higher. (Mountain heights also depend on how fast erosion takes place.) In a few tens of millions of

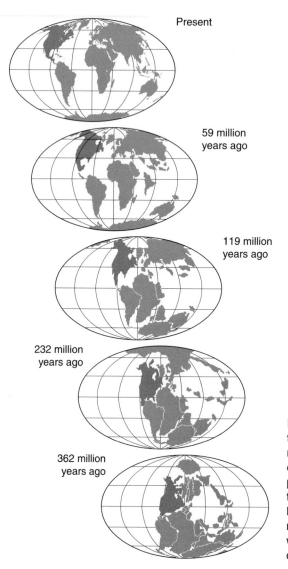

Present

59 million
years ago

119 million
years ago

232 million
years ago

362 million
years ago

Figure 8-12 Movement of the North American continent (shown in black) and other continents over the past 362 million years. Note that the North American landmass initially moved north and then moved westward as the Atlantic Ocean opened.

years, movement along the San Andreas Fault will carry Los Angeles northward to a position near San Francisco.

One of the important principles in geology is sometimes stated as the present is the key to the past. This means that events in the prehistoric past are likely to be similar to changes that you can observe. It also means that if scientists understand the geological processes that are happening today, they can better predict what will happen in the future.

TERMS TO KNOW

asthenosphere	meteorologist	polarity
convergent plate boundary	mid-ocean ridges	radiation
divergent plate boundary	ocean trenches	seafloor spreading
fluid	paradigm	subduction zone
hot spot	plastic	tectonics
island arc	plate tectonics	transform boundary
lithospheric plates		

CHAPTER REVIEW QUESTIONS

1. Remains of *Mesosaurus*, an extinct freshwater reptile, have been found in bedrock of similar ages at locations X and Y on the map below.

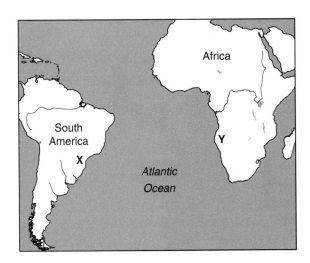

Which statement below represents the most logical conclusion to draw from this evidence?

(1) *Mesosaurus* migrated across the ocean from location X to location Y.
(2) *Mesosaurus* came into existence at widely separated locations at different times.
(3) South America and Africa were joined when *Mesosaurus* was alive.
(4) The present climates at locations X and Y are similar.

2. Which form of heat flow is responsible for the slow circulation within Earth's asthenosphere?

(1) insolation
(2) convection
(3) conduction
(4) radiation

3. Convection currents within Earth's mantle are making the Atlantic Ocean

(1) less salty.
(2) cooler.
(3) wider.
(4) narrower.

4. What is the approximate temperature at the boundary between the asthenosphere and the stiffer mantle?

(1) 600°C   (3) 2600°C
(2) 1000°C   (4) 3000°C

Use the information in the following map and data table to answer questions 5 and 6. The map shows the locations of volcanic islands and seamounts that erupted on the sea floor of the Pacific Plate as it moved northwest over a stationary hot spot beneath the lithosphere. The hot spot is currently under Kilauea.

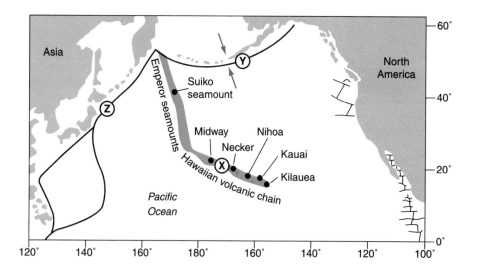

Data Table
Age of Volcanic Features

Volcanic Feature	Distance from Kilauea (km)	Age (millions of years)
Kauai	545	5.6
Nihoa	800	6.9
Necker	1070	10.4
Midway	2450	16.2
Suiko seamount	4950	41.0

5. Approximately how far has location X moved from its original position over the hot spot?

    (1) 3600 km
    (2) 2500 km
    (3) 1800 km
    (4) 20 km

6. According to the data table, what is the approximate speed at which the island of Kauai has been moving away from the mantle hot spot, in kilometers per million years?

    (1) 1
    (2) 10
    (3) 100
    (4) 1000

7. Which lithospheric plate boundary features are located at Y and Z?

    (1) trenches created by the subduction of the Pacific Plate
    (2) rift valleys created by seafloor spreading of the Pacific Plate
    (3) secondary plates created by volcanic activity within the Pacific Plate
    (4) mid-ocean ridges created by faulting below the Pacific Plate

8. The Himalaya Mountains are located along a portion of the southern boundary of the Eurasian Plate. Near the top of Mt. Everest (29,028 feet) in the Himalaya Mountains climbers have found fossilized marine shells

in the surface bedrock. From this observation, which statement is the best inference about the origin of the Himalaya Mountains?

(1) The Himalaya Mountains were formed by volcanic activity.
(2) Sea level has been lowered about 29,000 feet since the shells were fossilized.
(3) The bedrock containing the fossil shells is part of an uplifted seafloor.
(4) The Himalaya Mountains formed at a divergent plate boundary.

9. The diagram below shows land features that have been disrupted by an earthquake.

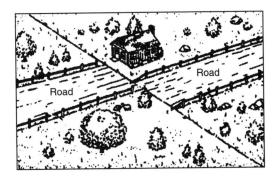

Which type of crustal movement most likely caused the displacement of features in this area?

(1) vertical lifting of surface rock   (3) downwarping of the crust
(2) folding of surface rock   (4) movement along a transform fault

Use the information in the diagram below to answer questions 10 and 11. The diagram shows the locations of deep-sea core drilling sites numbered 1–4.

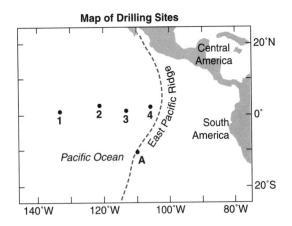

10. At point A, the East Pacific Ridge is the boundary between the

    (1) Cocos Plate and the North American Plate.
    (2) South American Plate and the Nazca Plate.
    (3) Pacific Plate and the South American Plate.
    (4) Pacific Plate and the Nazca Plate.

11. At which drilling site would the oldest bedrock most likely be found?

    (1) 1 (3) 3
    (2) 2 (4) 4

12. Compared with the thickness and density of the continental crust of South America, the oceanic crust of the Pacific floor is

    (1) thinner and less dense.
    (2) thinner and more dense.
    (3) thicker and less dense.
    (4) thicker and more dense.

Base your answers to questions 13–15 on the following diagram.

13. The Peru-Chile Trench marks the boundary between the

    (1) Pacific Plate and the Antarctic Plate.
    (2) Nazca Plate and the South American Plate.
    (3) North American Plate and the Cocos Plate.
    (4) Caribbean Plate and the Scotia Plate.

14. In which diagram do the arrows best represent the motion of Earth's crust at the Peru-Chile Trench?

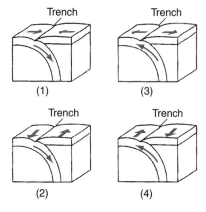

15. Which observation provides the best evidence for the pattern of crustal movement at the Peru-Chile Trench?

    (1) the direction of flow of warm ocean currents
    (2) the mineral composition of samples of mafic mantle rock
    (3) comparison of the rates of sediment deposition
    (4) the locations of shallow focus and deep focus earthquakes

Open-Ended Questions

16. What evidence indicates that lithospheric plates move over Earth's surface.

17. How is energy moved from one place to another in heat flow by convection?

18. Earthquakes are not evenly distributed over Earth. Why are earthquakes common along the west coast of the continental United States?

19. How can plates move apart at the mid-ocean ridges and not leave a deep gap in the lithosphere?

20. Classify each of the plate boundaries listed below as divergent, convergent, or transform.
    (a) East Pacific Ridge
    (b) Aleutian Trench
    (c) Western side of South American Plate
    (d) San Andreas Fault

Chapter 9

Geologic Hazards

 ## WHAT IS A GEOLOGIC HAZARD?

People who live near the coast of southern California enjoy a location with mild temperatures and beautiful mountain scenery. The climate is usually dry, but rain and snow in mountains to the east provide freshwater and recreational opportunities. However, other aspects of life in California are not as favorable. If it rains hard, there can be flooding and landslides. This is because the mountains are young with steep, unstable slopes. In addition, the dry summer climate results in relatively few plants to hold back surface water and keep soil in place. This area also has more destructive earthquakes than any other part of the United States except Alaska.

Most residents of this part of California do not realize that the mild climate and mountain scenery are a result of active geological forces. The mild climate is enhanced by winds off the ocean that are stopped by mountain barriers. The San Andreas Fault and other faults along which earthquakes occur are responsible for the mountains. If it were not for earthquakes, the mountains would not be there. If it were not for the mountains, the climate would not be as mild. It all relates to geology.

Most geologic events, such as uplift, weathering, and erosion, take place over a long period of time. Such slow events are seldom a danger to people. But, some geologic changes happen quickly, such as earthquakes, volcanic eruptions, landslides, and avalanches. These rapid changes can cause hazards to humans. A **hazard** is an event that places people in danger of injury, loss of life, or property damage.

Earthquakes

Some locations have more earthquakes than others. (See Figure 9-1.) The interaction between Earth's lithospheric plates makes plate boundaries the most active zones of change. Within the continental United States, our only plate boundary is the western edge of the North American Plate. The boundary between the Pacific and North American plates runs through California and then off the Pacific coast from Oregon to Alaska. Large earthquakes also can occur within the plates in places where the continental plates seem to be breaking apart. Two of the strongest seismic events in American history took place in the nineteenth century, one in Missouri and the other in South Carolina. Although large earthquakes are most common at plate boundaries, it appears that earthquakes can happen almost anywhere.

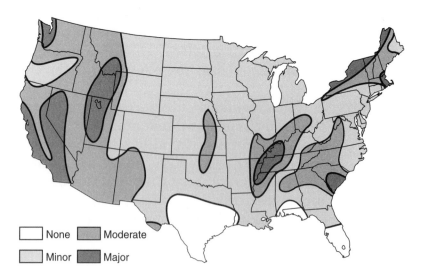

Figure 9-1 Most predictions of earthquake hazards are based on historical records. Although earthquakes are less common in the central and eastern United States than they are along the Pacific Coast, there have been major earthquakes in all three regions.

It is helpful to realize that when seismic waves pass from solid rock into loose sediment, the shaking is intensified. A major earthquake in Mexico in 1985 did more damage in Mexico City than in Acapulco, even though Acapulco was closer to the epicenter. Acapulco is built on or close to solid rock. On the other hand, Mexico City is built on sediments of an ancient lake bed. Not only do thick sediments amplify the shaking of an earthquake, they also provide a less secure foundation on which to construct large buildings.

Seismologists sometimes say, that Earthquakes do not kill people, buildings do. People are rarely hurt by motion of the ground. The collapse of buildings is the major cause of death and injury in most earthquakes. This is especially true if the buildings have no reinforcement to hold them together. Adobe is a building material made by mixing mud with straw, placing the mixture into molds to form blocks, and drying the blocks in the sun. Adobe blocks are stacked to make walls of the house and wooden rafters are placed across the tops of the walls to support a roof. This is a common form of construction in some third-world countries. Most adobe homes do not have any frame to hold the bricks together. Adobe structures are reasonably good at withstanding vertical motion, but horizontal motion can cause the walls and roof to collapse on anyone unfortunate enough to be inside. Wood frame houses are among the best buildings at withstanding earthquakes. In this kind of house, walls are held in place by adjoining walls. The foundation, floors, walls, and roof of the home can be secured with bolts. Wood frame structures absorb energy by bending and can still go back into their original shape. Steel frame buildings are also good absorbers of ground motion.

Fires often follow earthquakes in populated areas. The fire that followed the San Francisco earthquake of 1906 caused far more damage than was caused by the collapse of structures. The ground broke and shifted as much as 6 meters, causing gas lines to rupture. Damaged and sparking electrical lines ignited the highly flammable natural gas escaping from broken gas lines. Water pipes, which could have provided water to fight fires, were also broken, making it impossible to save burning buildings. The fire burned for several days, destroying most of the city.

Fortunately, builders have learned from their mistakes. Events in San Francisco and other cities that have experienced major earthquakes have helped engineers understand how to avoid damage. When the Alaska oil pipeline was constructed in the 1970s, special bends and joints were added where the pipeline crosses known fault lines. Should an earthquake occur causing movement along the fault, these bends should prevent the pipeline from breaking.

Earthquakes can also trigger ground failure. Near Anchorage, Alaska, 75 houses were lost in the 1964 earthquake. The houses were part of a development built on a high bluff of sediment overlooking an arm of the ocean. Shaking caused the sediments and the houses built on them to loosen and then collapse into the bay.

Sediments that hold groundwater pose another hazard. Saturated sediments can turn into a material like quicksand in a process known as **liquefaction**. This is caused by strong shaking that allows water to surround the particles of sediment, changing the sediment into a material with properties of a thick fluid. Buildings can sink where the ground is weakened by liquefaction.

Dam failure is also a hazard associated with earthquakes. The shaking of the ground or a landslide can break dams. If a reservoir of water is held back by the dam, people who live downstream will be in danger from flooding. Table 9-1 lists some of the world's greatest earthquakes as well as nearby seismic events.

Earthquakes sometimes cause a giant series of waves called a **tsunami** (sue-NAHM-ee). Although tsunamis are sometimes called tidal waves, these waves have nothing to do with the twice-daily rise and fall of ocean tides. That is why scientists prefer to use tsunami, a Japanese term that means harbor wave. The most destructive tsunamis are probably caused by a sudden motion of the ocean bottom or an underwater landslide released by an earthquake. In the open ocean, tsunamis may travel 1000 km/hr as a gentle swell that would be hard to notice on board a ship. At this speed, a tsunami can cross a major ocean in a few hours. When a tsunami moves into an open but shallow bay, its energy becomes more concentrated and the water can build high waves. Some tsunamis

TABLE 9-1. Selected Historic Seismic Events

Location	Date	Magnitude	Notes
Shensi, China	1556	Unknown	Worst natural disaster known; 830,000 deaths.
New Madrid, MO	1812	~8 (est.)	Few deaths due to sparse settlement. Flow of Mississippi River briefly reversed.
San Francisco, CA	1906	~8.3 (est.)	Most damage caused by uncontrolled fires. Water lines broken.
Massena, NY	1944	~6 (est.)	Largest in NY state. Chimneys destroyed; water lines broken.
Lebu, Chile	1960	9.6	Largest event measured with seismographs. Occurred in the Pacific Ocean.
Anchorage, AK	1964	9.3	Extensive tsunami. Damage to buildings extensive; 131 deaths.
Haicheng, China	1975	7.5	Predicted by scientists. City evacuated; only ~130 deaths.
Tengshan, China	1976	7.6	Prediction failed. Worst modern natural disaster; ~650,000 deaths.
Mexico City, Mexico	1985	8.1	Worst damage in city far from epicenter that was built on fill; ~9,000 deaths.
North Ridge, CA	1994	6.7	$10 billion in damages; 61 deaths.
Kobe, Japan	1995	7.2	Destroyed the port built on landfill; 5,500 deaths
Izmit, Turkey	1999	7.4	More than 12,000 dead and 34,000 injured. Lateral offsets of 2.5 m (9 feet). Largest event in a modern, industrialized area since San Francisco quake in 1906 and Tokyo quake in 1923.
Au Sable Forks, NY	2002	5.1	Felt throughout NY state. Caused damage to local roads.

grow to 10 to 20 meters high, sometimes higher. These waves may appear first as a giant wave or as a sudden drop in sea level. People have been drawn to the shore by the sudden withdrawal of the sea. They may not realize that they are observing a sign that a giant wave is coming that could sweep them away. Tsunamis are rare, but they cause great damage and loss of life in coastal locations. The city of Hilo in Hawaii has experienced several destructive tsunamis. The United

States now has a tsunami early warning system to alert people who live in coastal areas of approaching danger.

Earthquake Preparedness

Reducing the risk of injury or loss of property is especially important in places that have a history of damaging earthquakes. There are things people can do to prevent injuries and property loss. The following measures can be split into two categories: preparing for an earthquake and what to do during an earthquake. The following are examples of advanced preparations.

- Select a home built on or close to solid bedrock.
- Select a homesite that is not near a steep hill, open and shallow bay of the ocean, or downstream from a reservoir.
- Be sure your home meets local building codes.
- Know how to shut off gas, electricity, and water.
- Avoid storing heavy objects on high shelves.
- Store some food and freshwater in your home, as well as a battery-operated radio.
- Keep emergency telephone numbers in handy locations.
- Know where to find medical supplies and the location of the nearest doctor or hospital.
- Learn how to help any family member with special needs.
- Plan and rehearse what to do in case of an earthquake.

After a major earthquake, your help may be needed.

- Assist official emergency services as requested, but do not get in their way.
- Until help arrives, respond to the needs of people who are injured.

- Prevent further injuries by identifying unsafe structures, broken objects, chemical spills, and similar hazards.

- If possible, turn off gas, water, and electric supplies.

- Listen to the radio for information and instructions.

- Find safe shelter, undamaged food supplies, and clean water.

In a strong earthquake, your best protection depends on where you are. If you are outside, it is best to stay away from buildings or trees, since they may fall. If you are in a building, the best protection is probably under a strong object, such as a desk or a table, that can protect you from falling debris. Doorways are also good places in which to stay because the walls might be able to protect you if the ceiling collapses. While it would probably be safer to be outside, you may not have time to reach an exit before the earthquake is over. Most earthquakes last less than a minute. Do not use an elevator to leave a building during or after an earthquake. Elevators can be damaged by the original event or by aftershocks, trapping people inside. Stairways are a safer way to leave. Remember that aftershocks often occur following a major earthquake.

ACTIVITY 9-1 DEVISING AN EARTHQUAKE PREPAREDNESS PLAN

Work with your family or classmates to devise an emergency plan to use in the event of a damaging earthquake. Brainstorm with them to take into account unique characteristics of where you live and the needs of people around you. Create a list of things you can change to make your home more earthquake-safe.

PREDICTING EARTHQUAKES There have been many attempts to predict earthquakes. Such signs as bulging of the ground, unexplained changes in the water levels in wells, or even unusual animal behavior have been used to predict seismic events. Places along major faults that have not experienced ground movement as often as surrounding areas are considered likely places for future earthquakes. For example, seismologists might be able to predict that over the next 20 years there is a 60 percent chance of a magnitude 7 earthquake along a particular part of the San Andreas Fault.

Chinese scientists predicted a major earthquake and evacuated the city of Haicheng in 1975. A large earthquake did occur at that time and the death toll was remarkably light. However, a year later they missed a larger earthquake in the city of Tengshan where more than half a million people lost their lives.

In spite of a few successes, predictions of earthquakes that successfully specify a particular time and place are rare. Government agencies are reluctant to issue warnings when most earthquake warnings have been incorrect. If they issue many false warnings, the public will no longer respond to them.

 ## Volcanic Eruptions

Before it erupted in 1980, Mount St. Helens in the state of Washington was a nearly perfect volcanic cone almost 3000 meters high. The mountain was known to be an active volcano. In fact in 1975, scientists from the United States Geological Survey predicted that it would probably erupt before the turn of the century.

Signs of activity began in late March of 1980 with numerous small earthquakes generated by underground movement of magma. A bulge in the northern slope of the mountain grew to about 100 meters. On May 18, it broke loose and rushed down the northern slope of the mountain in a great, gas-charged cloud. The loss of pressure within the mountain resulted in a blast of hot ash and volcanic gases that filled Spirit Lake, a popular fishing resort on the north side of the mountain. About 1 cubic kilometer of the mountain was lost, and ash fell several centimeters deep hundreds of miles away. The mountain lost nearly a quarter of its height, and 60 people lost their lives. In fact, a number of the scientists who came to observe Mount St. Helens were burned to death or suffocated in

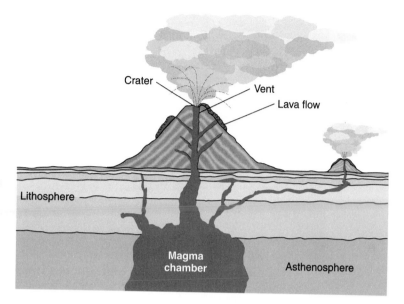

Figure 9-2 A volcano is a place where magma escapes from Earth's interior. Sometimes the eruption builds a mountain; other eruptions cover large, nearly flat areas with fluid lava. (Not drawn to scale.)

the ash cloud. Since the 1980 eruption, a small dome has grown inside the new crater. This dome could grow to the original height of the mountain, or it could lead to another eruption in the future.

A **volcano** is an opening in Earth's surface through which molten magma erupts. The source of magma is somewhere within the planet where the temperature is above the rocks' melting temperature. The temperature at which rock melts depends on its mineral composition and pressure on it. Felsic rocks melt at a lower temperature than mafic rocks. As pressure increases, so does the melting temperature. Fluid magma moves toward the surface through cracks or zones of weakness in the overlying rock. When the magma reaches the surface and releases gases into the atmosphere, it is called lava.

Figure 9-2 shows a cross section, or profile view, of a volcano. In this diagram you can see the magma chamber with vents that lead to the surface. Note the layering of ash and lava inside the mountain, which was built up by successive eruptions. Explosive eruptions of some volcanoes leave a bowl-shaped depression at the top of the mountain called a **crater**.

Some volcanoes erupt quietly. Kilauea on the island of Hawaii has continuously vented rivers of lava for several

Figure 9-3 The teardrop shape of this volcanic rock shows that it hardened from lava as it was thrown into the air. It is known as a volcanic bomb.

decades. The basaltic lava that feeds Kilauea is very hot and contains little gas, making its lava very fluid. But other volcanoes such as Mount St. Helens vent lava that is more felsic in composition. These lavas are recycled continental rocks. They contain more silicate minerals as well as dissolved gases such as water vapor, carbon dioxide, and sulfur dioxide. As felsic lava comes to the surface, decreasing pressure causes the gases to expand explosively like the soda in a bottle that has been shaken. Such volcanic eruptions release large quantities of volcanic ash and toss larger objects known as blocks and bombs into the air. These are considered the most dangerous eruptions. Figure 9-3 shows a volcanic bomb from a volcano in California.

TYPES OF VOLCANOES The cooling of magma at the surface builds volcanic features around the **vent** where the lava comes to the surface. Scientists recognize four types of volcanoes based on their shapes: shield volcanoes, cinder cones, composite volcanoes, and lava plateaus.

Repeated eruptions of hot, fluid, basaltic magma build a broad structure with gently sloping sides known as a shield volcano. The Hawaiian Islands contain shield volcanoes as much as 100 km across. Cinder cones are usually small fea-

Figure 9-4 Sunset Crater in Arizona is a cinder cone produced by an eruption approximately 1000 years ago.

tures built by cooler lava that was blown into the air, fell back to Earth, and hardened into a pile around the vent. (See Figure 9-4.) Composite volcanoes are mounds built up by alternating lava flows and layers of ash. Mount St. Helens is a good example of a composite volcano. The eastern part of the state of Washington is covered by hundreds of meters of flat layers of successive fluid lava flows that created lava plateaus. The lavas that formed these plateaus were so hot that they flowed over the surface almost like water before they hardened into basalt.

Some volcanoes form a **caldera**. This is a large bowl-shaped depression formed when the top of the volcano collapsed into the emptied magma chamber. Crater Lake in Oregon is a caldera that has filled with water, making a large, round lake where the top of the mountain used to be.

 ## Volcanoes as Hazards

How likely you are to suffer injury or loss of property from a volcanic event depends on where you are. If volcanoes erupted in your location in the past, you are probably in a place that could have future eruptions. The more recent the past eruptions, the higher the likelihood of future eruptions. Volcanoes are sometimes classified as active or dormant. If scientists see evidence of recent activity or if they see steam rising out of a volcano, it is considered active. However, dormant volcanoes can suddenly erupt, showing how difficult it is to classify them accurately.

| ACTIVITY 9-2 | ADOPT A VOLCANO |

Select a famous volcano. Be sure that your teacher approves the volcano you have chosen. Prepare a report about your volcano's activity. Please give your source(s) of information in the form of a bibliography.

Figure 9-5 shows that volcanoes, like earthquakes, tend to occur near plate boundaries. Notice the way that active volcanoes nearly surround the Pacific Ocean. That region is called the Ring of Fire. Many of these volcanoes occur inland near ocean trenches. Subduction zones are especially dangerous because this is where granitic (felsic) rocks are pulled into Earth's interior. With their low melting temperature, low-density, high-viscosity magma, and considerable gas content, subduction zone volcanoes are the most dangerous kind of volcano.

In the real estate business it is said that the value of a property depends on three factors: location, location, and location. The same factor(s) will determine your vulnerability to volcanic hazards. The first question is, Are there volcanoes in your area? If not, the likelihood of danger is low. If there is local evidence of volcanoes, the second location factor comes

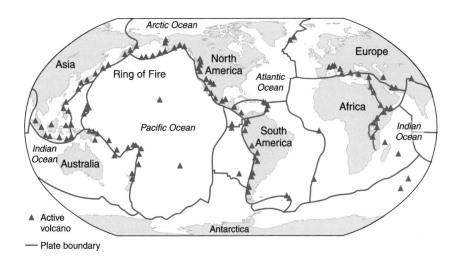

Figure 9-5 Most of the world's active volcanoes are at or near plate boundaries.

in. How close are you to a volcano that is or could become active? If the volcano is within a few tens of kilometers, the next question comes into play. Is your home, school, or place of work in a valley connected to the volcano?

Volcano damage can come in several ways. Sometimes, lava flows out of a volcano and runs downhill into valleys destroying anything in its path. However, most lava flows are slow enough that people can usually escape them. A greater threat is gas-charged flows of hot ash that can descend from a volcano at 100 km/hr or more. This was the kind of activity that killed observers and flattened forests around Mount St. Helens in 1980. Some volcanoes discharge poisonous gases that can suffocate people and animals in nearby lowland areas. Meltwater from the snow that covered Mount St. Helens before eruption was the greatest cause of damage in 1980 when the water quickly flowed into rivers already clogged with ash. Eruptions can also emit poisonous gases and choking dust, while they trigger landslides and mudflows that cause loss of life and property.

Volcanoes sometimes show signs of an impending eruption. The best way to protect yourself from their dangers is to move far enough away to avoid direct effects such as lava flows and gas clouds, and high enough to escape floodwaters.

 Mass Movement

Gravity is the force that pulls all matter toward Earth's center. As tectonic forces build mountains, erosion powered by gravity wears them down. Nearly all erosion starts with earth materials at a high elevation and moves them to a lower elevation. Wherever the ground is too steep for friction to hold rock and soil in place, there is danger that it will move downslope. **Mass movement** is the motion of soil or rock down a slope without the influence of running water, wind, or glaciers.

Some soil movement happens slowly such as the slow downhill creep that caused the trees in Figure 9-6 on page 234 to bend or the slumping of a block of soil shown in Figure 9-7 on page 234. These slow movements are most com-

Figure 9-6 Soil creep in Utah

mon in clay-rich soils, particularly when they become saturated with water. If roads or buildings are constructed along these slopes, the structures can lean and break apart as downslope motion stresses and carries them.

Sometimes there is the rapid, downslope movement of rock and soil known as a **landslide**. In unpopulated areas,

Figure 9-7 Slumping at Half Moon Bay in California

mass movement is of little concern. But each year downslope mass movement breaks up or covers roads, causes damage to property, and occasionally causes injury or loss of life.

Landslides are often triggered by water seeping into the ground. Clay minerals in soil can absorb many times their dry weight in water. Clay also offers little resistance to gravitational force. If the soil is composed primarily of moist clay, it can even slide down relatively gentle slopes. A 55-acre area near the bottom of a valley south of Syracuse, New York, slid downhill, covering a road and damaging three homes in the spring of 1993. The slope above the slide was steep, but held in place by bedrock. However, the valley had been the site of a large lake that left deposits of soft clay. When the clay was saturated by groundwater it moved downslope.

An **avalanche** is the rapid, downhill movement of snow, similar to a landslide, that occurs on a steep slope. Large quantities of rock can be carried down by avalanches. Some mountain valleys show evidence of avalanches where trees have been uprooted by slides in the past

Like volcanoes, landslides usually occur in places where they have happened in the past. The best way to protect yourself and your property is to be aware of where landslides have occurred previously. Avoid building on or below steep or unstable land. Be aware of avalanche dangers when skiing or traveling in mountain areas in the winter or early spring. Figure 9-8 shows a structure built to protect a mountain road from avalanches and landslides.

Figure 9-8 This snow shed in the Rocky Mountains protects the road and people who travel on it from avalanches and landslides.

TERMS TO KNOW

avalanche	hazard	mass movement	vent
caldera	landslide	tsunami	volcano
crater	liquefaction		

CHAPTER REVIEW QUESTIONS

1. Which city is most likely to have a destructive earthquake in the next 100 years?

   (1) New York City   (3) Houston
   (2) Los Angeles   (4) Chicago

2. Sometimes one destructive natural event is caused by another event. Which of the following is most likely?

   (1) a thunderstorm caused by a landslide
   (2) a tsunami caused by an earthquake
   (3) a hurricane caused by a landslide
   (4) lightning and thunder caused by an earthquake

3. Some people refer to tsunamis as tidal waves. Why do scientists seldom use the term "tidal wave?"

   (1) Most scientists speak Japanese.
   (2) Tsunami is easier to spell
   (3) Tsunamis are not caused by tides.
   (4) Tsunamis occur more often than tides.

4. A homeowner removed all heavy objects from high shelves in her home as a safety measure. That action would probably be most important in preparing for which of the following events?

   (1) a landslide   (3) an earthquake
   (2) a hurricane   (4) a volcanic eruption

5. Where is the safest place to build a home if your area has many earthquakes?

   (1) on the bank of a river   (3) on a thick layer of sediment
   (2) at the base of a weathered cliff   (4) on solid bedrock

6. Why is a wood frame house a good type of building to live in if your region has large earthquakes?

(1) Wood cannot burn.
(2) Wood is more dense than stone or brick.
(3) Wood frame houses often hold together when shaken.
(4) Wood does not conduct heat energy as well as stone or brick.

7. How do cities and other communities now protect citizens from earthquake hazards?

(1) They can require buildings that resist damage from shaking.
(2) The can pass laws to make earthquakes illegal.
(3) They can reinforce the ground to prevent lithospheric plates from moving.
(4) They can tell people exactly when earthquakes will occur and make them leave the area.

8. How would the magma that produces a cinder cone differ from lava that contributes a new layer to a lava plateau?

(1) The cinder cone magma comes out at the surface more slowly.
(2) The lava plateau magma is cooler and more liquid.
(3) The cinder cone magma crystallizes rapidly.
(4) The lava plateau magma does not reach Earth's surface.

9. Which location is most likely to have volcanic eruptions?

(1) near the North Pole
(2) near a large lake
(3) near plate boundaries
(4) near the centers of continents

10. A volcano whose top was covered by large glaciers and snowfields erupted. Which of the following is likely to cause the most property loss, injury, and deaths?

(1) flooding
(2) thunder
(3) violent shaking of the ground
(4) people falling into cracks in the ground

11. What kind of rock is likely to be found on the Hawaiian Islands?

(1) sandstone
(2) limestone
(3) gneiss
(4) basalt

12. What ocean is surrounded by a zone of frequent earthquakes and active volcanoes?

    (1) Atlantic
    (2) Pacific

    (3) Indian
    (4) Arctic

13. Where and when is a landslide likely to happen?

    (1) on bedrock that is wet
    (2) on bedrock that is dry

    (3) on clay sediments that are wet
    (4) on clay sediments that are dry

14. Which natural event usually occurs without warning and usually lasts less than a minute?

    (1) a tsunami
    (2) an earthquake

    (3) a hurricane
    (4) a volcanic eruption

15. Which natural disaster is most likely to include solid objects falling from the sky?

    (1) floods
    (2) earthquakes

    (3) tsunamis
    (4) volcanoes

Open-Ended Questions

16. Students read an article in a local newspaper stating that a major earthquake can be expected to affect that region within the next year. The students plan to stay in the region. As a result, the students decide to help prepare their home and family for this expected earthquake.
    State three specific actions the students could take to increase safety or reduce injury or damage from an earthquake.

17. An Earth science class is creating a booklet on emergency preparedness. State one safety measure that should be taken to reduce injury or deaths during a nearby volcanic eruption.

18. What is one form of evidence that might indicate that your area is in danger from volcanic eruptions?

19. Volcanoes' shapes and surface features can often be used identify them. State one way in which a volcano is likely to look different from other mountains.

20. Why are earthquakes, volcanoes, and landslides considered hazards?

Chemical Weathering

Sometimes the weathering process does more than simply break the rocks apart. If you find a steel nail that has been exposed to the weather for a long time, it will probably be rusted. Rusting is a **chemical change**, which results in the formation of a new substance. Iron, the major component of steel, can combine chemically with oxygen in the atmosphere to form rust (iron oxide). The presence of moisture accelerates the rusting process. This is a form of chemical weathering because a new substance (rust) is formed. **Chemical weathering** is a natural process that occurs under conditions at Earth's surface, forming new compounds. Although steel is not found in nature, many minerals do contain iron. Iron is often one of the first components to weather. When iron combines with oxygen in the atmosphere it forms iron oxide, which gives rock a rusty red to brown color. The chemical equation for this change is (Fe is the chemical symbol for iron)

$$4Fe + 3O_2 \rightarrow 2Fe_2O_3$$
iron + oxygen → iron oxide

Calcite, the principal mineral in limestone and marble, is chemically weathered by water that is acidic. The chemical formula for limestone is $CaCO_3$ (calcium carbonate). Rainwater absorbs carbon dioxide as it falls through the atmosphere, making rain a mild acid. This is not strong enough to hurt you or your clothing, but it can slowly break down limestone. When rainwater infiltrates the ground, it picks up more carbon dioxide from decaying plant remains. The acid (represented by H^+) can then react with limestone. The chemical equation for this change is written as

$$CaCO_3 + 2H^+ \rightarrow Ca^{2+} + H_2O + CO_2$$
calcium carbonate + acid → calcium ions + water + carbon dioxide

This process forms limestone caverns, such as Howe Caverns and Secret Caverns near Cobleskill, New York. Although

Figure 10-6 This sample of limestone has been shaped by chemical weathering as natural acids in rainfall have partly dissolved the calcite.

the longest limestone caverns in New York State have been explored to about 10 km, Mammoth Caves in Kentucky are known to have more than 500 km of connected underground passages. Figure 10-6 shows a sample of limestone that has been weathered by natural acids in rainwater.

Limestone and marble make excellent building stones, although the calcite in them has a hardness of only 3 on Mohs' scale. These rocks are soft enough to be cut into blocks but strong enough to support the weight of a large building. Limestone and marble are also relatively easy to shape into sculptures and ornamentation. Many of our historic buildings are built with limestone and marble. Unfortunately, as cities have developed into centers of industry and commerce, air pollution has weathered the surface of some of these buildings.

Sulfur dioxide primarily from the burning of fossil fuels is a source of sulfuric acid when it combines with moisture in the atmosphere. When combined with moisture in the atmosphere, nitrogen oxides from motor vehicles and electrical power plants form nitric acid. When acid precipitation falls on limestone and marble it changes the mineral calcite into a chalky powder. Many historic buildings and outdoor statues in Europe and North America have been damaged by acid weathering. This is a major reason there are laws to limit acid pollution. Although these measures cannot restore damaged structures, they have slowed the further chemical weathering of buildings and monuments made of limestone and marble.

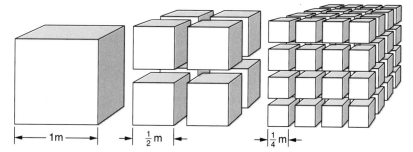

Figure 10-7 As solid rock is broken into smaller pieces, its total surface area increases. Additional surface area can increase the rate of weathering.

SURFACE AREA Figure 10-7 shows that breaking a rock into smaller fragments increases the surface area of the material. Weathering is active on the surfaces, breaking up a rock exposes more surface area, which accelerates the rate of weathering.

| ACTIVITY 10-2 | CALCULATING SURFACE AREA |

Figure 10-7 shows a single cube of rock 1 meter on each side (A) that is divided into progressively smaller pieces. Calculate the total surface area of the samples in parts A, B, and C of the diagram. Show your work. Start with an algebraic formula, substitute numbers and units, and show the mathematical steps to each solution.

| TEACHER DEMONSTRATION | REACTION RATE AND SURFACE AREA |

(**Caution:** Acids can cause skin burns and damage to clothing. Always handle acids with care.)

Materials: 2 small beakers (50–100 mL), natural chalk, a mortar and pestle, about 25 mL of 1 molar hydrochloric acid.

Break off two equal sized lengths (about 1 cm) of natural chalk. Place the first piece of chalk in a beaker of acid and watch the reaction. Use the mortar and pestle to crush the second piece of chalk into a fine powder. Add the powder to a second beaker containing acid. Why do the reactions in the two beakers differ?

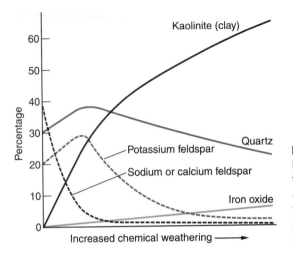

Figure 10-8 Fresh granite is composed primarily of feldspar and quartz. Through a long process of chemical weathering, the primary components change to clay, quartz, and iron oxide."

Feldspar is the most common mineral in rocks at or near Earth's surface. But feldspar is not stable when it is exposed to the atmosphere over very long periods of time. Feldspar weathers to a softer material composed primarily of clay and silica.

Figure 10-8 shows the changing mineral composition of granite as chemical weathering takes place. The unweathered rock is composed primarily of quartz and feldspar. After a long period of weathering, the sediments are mostly clay, quartz, and iron oxide. Of the original minerals, the only one to remain abundant is quartz. This shows that quartz is stable over a wide range of environmental conditions.

ACTIVITY 10-3 CHEMICAL WEATHERING AND TEMPERATURE

Materials: small beakers (100–250 mL), hot and cold running water, thermometers, three to four small pieces of antacid tablets.

In this activity, you will devise your own laboratory procedure. The objective is to find out how temperature affects the rate of a chemical reaction. Once you have planned and written down your procedure, check it with your teacher. When your procedure is approved, perform the experiment and create a data table. Finally, record your conclusion about the effect of temperature on this chemical reaction.

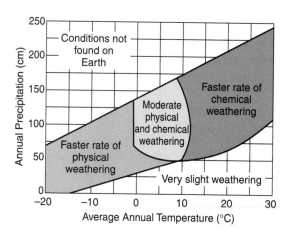

Figure 10-9 Cold climates favor physical weathering, especially frost action. Chemical weathering dominates under conditions of warm temperatures and abundant rainfall.

Factors That Affect Weathering

The amount and kind of weathering that takes place depends on three factors. You have read that the harder a rock is, the more it resists physical weathering. The more chemically stable its minerals are, the better a rock resists chemical weathering. The final factor is climate. Figure 10-9 shows that cold climates tend to favor physical weathering. Daily cycles of temperatures above and below freezing promote frost action in cold climates. Warm and moist climates accelerate chemical changes. For this reason chemical weathering is especially active in tropical locations.

HOW DOES SOIL FORM?

To this point in the chapter, weathering has been considered a destructive process that loosens rock and wears down the land. But weathering is responsible for one of our most important natural resources—soil. **Soil** is a mixture of weathered rock and the remains of living organisms in which plants can grow.

Figure 10-10 on page 250 shows the development of soil on a solid rock surface. In the first column in the diagram, the bedrock is unbroken, but it is exposed to the atmosphere and

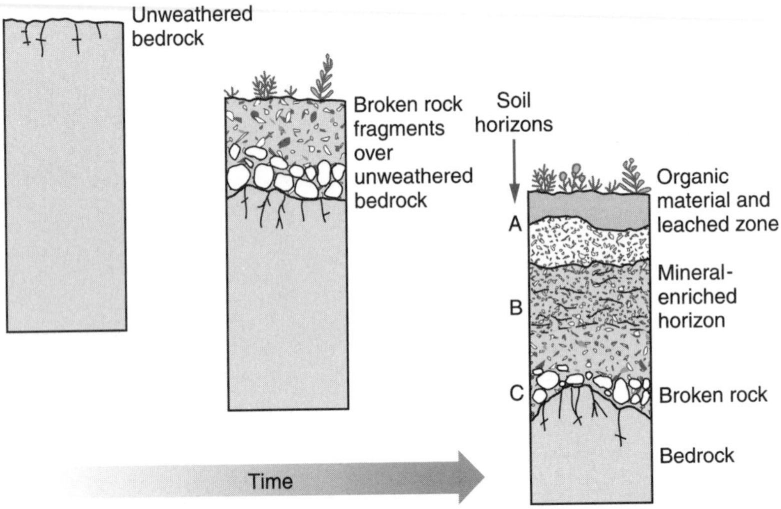

Figure 10-10 Weathering, infiltration of water, burrowing by animals, plant growth, and decay of organic remains contribute to the formation of soil. A mature soil usually develops layering called soil horizons.

weather. The weathering process begins as rainwater reacts with the rock surface and water infiltrates cracks in the rock. Water that seeps into crevices and fissures may change to ice and push the rock apart. Minerals soften and some minerals expand as they react with rainwater and groundwater. The second column shows fragments of broken rock covering solid bedrock. The third column shows a complete soil in which organic remains, mostly dead plant material, have been mixed into the topsoil. **Infiltration** (water seeping into the ground) has carried some water deeper into the soil. The mature soil in the third column shows layering called **soil horizons** that are typical of well-developed soils. The topsoil is usually enriched with organic remains but may lack some soluble minerals that water carried deeper into the soil. As a result, the soil below is enriched in soluble minerals. At the bottom of the soil profile, a layer of broken rock overlies the solid bedrock from which the soil may have formed.

The soil formed at any location depends on the composition of local bedrock, the climate, and the time for development. Warm and moist climates favor chemical weathering and usually produce thick soils, although the movement of groundwater through the soil may wash away important nutrients. Polar locations more often have thin, rocky soils with little chemical weathering. Animals take part in soil forma-

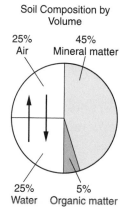

Soil Composition by Volume

25% Air

45% Mineral matter

25% Water

5% Organic matter

Figure 10-11 The best soils for the growth of most plants contain a mixture of weathered minerals and organic remains. The proportions of air and water depend upon where the soil is located and recent weather conditions.

tion as they burrow by mixing the components of soil (minerals in various states of weathering and organic remains), by loosening soil, and by allowing air and water to circulate.

Active volcanoes can be dangerous, but the soils that result from the weathering of volcanic rocks are usually very fertile. At least two cities were destroyed by the eruption of Mount Vesuvius in southern Italy in 79 CE. In spite of the danger, people soon moved back to the slopes of Vesuvius. The volcano is still active, and another major eruption is possible at any time. In spite of this, farmers are drawn back to the slopes of the mountain by the rich soil.

Figure 10-11 shows the composition of a well-balanced soil. The mineral content provides important nutrients and support for plants. Organic material retains water in the soil and holds the soil together. Water is an essential component for plant growth, but air is also important. Many plants cannot thrive if their roots are submerged in water all the time.

Soil that is formed in place and remains there is called a **residual soil**. Residual soils develop through the processes of weathering over hundreds or even thousands of years. **Transported soil** is formed in one location and moved to another location. In most areas, including New York State, transported soils are more common than residual soils. Continental glaciers that repeatedly formed in Canada and moved southward pushed, carried, and dragged most of our sediment and soil from the place where it formed to New York State. The absence of a layer of broken bedrock that grades into solid rock in most New York locations is evidence of this transportation of soil.

TERMS TO KNOW

abrasion	frost wedging	sediment
bedrock	infiltration	soil
biological activity	mechanical weathering	soil horizon
chemical change	physical weathering	transported soil
chemical weathering	residual soil	weathering

CHAPTER REVIEW QUESTIONS

1. How does weathering affect rocks?

   (1) Weathering causes the mineral grains to increase in size.
   (2) Weathering makes rock harder.
   (3) Weathering occurs when sediment changes to sedimentary rock.
   (4) Weathering weakens rock so it can be carried away by erosion.

2. A tree growing on bedrock extends its root into a crack in the rock and splits the rock. The action of the root splitting the bedrock is an example of

   (1) chemical weathering.   (3) erosion.
   (2) deposition.   (4) physical weathering.

3. Which statement best describes physical weathering that occurs when ice forms within cracks in rock?

   (1) Physical weathering occurs only in bedrock composed of granite.
   (2) Enlargement of the cracks occurs because water expands as it freezes.
   (3) The cracks become wider only because of chemical reactions between water and rock.
   (4) This type of weathering is most common in regions with warm, humid climates.

4. Two different kinds of minerals, A and B, were placed in the same container and shaken for 15 minutes. The diagrams below represent the size and shape of the various pieces of mineral before and after shaking. What caused the resulting differences in shapes and sizes of the minerals?

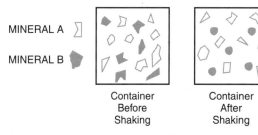

   (1) Mineral B was shaken harder.
   (2) Mineral B had a glossy luster.
   (3) Mineral A was more resistant to abrasion.
   (4) Mineral A consisted of smaller pieces before shaking began.

5. Which of the following is probably bedrock?

   (1) a large boulder transported by a glacier
   (2) the solid rock walls of a deep canyon
   (3) minerals in a granite block that is part of a new building
   (4) a statue of a Greek hero displayed in a museum

6. Which kind of rock is most likely to form new compounds when it is exposed to air polluted with acids?

   (1) gneiss   (3) granite
   (2) limestone   (4) schist

7. Which geologic feature is caused primarily by chemical weathering?

   (1) large caverns in limestone bedrock
   (2) a pattern of parallel cracks in a granite mountain
   (3) blocks of basalt at the base of a steep slope
   (4) the smooth, polished surface of a rock in a dry, sandy area

8. As rock is broken apart by physical weathering processes,

   (1) its total surface area decreases.
   (2) its total surface area increases.
   (3) new minerals form in the rock material.
   (4) the mass of the rock increases.

9. Which factor has the greatest influence on the weathering rate of bedrock at Earth's surface?

   (1) local air pressure   (3) local weather and climate
   (2) position of the sun in the sky (4) age of the bedrock

10. Marble is a metamorphic rock composed primarily of the mineral calcite. One hundred grams of marble is added to each of two identical beakers of hydrochloric acid. One marble sample is added as coarse marble chips. The second sample is added as a finely ground powder of marble. Why does the fine powder react more quickly with the acid?

    (1) Grinding changes the chemical composition of marble.
    (2) Fine particles of marble are less dense than coarser particles.
    (3) The finely ground powder has a greater total surface area.
    (4) The coarse marble chips have a greater total surface area.

11. Assuming that rainfall and other precipitation was constant, which graph below best shows how the amount of chemical weathering changes through the calendar year in New York State?

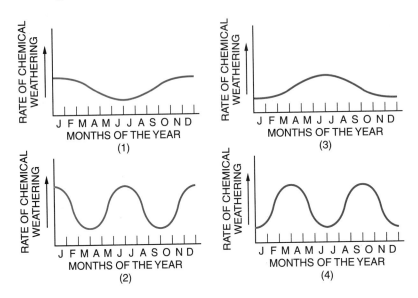

12. Which of the following changes does *not* directly contribute to the formation of soil?
    (1) melting of rock to make molten magma
    (2) plant roots growing into cracks in the ground
    (3) acidic rainfall reacting with the mineral calcite
    (4) rocks split apart by water freezing in cracks

13. A residual soil forms by
    (1) cooling of magma.
    (2) erosion of weathered rock.
    (3) cementing of mineral grains.
    (4) physical and chemical weathering.

14. The lowest horizon of a residual soil is composed primarily of
    (1) organic remains.
    (2) solid bedrock
    (3) broken bedrock.
    (4) products of intense chemical weathering.

15. Two very old tombstones of the same age sit next to each other in a cemetery. Both of them face south. One was cut from granite and the other was cut from marble. The carved writing on the granite stone is sharp and clear. But similar writing carved in the marble is now hard to read. Why is the writing in granite so much easier to read?

(1) The marble was exposed to greater changes in temperature.
(2) Marble is made of minerals that are less resistant to weathering.
(3) Granite formed from molten magma before the marble was metamorphosed.
(4) Granite is relatively soft because it contains large crystals of quartz.

Open-Ended Questions

Base your answers to questions 16 to 19 on the reading passage below and Map I, which shows the location of major producers of nitrogen oxides and sulfur dioxide and Map II, which shows the average pH of precipitation in the continental United States.

Acid Rain

Acid deposition consists of acidic substances that fall to Earth. The most destructive type of acid deposition is rain containing nitric acid and sulfuric acid. Acid rain forms when nitrogen oxides and sulfur dioxide gases combine with water and oxygen in the atmosphere.

Human-generated sulfur dioxide results primarily from coal-burning electric utility plants and industrial plants. Human-generated nitrogen oxide results primarily from burning fossil fuels in motor vehicles and electric utility plants.

Natural events, such as volcanic eruptions, forest fires, hot springs, and geysers, also produce nitrogen oxide and sulfur dioxide.

Acid rain affects trees, human-made structures, and surface water. Acid damages tree leaves and decreases the tree's ability to carry on photosynthesis. Acid also damages tree bark and exposes trees to insects and disease. Many statues and buildings are composed of rocks containing the mineral calcite, which reacts with acid and chemically weathers more rapidly than other common minerals. Acid deposition lowers the pH of surface water. Much of the surface water in the Adirondack region has pH values too acidic for plants and animals to survive.

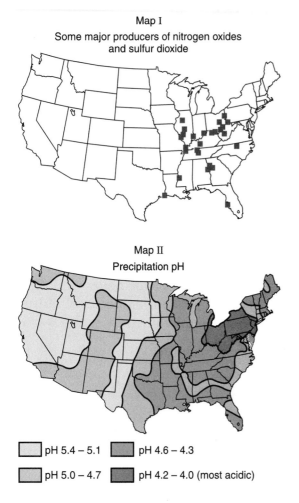

Map I
Some major producers of nitrogen oxides
and sulfur dioxide

Map II
Precipitation pH

pH 5.4 – 5.1   pH 4.6 – 4.3

pH 5.0 – 4.7   pH 4.2 – 4.0 (most acidic)

16. State one reason why the northeastern part of the United States has more acid deposition than other parts of the country.

17. Name one sedimentary or one metamorphic rock that is most chemically weathered by acid rain.

18. Describe one law that could be passed by the government to prevent some of the problems of acid deposition.

19. Explain why completely eliminating human-generated nitrogen oxides and sulfur dioxide will not completely eliminate acid deposition.

20. What is the major cause of physical weathering of big rocks transported along a large stream that has a steep gradient.

Chapter 11

Erosion and Deposition

WHAT IS EROSION?

Have you ever visited the Grand Canyon in Arizona? Each year millions of people come to see and experience one of the most spectacular natural wonders of the world. The canyon is 10 to 20 km wide and 1.5 km deep. It is one of Earth's most inspiring monuments to erosion and deposition. The walls of the canyon are mostly sedimentary rocks representing millions of years of deposition. The Colorado River eroded the canyon in just a few million years as tectonic forces pushed up the Colorado Plateau more than 1000 meters.

New York State also has striking, if smaller, erosional features such as Letchworth Gorge on the Genesee River. This gorge is sometimes called the "Grand Canyon of the East." It has sheer cliffs nearly 200 meters high on both sides. The Genesee River created Letchworth Gorge when deposits from the most recent advance of continental glaciers blocked the river. In the past 15,000 years, the river has cut a new route through thick layers of sedimentary rock. Watkins Glen and Enfield Glen in the Finger Lakes region are smaller gorges that feature trails following streams with numerous waterfalls and potholes. Ausable Chasm south of Plattsburgh is nearly 50 meters deep and also has sheer sides and waterfalls.

The Niagara River drops over Niagara Falls and follows a narrow gorge for about 10 km before it opens into the plain of Lake Ontario.

In the previous chapter, you learned about one of two processes that wear down the land. Weathering weakens and breaks up solid rock transforming bedrock into sediments, so they can be carried away by erosion. **Erosion** is the transportation of sediments by water, air, glacier, or by gravity acting alone. If erosion did not occur, sediment and soil would form on bedrock without any process to wear down the land. There would be no stream valleys, no canyons, or waterfalls. Mountains pushed up by tectonic forces within Earth would become higher and higher as long as uplift continued. However, scientists know that the processes of weathering and erosion down wear mountains and balance mountain-building the forces within our planet.

 ## Mass Movement

The force of gravity drives all erosional processes. Sometimes gravity alone transports earth materials. If weathering weakens rocks near the top of a cliff, the force of gravity may be greater than the strength of the rock holding it in place. When this happens, the rocks fall to a lower elevation. In Chapter 9, you read about the dangers to people and property caused by mass movement of rock and sediment. The position of rocks at the bottom of a landslide have little or no relationship to their position or organization before the landslide occurred. Landslides are most common in areas where tectonic forces within Earth are building high mountains. Figure 11-1 shows blocks of rock that have fallen to the bottom of a cliff.

Other forms of mass movement occur more slowly. If clay-rich sediments along a slope become saturated with rainwater or snowmelt, they may move downslope as a mudflow. Sometimes blocks of sediment slide down steep slopes along weakened layers, but the internal structure of the blocks of sediment remains unchanged. This kind of erosion is common where streams or ocean waves cut cliffs or bluffs into thick layers of sediment.

Figure 11-1 At Devil's Post-pile in California, gravity has caused blocks of fractured basalt to fall to the bottom of the cliff.

Rocks eroded by gravity without being transported by water, wind, or glaciers are usually angular and rough. Recently broken surfaces show fewer signs of weathering than parts of the rock that have been exposed to weather for a longer time.

 Erosion by Water

If the motion of water, wind, or ice causes erosion, these substances are called **agents of erosion**. Running water is the most important agent of erosion because it carries more sediment than any other agent of erosion does. Each year streams and rivers carry millions of tons of sediment into lakes and oceans. The Mississippi River alone is estimated to carry about 6 tons of sediment per second into the Gulf of Mexico.

METHODS OF TRANSPORT Streams carry sediment in several ways. The smallest sediments are dissolved in water and are carried in **solution**. For example, when a stream flows over rock or sediment rich in the mineral halite (sodium chloride, or rock salt), the halite dissolves in water to form a solution. The sodium and chlorine enter the solution as ions, or individual atoms with an electrical charge. These ions are so small that they cannot be seen or separated from water by filtration. A filter fine enough to catch the ions would not allow the molecules of water to pass through. Materials carried in solution can give water a color, but a solution is transparent. Natural water always has some substances in solution, even if they are present in very small amounts.

Sediments carried in **suspension** are small enough that they settle out of the water very slowly. Suspended sediments can be removed by passing the water through a filter. Most streams are turbulent. That is, the water does not flow smoothly downstream. If mixing and tumbling currents are faster than the rate at which suspended particles settle, the suspended load of a river can be carried indefinitely. Silt and clay in suspension give streams a muddy appearance. However, not all suspensions are in water. Clouds in the sky are actually accumulations of tiny ice crystals or water droplets so small they can remain suspended in air.

Some of the sediment load of a stream is too large to be carried in solution or suspension. Larger particles that are less dense than water float to the surface. Fresh leaves and other organic remains may be carried downstream by **flotation**. Large particles that are more dense than water settle to the bottom of a stream. If the stream is flowing quickly enough, these sediments will roll or bounce along the bottom of the stream as **bed load**.

SIZE AND VELOCITY Figure 11-2 from the *Earth Science Reference Tables* shows the relationship between the size of rock particles and the stream velocity needed to transport them. Boulders are rocks greater than 25.6 cm in diameter. The graph shows that the smallest boulders require a stream velocity of about 300 cm/second to keep them moving.

To read this graph with care, it will be helpful to use the 90° corner of a sheet of paper. Carefully align one edge of the paper with the horizontal axis. Then slide the vertical edge of the paper to intercept the graph line at the desired value. (See Figure 11-3.)

Several important factors that influence the relationship between particle size and stream velocity are not accounted for on this graph. For example, the speed of current needed to start particles of sediment moving is greater than the speed

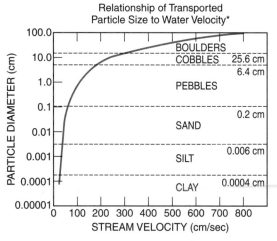

Relationship of Transported
Particle Size to Water Velocity*

Figure 11-2 There is a direct relationship between the velocity of a stream and the size of particles it can transport. The larger the grains of sediment, the faster a stream must move to keep them in motion. This graph shows stream velocity needed to maintain particles in motion, but not start them moving.

*This generalized graph shows the water velocity needed to maintain, but not start, movement. Variations occur due to differences in particle density and shape.

needed to keep them in motion. This graph applies only to particles already in motion. A second issue is shape of the particles. The values obtained from the graph refer to rocks that are neither spherical nor flat, but some average shape. Another factor is density. The denser a rock, the harder it is to move. A final factor is the nature of the bottom of the stream.

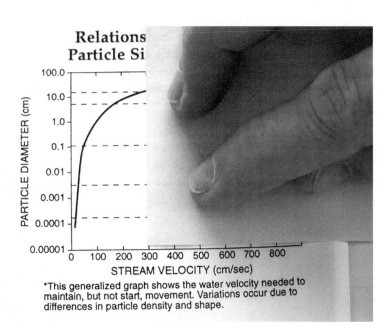

Figure 11-3 Holding the edge of a sheet of paper along the horizontal (or vertical) scale can help you read line graphs with increased accuracy.

*This generalized graph shows the water velocity needed to maintain, but not start, movement. Variations occur due to differences in particle density and shape.

The smoother and harder the bottom of the stream, the easier it is to keep sediments in motion. Therefore, this graph represents some typical characteristics of shape and density of moving rock particles, as well as typical conditions of the bottom of the stream.

Notice that Figure 11-2 gives the particle size for the various kinds of sediment: clay, silt, sand, pebbles, cobbles, and bounders. You can read particle size (diameter) on the vertical axis on the left side of the graph or you can read the numbers on the dotted lines that separate each particle from larger grains above or smaller grains below. As an example, sand is defined as particles between 0.006 and 0.2 cm in diameter. Particle diameter is the typical distance across the particle. Once again, there is an assumption that the particle is partly rounded.

The term clay has two different meanings as geologists use it. Sometimes clay refers to a group of minerals. In earlier chapters, you read about weathering of such minerals as feldspar and mica into clay. Clay minerals readily absorb and release water. They also tend to be plastic and soft. But the term clay as it is used here refers to any particles of sediment smaller than 0.0004 cm. Clay-sized particles are often the minerals known as clay, but other minerals can be reduced to the size geologists call clay.

How fast must a stream be moving to keep all sand-size particles in motion? The largest sand grains are 0.2 cm in diameter. That is the size that requires the fastest stream velocity. The line on the graph crosses the 0.2-cm sand-pebbles interface at a velocity of approximately 50 cm/second.

One method to determine the relative velocity of two nearby parts of a stream is to observe the size of sediments in the stream. In the faster parts of the stream, only larger particles can settle. In the slower sections smaller particles settle out. Therefore, the speed of a stream determines the size of particles to be found along the streambed. The larger the sediments, the faster the water velocity.

This principle should be applied with care. For example, a small stream may transport large rocks only when the stream is in flood. Large rivers often flow quickly near the surface. However, they flow more slowly near the bottom where they transport only small particles. In spite of the rapid surface current, only fine sediments are carried along the bottom of the river.

Particles of sediment transported by running water become rounded by abrasion as they are transported downstream. The farther rocks are carried downstream, the more rounded they become. Round, smooth rocks are often a sign of transport by running water.

Erosion by Wind

Strong winds can also transport sediments. However, because air is less dense than water, wind is generally unable to move particles larger than sand. As with stream water, the faster the wind, the larger the particles it can carry.

Wind erosion is most active in places such as deserts and beaches. Here there are few plants to slow the wind and hold soil in place. Wind-blown particles cause weathering by abrasion. Softer minerals in a rock are worn away, which may give wind-abraded rocks a pitted look. Ventifacts are wind-worn rocks that have flat surfaces (facets) like those of a regular geometric solid. These facets are created when the rock is partly buried in sand and abrasion wears the exposed face to a flat surface level with the top of the sand. Other flattened surfaces form when the rock is moved and another surface is exposed to the wind.

Rocky deserts are more common than sandy deserts. Figure 11-4 on page 264 shows a wind-blasted desert surface in Arizona. This type of surface is called desert pavement. Without plants to hold the soil, fine sediment is blown away, leaving a surface covered by rocks that are too large to be carried away by the wind. Many of the remaining rocks have flat faces and straight edges, a shape typical of wind-eroded rocks.

Erosion by Glaciers

Geologists know that the climate of New York State has not always been as temperate as it is now. Thousands of years ago, the climate was cold enough that winter snow did not melt during the brief summers. In eastern Canada, layers of snow

Figure 11-4 Desert pavement is common where wind has carried away fine sediments and left behind rocks too large to be moved.

built up and compressed the layers beneath them into ice. Eventually, the ice became thick enough that it flowed southward into New York under the influence of gravity. A **glacier** is a large mass of ice that flows over land due to gravity.

As the ice in a glacier flows downhill, it pushes, drags, and carries rocks, soil, and sediment. Rocks along the bottom of the glacier are worn down into partly rounded shapes. Glaciated rocks often have parallel scratches known as striations, which are caused by a glacier dragging rocks along the surface of other rocks. Figure 11-5 shows rocks with shapes characteristic of different kinds of erosion.

WHAT IS DEPOSITION?

Have you heard people talking about depositing money in a bank? Making a deposit is the act of placing something where it will stay and be available. In the same way that people deposit money, agents of erosion deposit sediments. Therefore, agents of erosion are also agents of deposition. **Deposition** is the settling, or release, of sediments carried by an agent of erosion. Gravity and glaciers transport sediment without regard to the sizes of the particles. Everything gets transported

Figure 11-5 Each agent of erosion produces a characteristic shape and texture in rocks. Rock A was eroded by windblown sand so it has flat faces. B was eroded by gravity alone so it is angular with a fresh face where it was recently broken. C is partly rounded with scratches, called striations, produced by movement in a glacier. Rock D is round and smooth, characteristics of erosion by water.

and dumped together. There is no organization in sediments deposited by gravity or by glaciers. But deposition by running water and wind is more selective. The size, shape, and density of the particles of sediment affect the rate at which they are deposited.

The Effect of Particle Size

The size of particles transported by wind or running water determines how quickly they settle out of their transporting medium. Suppose that a flooded river is moving very fast and transporting a very wide range of particle sizes. When the river slows down, sediments will be deposited. According to Figure 11-2 on page 261, the first group of sediments to settle out will be the largest particles: boulders. By the time the current slows to about 300 cm/second, all the boulder-sized particles should be deposited and cobbles will start to settle. If the current continues to decrease, you would see pebbles, sand, silt, and finally clay particles deposited. Because the

Figure 11-6 When a jar containing water and sediments of mixed sizes is shaken, sediments will settle in size order from the largest particles at the bottom to the smallest particles at the top. This progression of sediment sizes is called graded bedding.

largest particles settle first, they are on the bottom. This pattern is called graded bedding.

Figure 11-6 shows a procedure you can try yourself to observe how particles are deposited. When a jar of water containing particles of mixed sizes is shaken, all the particles mix with the water. When the shaking stops, the larger particles settle first followed by smaller and smaller particles. The result will be sorted sediments with the largest particles at the bottom and a gradual change to smaller particles toward the top of the sediment.

The Effect of Particle Shape

The shape of particles of sediment affects how quickly they settle. Spherical particles settle fastest because they are streamlined. Flat and irregular particles must push more water out of the way as they settle, which slows them. Friction is also greater for flat and irregular particles that have more surface area. So flat or irregular particles fall through water at a slower rate than do spherical particles.

The Effect of Particle Density

Among particles of the same size and shape, the most dense particles settle first. How quickly sediments settle depends on the balance between resistance and weight. Resistance, which is determined by size and shape of the sediments, holds them back. The force of gravity (the weight of the particles) causes them to sink. Therefore if the weight (or density) of a particle increases with no change in resistance, the particle will settle more quickly.

Running Water Sorts Sediments

Sorting is the separation of sediments by their shape, density, or size. Among particles of sediment transported by streams, the shape and density of particles seldom changes enough to make separation by shape or density apparent. Sediments are often sorted by their sizes. How does this occur?

If a landslide releases a large mass of sediment into deep water or if an underwater slide occurs along a steep slope, particles in a wide range of sizes begin to settle at the same time. The largest particle will settle to the bottom first, followed by progressively smaller particles. Each such event results in a layer of sediment. Within the layer, the largest particles are on the bottom and the size of particles decreases toward the top. Figure 11-7 shows a vertical cross section of layers of sediment produced in this way. Notice that within each layer there is a gradual change in sediment size from the bottom to the top showing the order in which

Figure 11-7 Layers of graded bedding represent pulses of deposition in which each release of sediments is represented by a single layer. The particle size in each layer gradually changes from large particles at the bottom to smaller particles toward the top.

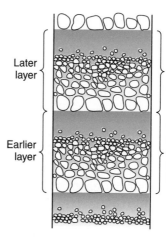

Later layer

Earlier layer

Each layer of graded bedding represents a single event of deposition.

Within a single layer, the particles gradually change from largest at the bottom to smallest at the top.

the particles settle to the bottom. This is called **vertical sorting**, or **graded bedding**. Please note that you are not looking at an alternation of layers of fine sediments and layers of coarse sediments. A single layer shows two features. Each layer has a wide range of sediment sizes and a gradual change in the size of particles within a single layer.

ACTIVITY 11-1   GRADED BEDDING

Obtain a transparent plastic tube about 5 cm in diameter and about 1.5 to 2 meters long. Securely seal one end of the tube. Fill the tube about half way with water. In a container, combine portions of sediments of mixed sizes from gravel to fine sand. Individually (or in lab groups) use a 100-mL beaker to quickly pour about 50 mL of the sediment into the cylinder. Watch the particles settle and observe which particles settle fastest and which settle more slowly. Pour in another 50 mL of the mixed sediment. Can you see the pattern that forms each time sediments are added? Each addition of sediments should produce a new layer of graded bedding. Draw two successive layers of graded bedding.

Another kind of sorting occurs when a fast-flowing river enters the relatively still water of a lake or an ocean. The fast current is capable of transporting sediments in a range of sizes. As the river enters the larger body of water and slows down, there is a progression of settling. The largest, most dense, and roundest sediments settle out first just as the river water starts to slow down. Flatter, less dense, and smaller particles are carried farther and settle more slowly. The pattern that results from the decrease of velocity of the current is not a vertical separation, as we saw in graded bedding, but a change in particle characteristics with distance. **Horizontal sorting** is a decrease in the size of sediment particles with increasing distance from the shore that is produced as a stream enters calm water. Figure 11-8 illustrates this form of horizontal sorting.

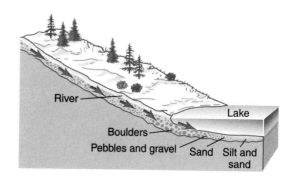

Figure 11-8 Horizontal sorting occurs as a fast-moving stream enters still water. The largest particles are usually deposited near shore. The finest sediments are carried the greatest distance into calm water.

River

Lake

Boulders

Pebbles and gravel   Sand   Silt and sand

Deposition by Wind

Like running water, wind sorts sediments. The primary difference is that wind cannot pick up large rocks. Sediments deposited by wind tend to be sand size (0.006 cm to 0.02 cm) and smaller. In areas where wind is the primary agent of erosion and deposition, wind blows sand into hills or ridges of sediment called **dunes**. The wind blows particles of sand up the windward side of sand dunes and deposits them on the protected downwind face. Dunes are found in some desert regions where there are few plants to hold the sand in place and some locations along lakes and oceans where sand is plentiful. Figure 11-9 shows a sand dune in a desert area of eastern California.

Figure 11-9 Sand is blown up the left side of this dune and settles on the down-wind side where the wind speed drops off.

Wind-blown deposits are sorted by size. Large rocks may be present under or around the dunes, but rocks are not picked up and moved over the dunes the way sand is. Particles smaller than sand are transported out of the dune area by the wind.

Have you been to a place along the ocean where stable land areas are separated from the water by sand dunes? The line of dunes protects inland areas from storms and erosion by waves. Plant cover on the dunes may be scarce or absent because sand lacks important nutrients; the sand is unable to hold onto water, and the sand has a loose consistency. Sand dunes are fragile ecological features that play an important function in the beach environment. Only recently have scientists become aware of the importance of protecting dunes from motor vehicle and foot traffic as well as residential or commercial development.

 ## Deposition by Ice

In some ways glaciers are like running water. They flow from higher areas to lower areas. Glaciers often occupy valleys that they form by erosion. But the rate of flow of a glacier is usually a meter or less per day. Furthermore, glaciers are not able to separate different sizes of particles. Ice is not fluid enough to allow sediments to settle through the ice. As a result, sediments carried by moving ice are not deposited until the glacier melts. Giant boulders, fine clay, and every sediment size in between are deposited in irregular mounds with no separation or sorting.

Sorted and layered sediments can be found in areas where glaciers are or were active. But streams running out of the glaciers deposited these sorted sediments. Sediments deposited directly by ice are unsorted.

ACTIVITY 11-2   WHAT'S IN SEDIMENT?

Materials: containers of sand and other sediments, magnifiers, metric rulers

Your teacher will set out containers of sediments collected from a variety of locations. Compare the sediments by overall color, colors of the grains, shapes of the grains, average particle size, sorting, range of sizes, and any other unusual characteristics you may observe.

EQUILIBRIUM OF EROSION AND DEPOSITION

Consider a section of a river that is carrying sediment. You may observe that the river looks about the same over a long period of time. If the riverbed is not filling with sediment and it is not cutting deeper into its bed, the river is in equilibrium. **Equilibrium** is any state of balance. In the case of the river, the sediment washed into this stretch of river must be equal to the sediment that is carried away.

There are two ways in which this can occur. It is possible that the sediment entering this part of the river is carried along without any erosion or deposition. The particles of sediment that are carried in are the same particles that are carried out. The river does not change and the sediment that makes up the bottom of the river does not change. In this condition there is no erosion and no deposition.

It is more likely, however, that some new sediment is deposited, and some sediment from the bottom of the river channel is carried away by erosion. This is especially likely if the volume of water and the velocity of the water change through time, as they do in nearly all rivers. Flooding causes erosion of the streambed and deposition occurs when the flow is reduced. If equilibrium is reached over the course of the year, erosion and deposition are equal. However, some sediment from the bed of the river has been carried away, while some new sediment has been deposited.

If the appearance of the river has remained about the same even though some of the sediments on the river bottom have changed, the river is said to be in a state of **dynamic equilibrium**. In a dynamic equilibrium, opposing processes

Figure 11-10 A stream is in dynamic equilibrium if the amount of erosion is equal to the amount of deposition.

are taking place, but they balance out because they take place at the same rate. Figure 11-10 shows a stream in dynamic equilibrium.

 TERMS TO KNOW

agent of erosion	**erosion**	**landslide**
bed load	**flotation**	**mass movement**
deposition	**glacier**	**solution**
dune	**graded bedding**	**sorting**
dynamic equilibrium	**horizontal sorting**	**suspension**
equilibrium		

CHAPTER REVIEW QUESTIONS

1. The cross section below shows sedimentary rocks being eroded by water at a waterfall.

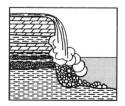

The sedimentary rock layers are being eroded at different rates primarily because the rock layers

(1) formed during different periods.   (3) have different compositions.
(2) contain different fossils.   (4) are horizontal.

2. What is the largest particle that can be kept in motion by a stream that has a velocity of 100 cm/second?

(1) silt   (3) pebble
(2) sand   (4) cobble

3. A stream with a velocity of 300 cm/second decreases to a velocity of 200 cm/second. Which sediment size would most likely be deposited?

(1) pebbles   (3) boulders
(2) sand   (4) cobbles

4. The diagram below shows four identical columns of water. Four sizes of spherical object made of the same uniform material are dropped into the columns where they settle to the bottom as shown below.

Which graph best shows the relative settling times of the four objects?

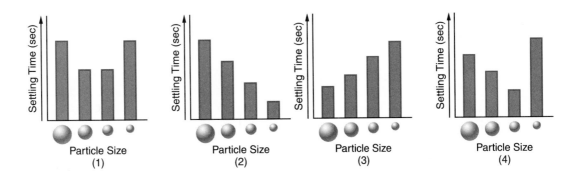

5. When small particles settle through water faster than large particles, the small particles are probably

(1) lighter in color.
(2) less rounded.
(3) better sorted.
(4) more dense.

6. The table below shows the density of four mineral samples.

Mineral	Density (g/cm³)
Cinnabar	8.2
Magnetite	5.2
Quartz	2.7
Siderite	3.9

If the shape and size of the four mineral samples are the same, which mineral will settle most *slowly* in water?

(1) cinnabar
(2) magnetite
(3) quartz
(4) siderite

7. The diagram below shows a cylinder filled with clean water. At the left of the cylinder is a light source, and at the right of the cylinder is a meter that measures intensity (brightness) of light as it passes through the water. One minute after the light is turned on, a mixture of sand, silt, and clay is poured into the cylinder.

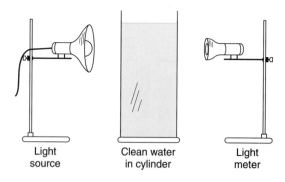

Light source   Clean water in cylinder   Light meter

Which graph below shows the probable change in light intensity (brightness) recorded during the 6-minute period after the light is turned on?

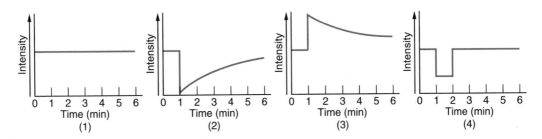

8. Where the stream velocity decreases from 300 to 200 cm/second, which size sediment will be deposited?

(1) cobbles   (2) sand   (3) silt   (4) clay

9. The particles of sediment at the top of a layer of graded bedding are usually those that are

(1) most dense, best rounded, and smallest.
(2) least dense, best rounded, and largest.
(3) most dense, most flattened, and smallest.
(4) least dense, most flattened, and smallest.

10. The table below shows the amount of each size of sediment in four samples of sediment.

Mass Composition of Four 100-gram Samples of Sediment

Sample	A	B	C	D
Clay	10	18	0	20
Silt	15	20	0	20
Sand	20	21	5	20
Gravel	25	22	90	20
Cobbles	30	19	5	20

Which sample is best sorted by sediment size?

(1) A   (2) B   (3) C   (4) D

11. The diagram below shows sediments along the bottom of a stream.

Water level

Sand   Gravel

What can we say about the water velocity in this stream at the time these sediments were deposited?

(1) The stream velocity increased to the right.
(2) The stream velocity increased to the left.
(3) The stream velocity was constant in this section of the stream.
(4) There is no relationship between sediment size and stream velocity.

12. Where is the most deposition likely to occur?
(1) at the mouth of a river where it enters an ocean
(2) on the side of a sand dune facing the wind
(3) at a place where glacial ice scrapes bedrock
(4) at the top of a high mountain

13. Long, sandy islands along the south shore of Long Island are composed mostly of sand and rounded pebbles arranged in sorted layers. The agent of erosion that most likely shaped and sorted the sand and pebbles while transporting them to their present location was
(1) glaciers   (3) wind
(2) landslides   (4) ocean waves

14. The table below shows the rate of erosion and the rate of deposition at four stream locations.

Location	Rate of Erosion (tons/year)	Rate of Deposition (tons/year)
A	3.00	3.25
B	4.00	4.00
C	4.50	4.65
D	5.60	5.20

A state of dynamic equilibrium exists at location

(1) A.   (3) C.
(2) B.   (4) D.

15. Which statement below best describes a dynamic equilibrium between erosion and deposition?

 (1) There is more erosion than deposition.
 (2) There is more deposition than erosion.
 (3) There is no erosion and there is no deposition.
 (4) The amounts or erosion and deposition are equal.

Open-Ended Questions

16. What is the range of sediment sizes that are classified as pebbles?

17. Which agent of erosion deposits irregular mounds of unsorted sediment with parallel scratches on rounded particles?

18. A cloud of volcanic ash from the eruption of El Chichon in Mexico was blown to the west by wind currents. As the ash cloud moved away from El Chichon, some ash particles fell back to Earth. Describe how the size of the particles affected the pattern of deposition.

19. Describe how the density of particles from El Chichon (question 18) affected the pattern of deposition.

20. A brick factory was constructed along a river. Stacks of identical bricks were left outside the building for shipment. A major flood destroyed the factory and washed away the bricks. Some bricks were carried several miles downstream; others were deposited close to the factory site. State two characteristics of the brick recovered far downstream that are likely to be different from the brick found near the factory site.

Chapter 12

River Systems

 WHAT IS A RIVER SYSTEM?

Rivers have always been important to us. The trade routes of native Americans as well as the routes of exploration by Europeans followed rivers to reach the interior of North America. Travel by water was easier than travel by land, especially where vegetation was thick, and travelers had to carry food and supplies. Before roads were built, rivers were the avenues of travel, trade, and communications. The earliest European settlements were along the coastline, where there was access to shipping and manufactured goods. From coastal cities, settlement and commerce followed waterways to the interior. Today, highways, railroads, and air routes have replaced streams as our primary arteries for travel, settlement, and commerce. Rivers and lakes still carry some commerce while they supply freshwater and serve as recreational areas.

A system is a collection of parts that contribute to a single function. A river system, or **stream system**, consists of all the streams that drain a particular geographic area. A **stream** is any flowing water, such as a brook, river, or even an ocean current. The function of a river is to transport water and sediments from a specific land area to an ocean or a lake. Water

Figure 12-1 This braided stream in Alaska is transporting so much sediment that the stream spreads over its own sediment load.

and sediments have potential energy at the beginning of their journey. The amount of energy depends on how high they are above the end, or mouth, of the stream. As water and sediment flow downhill, potential energy is changed to kinetic energy. At the end of the stream where water flows into the calm water of a lake or ocean, potential energy has decreased because water and sediments are at their lowest elevation. Kinetic energy decreases as these materials stop moving. For these materials, the transporting function of the stream system has been accomplished. Figure 12-1 shows a stream that is transporting a large load of sediment.

 ## Watershed

The geographic area drained by a particular river or stream is its **watershed**, or drainage basin. All the rain, snow, and other precipitation that falls into the watershed and does not escape by infiltration, evaporation, or transpiration must exit the watershed through its principal river, stream, or other body of water. (Sometimes a lake or an ocean rather than a river is used to define a watershed.) **Drainage divides** separate one watershed from the next. These are high ridges from which water drains in opposite directions. You can identify watershed boundaries and trace the perimeter of

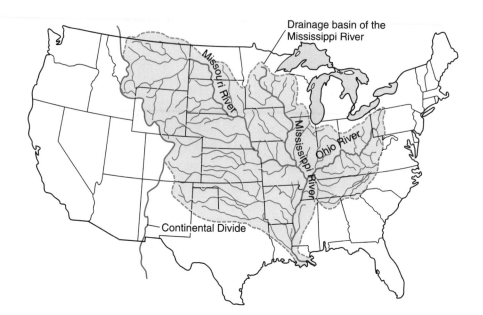

Figure 12-2 The Mississippi River and its tributaries drain about half of the surface area of the United States.

a watershed by drawing a line that separates all the streams draining into one watershed, from streams that flow into adjacent watersheds. Drainage divides never cross streams or rivers. At its lowest level, the drainage divide of a particular stream ends only where the stream flows into another body of water. Figure 12-2 shows the Mississippi River's watershed.

Watersheds are important because they show the region drained by a particular stream system. Figure 12-3 shows a small watershed in Utah. Communities that draw water from a nearby stream or river depend on rain and snow that falls within the watershed. The availability of water for such a community depends only on the amount and the quality of

Figure 12-3 The water that flows to the lowest point in this closed basin in Utah evaporates during dry summer weather.

precipitation in the watershed upstream from the municipal intake. When a water-soluble form of pollution is released into the environment, it flows downhill, and is carried into the nearest stream. As smaller streams join larger streams, the pollution affects only downstream locations in the watershed. Like water pollution, flooding is also confined to a particular watershed.

Most of the precipitation that falls over the continents falls on solid ground. If this water does not infiltrate the ground or evaporate, it must flow downhill under the influence of gravity as **overland flow**, or **runoff**. The amount of runoff depends on the slope of the land, the permeability of the surface, and the amount of precipitation. The steeper the slope, the greater the runoff. More water runs off a hard surface, such as pavement, than off a permeable surface, such as soil. The presence of grasses and shrubs in the soil also decreases runoff because they absorb water. The greater the amount of precipitation, rain or snow, the greater the runoff. Overland flow continues until the water reaches a stream.

Names such as brook or creek are often used to label small streams that flow into larger streams such as rivers. A stream that flows into another larger stream is called a **tributary**. In large watersheds, small tributaries join to form larger tributaries which themselves may be a tributary of even larger streams. The Bedrock Geology of New York State map in the *Earth Science Reference Tables* shows some of the major rivers of New York State.

| ACTIVITY 12-1 | DRAINAGE OF THE SCHOOL GROUNDS |

Make a map of your school grounds to determine how water drains off different parts of the property. On the map, identify potential sources of water pollution and show what parts of the grounds would most likely be affected by these sources of pollution. Also show where runoff could cause erosion problems and suggest ways to prevent these problems.

Features of Streams

As most streams flow from their source to their mouth, the *slope*, or *gradient*, of the stream decreases and the shape of the valley becomes broader. Streams that form steep, V-shaped valleys in mountain areas emerge into regions where the land is less steep. Here the valleys become broad with floodplains. (See Figure 12-4.)

FLOODPLAIN   Most of the time, the stream is confined to a relatively narrow and winding path along the bottom of the valley. But in times of flood, streams overflow their banks and spread over a flood plain. A **floodplain** is a flat region next to a stream or river that may be covered by water in times of flood. Floodplains are valued as agricultural land because sediments brought by periodic floods enrich the soil with important minerals and nutrients for plant growth. Figure 12-5 shows a stream as it changes through time from a narrow, steep-sided valley to a broad valley in which the stream can wander over its floodplain.

The occasional flooding of agricultural land is considered an acceptable risk. However, when local governments allow

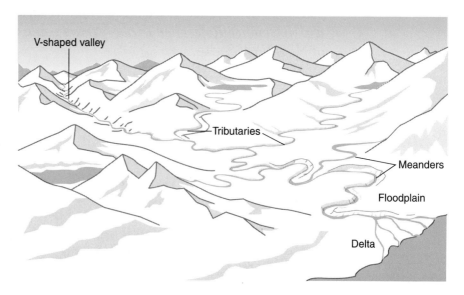

Figure 12-4 As a stream flows out of mountains and tributaries add water, V- shaped valleys change to broader valleys with flood plains. Deltas are deposited where the stream slows upon entering a lake or ocean.

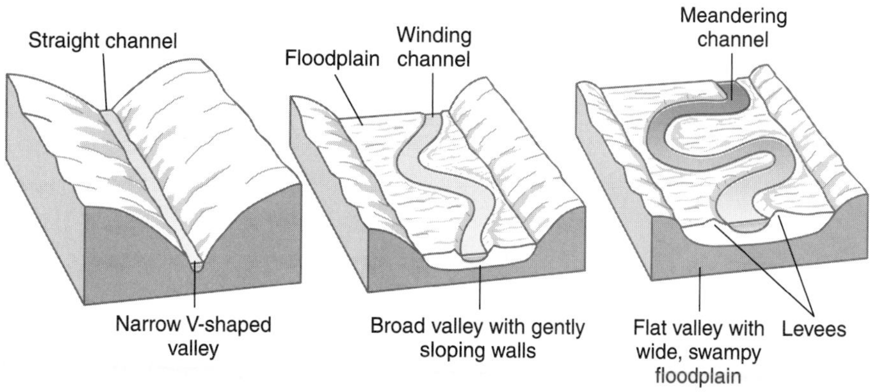

Figure 12-5
Through time, rivers erode their valleys changing from a narrow, steep sided, V–shaped valley to a valley with a broad U–shaped bottom and a floodplain.

houses and other buildings to be constructed on floodplains, periodic flooding can cause considerable losses. This is why zoning laws are important to protect citizens from property loss and even loss of life in times of flood.

DELTA   As most rivers continue downstream, they empty into the calm water of a lake or ocean. With loss of velocity, the water also loses its ability to transport sediment. Deposition often forms a delta at the end of the stream. A **delta** is a region at the end (mouth) of a stream or river that consists of sediments deposited as the velocity of the stream decreases. The name *delta* comes from the Greek letter Δ (delta). Sometimes delta deposits have this shape. Look again at Figure 12-4, which shows stream features including V-shaped valleys, tributaries, a floodplain, and a delta.

MEANDERS   As a stream flows over relatively flat land, its path develops curves called **meanders**. Builders of irrigation canals have discovered that when the channel has a soft streambed and little slope, the path of the canal tends to meander. Even when the path of the water is initially straight, meanders develop, through time, unless the banks are lined with a hard material such as concrete. The curves of a meandering stream are the natural shape of steams and rivers that have a low gradient and flow over a broad valley.

Once meanders have formed, they do not remain stationary through time. If you could see a greatly speeded-up view

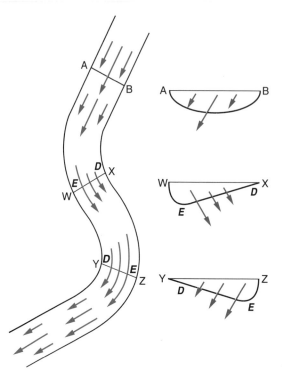

Figure 12-6 In a straight section of a stream (A–B) the fastest current is at the center of the stream. But in a meander (W–X and Y–Z) the fastest current swings to the outside of the curve. Erosion (**E**) occurs on the outsides of meanders where the current is fastest and deposition (**D**) takes place on the inside of meanders where water flows slowest. Erosion and deposition cause streams to change their paths through time.

of a meandering stream over many years, the meanders would shift like the slithering motion of a snake. How do they do this? Streams change their course as a result of erosion and deposition. Erosion occurs where the water flows fastest and deposition takes place where the water slows down.

Figure 12-6 is a diagram of a section of a stream. The length of the arrows indicate the water's speed at various places in the stream. Where the stream is straight, the fastest current is near the center. But when the water flows through a meander, the fastest current swings to the outside of the curve. Inertia is responsible for the changing position of the fastest water. This is the same force you feel when you quickly round a corner in a car. Everyone in the car feels a force to one side of the car. In the same way, the water in the stream travels in a straight path until water building up at the side of the stream forces the water to turn.

This pattern of water velocity with continuing erosion and deposition change the course of a meandering stream. For ex-

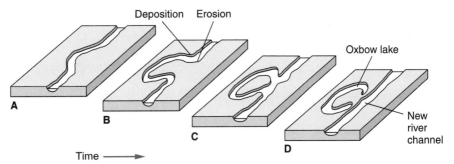

Figure 12-7 Meanders form where the stream gradient is low and there is a broad floodplain. Diagram A shows the stream starting to form a meander. In B erosion and deposition are starting to cut through the meander. The cutoff is complete in C. Deposition along the edge of the river leaves the meander isolated as an oxbow lake in D.

ample, in Figure 12-6 along profile A–B, the shape of the stream is not changing. Therefore erosion and deposition are in equilibrium. But at a meander, the faster current swings to the outside of the curve and causes erosion. On the inside of the bend where the water is slow, deposition dominates. If a stream has a low gradient, a straight stream will tend to form meanders and then cut off its meanders as shown in Figure 12-7.

LEVEES Streams in broad valleys sometimes flood and leave deposits of sand and silt on the land bordering and parallel to the streams. These ridges are called levees. In the area of New Orleans, Louisiana, the first homes were built on the natural levees along the Mississippi River. These areas were the highest and driest locations in the flat delta region. The land away from the riverbanks is often low and swampy. These natural levees are often used as foundation for tall, human-made levees, which are built to keep the river confined to its channel in times of flood.

However, when people alter natural systems, untended results often occur. Tall levees may force the river to deposit its sediment load in the river channel. In some locations, the Mississippi River is now higher than the surrounding land. This increases the danger of flooding, and may require the artificial levees to be built higher and higher through time.

ACTIVITY 12-2   MODELING A STREAM SYSTEM

Materials: stream table, fine sand, or coarse silt

Observe and list characteristics and common features of streams that develop on a stream table. How does the path of the stream change through time? Where do erosion and deposition occur?

HOW DO WE MEASURE STREAMS?

Two aspects of streams that scientists measure are size and velocity. Size is more than just the length of the stream. Several factors affect the velocity of a stream.

 Stream Size

The size of a stream can be measured in several ways. One measure is the area of its watershed. In general, the larger the drainage basin, the larger the stream. However, some locations receive more precipitation than others. In a dry region, a large watershed may only supply water to streams that are dry most of the year. Some watersheds receive so little rain and snow that none of the water in the stream flows out of the watershed. Streams in these areas run into bodies of water that lose their water by evaporation, or the stream water may seep into the ground before the stream reaches its lowest level.

Stream size is more often measured by finding discharge. **Discharge** is the amount of water flowing in a stream past a particular place in a specified time. For example, a small stream may have discharge of a fraction of a cubic meter per second. However, the Amazon River in South America, which has the greatest discharge of any river on Earth, discharges about 200,000 m^3 of freshwater into the Atlantic Ocean each second.

To measure the discharge of a stream, you need to measure the area of its cross section at a particular location, then multiply that value by the velocity of the stream. You can determine the area of the cross section by multiplying the average depth of the stream by its width at that point. Area is in units such as square meters. Velocity can be expressed in meters per second. The product of these values is therefore cubic meters per second. The formula below shows how to calculate discharge volume:

$$\text{Discharge} = \text{area of cross section} \times \text{stream velocity}$$

ACTIVITY 12-3   MEASURING STREAM DISCHARGE

Materials: Device to measure distance, watch, or timer

Use the method in the text above to measure the discharge of a stream near your school or home. Can you devise a different method to find the discharge in order to verify the first value that you obtained?

Streams respond differently to rainfall depending on their sizes. As you can see in Figure 12-8, large streams are slower to respond than small streams. This is because the water in a large drainage basin has a greater distance to flow before it reaches the major stream. Streams in smaller watersheds

Figure 12-8 Maximum stream discharge occurs after the time of maximum rainfall. In general, small streams respond more quickly than do larger streams because small streams have smaller watersheds.

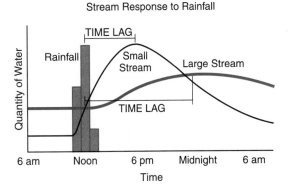

generally respond quickly because the precipitation does not flow far to reach a small stream.

 Stream Velocity

How quickly water flows in a stream is a function of three factors: shape of the stream channel, gradient (slope), and volume of water. If the stream channel is straight and smooth, water can flow quickly. But if the stream is flowing through large rocks, the rocks slow the water as it bounces from rock to rock. Many mountain streams are filled with coarse sediment such as large cobbles and boulders that slow the stream velocity.

The gradient of a stream also determines how quickly water moves. The force of gravity maintains flow of water. Other factors being equal, the steeper a stream channel, the faster water will flow.

The third factor is the volume of water flowing in a stream. As the quantity of water in a stream increases, so does the pull of gravity. If the discharge of a stream increases, usually the weight of water increases more quickly than the resistance of the stream channel. Most streams are steepest at the beginning of the stream and the gradient decreases as they flow downstream. But many rivers flow faster as they move downstream because the increase in the volume of the stream dominates over the reduction in slope.

Figure 12-9 shows profiles along the length of the Hudson River and the Colorado River. The concave shape of these profiles is typical of many streams: steeper near their sources than they are near the end of the stream.

ACTIVITY OR DEMONSTRATION 12-4	**WATER VELOCITY**

Materials: Stream table or running water and tilted trough to represent a stream bed, meter stick, stop watch, or timer

Determine the influence of gradient and stream discharge on stream velocity. Measure the velocity of the water as the gradient and the discharge are changed. Use your data to explain how gradient and discharge affect stream velocity.

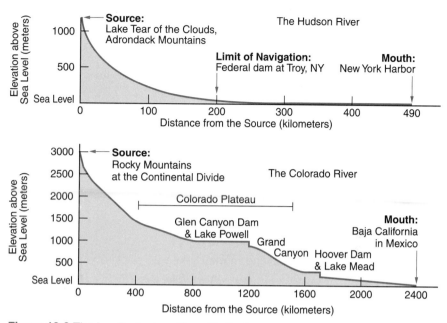

Figure 12-9 The longitudinal profiles of both rivers are concave as their slope decreases downstream. The double concave shape of the Colorado River profile is evidence of uplift of the Colorado Plateau.

You can measure the speed of a stream by selecting a relatively straight section of a small stream. You will need a device to measure distance, such as a meter stick, a timing device such as a watch with a second hand, or a stopwatch, and an object to float downstream. (In case there is wind, it may be best to use an object that floats, but is mostly submerged.) Measure the length of the stream section in units such as meters. Then place the floating object in the water above the measured section and time how long it takes for the object to float through the measured distance. The stream velocity can be calculated using the formula.

$$\text{Velocity} = \frac{\text{distance}}{\text{time}}$$

SAMPLE PROBLEM

Problem Two students stand 53 m apart along a straight portion of a small stream. One student places a floating marker in the stream and immediately begins timing it with a stopwatch. If the marker passes the second student in 15 s, what is the average velocity of this section of the stream?

Solution

$$Velocity = \frac{distance}{time}$$

$$= \frac{53\,m}{5\,s}$$

$$= 3.5\,m/s$$

Notice that the solution is expressed to two digits, as are the values of time and distance.

 Practice

1. A floating marker takes 21 s to travel 88 m along a straight portion of a stream. What is the average velocity of this section of the stream?
2. The average velocity of a stream is 2.7 m/s. How far will a marker travel in 16 s?

| ACTIVITY 12-5 | MEASURING STREAM VELOCITY |

Materials: Device to measure distance, watch, or timer

Use the method in the text above to measure the velocity of a small stream near your school or home. (**Note:** This should be done under adult supervision to ensure safety.)

WHAT IS A DRAINAGE PATTERN?

Streams seek the lowest path as they move downhill, and they tend to erode their beds in places where the ground is weak. Therefore, both topography and geologic structure influence the path streams follow through an area, which we call the **drainage pattern**. By looking at a map view of a stream, you can often infer the underlying bedrock structures. Figure 12-10 shows the relationship between stream pattern and rock structure.

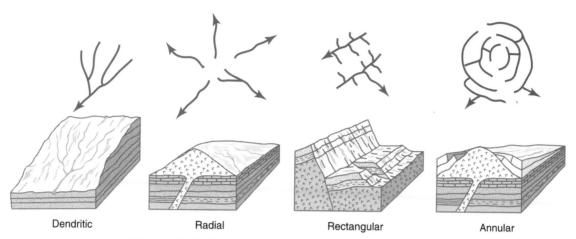

Dendritic   Radial   Rectangular   Annular

Figure 12-10 Underlying rock features influence the drainage patterns we see on map views of streams. Streams follow low areas, weak rock types, and fractured rocks.

The most common stream pattern is a dendritic drainage. Dendritic streams flow downhill in the same general direction and they join to make larger streams. As a result, they have a branching appearance. This pattern is common where the bedrock is uniform, without faults, folds, or other major structures or zones of weakness to capture the streams. Dendritic drainage is also common where the rock layers are horizontal. Much of the region of western New York State north of the Pennsylvania border has dendritic drainage because rock layers are flat and there are few faults or folds to divert streams.

A region that has prominent parallel and perpendicular faults, repeated folds, or a strong rectangular jointing pattern will display a rectangular drainage pattern. (Joints are cracks in bedrock along which no significant movement has occurred. They may be related to expansion or regional forces acting on bedrock.) Streams seek the lowest areas of folds, fractured rocks along faults, or the weakest surface bedrock locations.

Annular drainage is a pattern of concentric circles that are connected by short radial stream segments. This type of drainage occurs in an eroded dome

A radial drainage pattern resembles the spokes of a wheel. Streams flow away from a high point at the center of the pattern. Radial drainage may develop on a smooth dome or a volcanic cone. The Adirondack Mountain region of New York displays radial drainage, although rock structures such as faults and folds in the Adirondacks alter the regional pattern and may rake radial drainage hard to observe.

The important point is that the underlying rock types and geologic structures influence streams, and that different structural features produce different patterns of drainage.

TERMS TO KNOW

delta	floodplain	stream system
discharge	meander	tributary
drainage divide	overland flow	watershed
drainage pattern	runoff	

CHAPTER REVIEW QUESTIONS

1. Why are streams and rivers with the features of their watersheds called systems?

   (1) All streams and rivers are the same size.
   (2) Streams and rivers form straight channels.
   (3) Different parts contribute the same outcome.
   (4) The features of one stream system are not found in other stream systems

2. What name is applied to the outside boundaries of a watershed?

   (1) stream channel
   (2) meander
   (3) drainage divide
   (4) tributary

3. Which river is a tributary of the Hudson River?

   (1) Delaware River
   (2) Susquehanna River
   (3) Mohawk River
   (4) Genesee River

4. Which condition would cause surface runoff to increase in a particular location?

    (1) covering a dirt road with pavement
    (2) reducing the gradient of a steep hill
    (3) planting grasses and shrubs on a hillside
    (4) having a decrease in the annual rainfall

Base your answers to questions 5 and 6 on the diagram below that shows a stream meander.

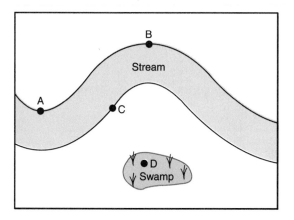

5. At which point is erosion greatest?

    (1) A   (2) B   (3) C   (4) D

6. Where is deposition most likely to occur?

    (1) A   (2) B   (3) C   (4) D

7. The map below shows a meandering stream. A–A′ is the location of a cross section. The arrows show the direction of the stream's flow.

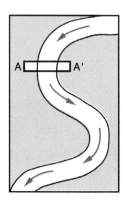

Which diagram below best represents the shape of the river bottom at A–A′?

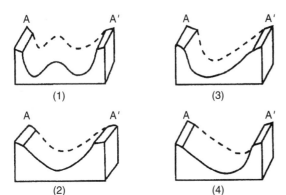

8. Which location along a river is the most likely place for deposition to dominate over erosion?

   (1) the base of a waterfall
   (2) the outside of a meander
   (3) a section of the river where the gradient is steep
   (4) the delta where the river enters the calm water of an ocean

9. Which of the following changes will cause an increase in the amount of erosion in a stream?

   (1) an increase in the hardness of bedrock
   (2) an increase in the stream discharge
   (3) a decrease in the gradient of the stream
   (4) a decrease in the average temperature

10. Which change would cause an increase in stream velocity?

    (1) an increase in the concentration of dissolved solids in the water
    (2) an increase in the volume of water flowing in the stream
    (3) a decrease in the slope of the stream
    (4) a decrease in the temperature of the water

11. Which change would be most likely to increase velocity of water flowing in a river?

    (1) a decrease in gradient   (3) an increase in latitude
    (2) a decrease in rainfall   (4) an increase in discharge

12. The diagram below is a map showing the stream drainage pattern for an area of Earth's crust.

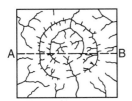

Which geologic cross section shows the most probable underlying rock structure and surface for this area along line A–B?

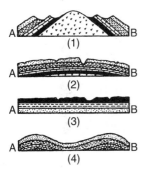

13. Which type of stream drainage pattern is most likely in a location where a constant slope dips gently to the north and the bedrock is uniform granite without folding, faulting, or other geologic structures?

(1) a rectangular drainage with streams aligned at right angles
(2) a branching drainage in which streams join as they flow north
(3) streams flowing away from a central location like spokes on a wheel
(4) a drainage pattern that contains many concentric circles connected by short radial stream segments

14. Which of the following factors is *least* likely to influence the drainage pattern of streams in a watershed?

(1) differences in the hardness of bedrock
(2) size of grains in the bedrock
(3) folding of the bedrock
(4) faulting of the bedrock

15. Mercury was found in water wells of a small community. What was the most probable source area of the mercury?

    (1) rainfall in a different watershed
    (2) a waste dump in a different watershed
    (3) a factory upstream in the same watershed
    (4) a housing development downstream in the same watershed

Open-Ended Questions

16. What are the principal functions of any stream system?

17. Why are low areas near rivers better suited for farmland and growing crops than they are for use as home sites.

18. A student decided to measure the speed of a stream by floating apples down a straight section of the stream. Describe the steps the student must take to determine the streams surface rate of movement (speed) by using a stopwatch, a 4-m rope, and several apples. Included the equation for calculating rate.

19. The diagram below shows a stream with a constant flow running through an area where the land around the stream is made of uniform sand and silt. Show the future path of this stream resulting from erosion and deposition as it usually occurs along a meander by drawing two lines to show the future stream banks.

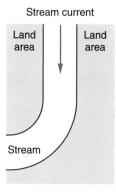

back to Earth as **precipitation**: rain, snow, sleet, or hail. Most precipitation falls into the ocean where it completes the water cycle. But precipitation that falls onto land can accumulate on the surface, it can run overland into streams, or it can infiltrate the ground.

Infiltration is the process by which water soaks into the ground under the influence of gravity. Whether water sits on the surface, runs off, or infiltrates depends on many factors. If the ground is saturated, it cannot hold any more water. Frozen ground also stops infiltration. Precipitation that falls as snow may remain where it fell until it melts. Infiltration is more likely if precipitation is not too rapid. The ground can absorb more water when rainfall comes as a long, steady rain. Rapid, intense rainfall may be more likely to run into streams than seep into soil. Slope is also important. The steeper the surface gradient, the less likely water is to infiltrate and more likely it is to run into streams. Water is more likely to infiltrate a permeable soil with large pores.

Land use is also important. Grass, trees, and other natural vegetation slow runoff and give surface water more time to soak in. However, when the ground is covered with pavement and buildings, water runoff is rapid and infiltration may not take place at all.

Permeability is the ability of soil or sediment to allow water to flow down through it. A loose or sandy soil is more permeable than clay or a soil with mineral deposits that block infiltration.

GROUNDWATER ZONES

Gravity pulls water into the ground until it reaches the **zone of saturation**, the part of the rock and soil where all available spaces are filled with water. Below the zone of saturation, the layers or rock or other material (such as clay) are compact enough that water cannot penetrate. (These layers are impermeable.) Above this zone of saturation is the **zone of aeration**, the region in which air fills most of the available

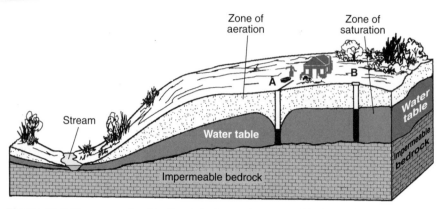

Figure 13-3 The zone of saturation is between impermeable layers below and the zone of aeration above. The top surface of the zone of saturation is the water table. The height of the water table depends on the balance between water entering as infiltration and water leaving as underground flow and usage from wells. Well A is active and it has depressed the water table nearby. The well at B does not affect the water table because it is not in use.

spaces in the rock and soil. As groundwater infiltrates, it moves through the zone of aeration and enters the zone of saturation.

The upper limit of the zone of saturation is known as the **water table** as shown in Figure 13-3. Therefore the zone of saturation extends from impermeable layers at the bottom to the water table at the top. The height of the water table changes depending upon infiltration, horizontal flow of groundwater, and usage. Sometimes there is a period of plentiful precipitation. At this time, infiltration is greater than the amount of water taken from wells or that flows out through the ground. In response, the water table will rise and groundwater will move closer to the surface. However, a dry spell is likely to result in a drop of the water table, as inflow is less than usage from wells and water that flows away.

The depth of the water table is critical to anyone who uses a well to obtain groundwater. Unless the well penetrates below the water table, water will not flow into the well. As water is drawn from a well, the use of groundwater causes the water table to fall near the well as shown in Figure 13-3. It is important to limit the usage of water from a well so that the amount of water taken out is not greater than the ability of water to flow into the well.

In moist years when there is plentiful rainfall, the water table is likely to rise. Well water will be plentiful in this situation. In dry spells, the water table can fall. If the water table falls below the bottom of the well, the well is said to run dry. Unfortunately, a well is most likely to run dry when there is a lack of surface water and the need for well water is critical.

There are several alternatives to correct this problem: reduce the pumping of groundwater, allowing the water table to rise back into the well; find other sources of water to reduce the use of well water; and dig the well deeper.

The water table is often an irregular surface that has high and low places where there are hills and valleys on the land. However, the water table usually has less relief, or change in elevation. The water table is usually deeper below the surface at hilltops, and the water table may come to the surface in valleys. If the water table comes to the surface, it can feed ponds or lakes or it may flow out of the ground onto the surface. A **spring** is a place where groundwater flows onto the surface of the ground.

In moist areas where streams run even in the periods between rainstorms, water flows through the ground and into streams through the streambed. Figure 13-4 shows two

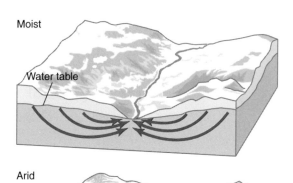

Figure 13-4 In moist regions, groundwater feeds into streams to maintain flow between rainstorms. But in arid climates, streams quickly run dry as they lose water into the ground through their streambed.

streams in different climatic regions. The top stream is in a moist location where it is fed by groundwater. The other stream is in a dry area. It loses water through its bottom into the ground.

ACTIVITY 13-1 | **GROUNDWATER MODEL**

You can construct a model of groundwater zones in a watertight container such as a fish tank. The bottom of the tank can represent the impermeable zone, although you may be able to place a layer of clay in the bottom of the tank to represent this barrier. Well-sorted sand, such as beach sand, is a good choice to represent the part of the ground in which water can circulate. Soda straws can be used to represent wells of various depths. Note the flow of groundwater as well as changes in the position of the water table. Make a list of your observations as water infiltrates the model and as water is drawn from the wells.

HOW DOES GROUNDWATER MOVE?

The ability of soil and bedrock to hold and transfer groundwater differs from place to place. In some locations, groundwater is plentiful while other places have little or no groundwater. Even if groundwater is present, it is easier to extract water from some materials than from others. Several factors affect where groundwater can occur and how it moves.

Permeability

If the ground is very permeable, infiltration will be rapid, and water can flow freely within it. Therefore, permeable ground prevents or reduces flooding. Permeable soils allow water to infiltrate before it reaches streams. Permeability affects the recharging, or replacement, of groundwater. The more permeable the soil, sediment, and bedrock are, the more readily water can flow down into the zone of saturation to replace groundwater lost to outflow or pumping from wells.

Large openings that are connected make rocks or soil very permeable. Even bedrock can be permeable if it is composed of particles with spaces between them, such as well-sorted sandstone. Bedrock can also be permeable if it has large and connected cracks, such as a fractured granite, or limestone that has underground passageways. Figure 13-5 shows the permeability of several kinds of ground materials.

Among sediments, the most permeable are well-sorted sediments with the largest, roundest particles. Soils made of uniform large particles have large spaces between the particles. This is especially true for rounded grains, which cannot be packed as tightly as flat or rectangular particles. If smaller particles are mixed with the large particles, the smaller grains fill in spaces between the large particles and reduce the permeability.

Even though a fine-grained soil may have the same total space between the grains as a coarse-grained soil, water cannot flow as quickly through these smaller spaces. Water clings to the surfaces of grains of sediment. Because fine-grained sediments have more total surface area, water has difficulty

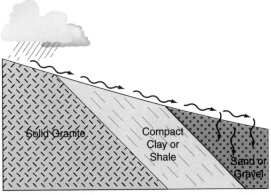

Figure 13-5 Solid, unbroken bedrock and compact clay can be impermeable. But sand, gravel, or fractured bedrock usually allow rapid infiltration.

passing through fine-grained sediments. Therefore materials like silt and clay have low permeability. Compact clay can be nearly as impermeable as solid bedrock.

Porosity

Porosity is the ability of a material to hold water in open spaces, or pores. It is an important property of a soil, sediment, or bedrock because porosity determines how much water the ground can store. If the porosity is low, there cannot be much groundwater stored. Rock and sediment with a high porosity generally result in an abundant supply of groundwater. Porosity can be calculated using the following formula:

$$\text{Porosity} = \frac{\text{volume of pore space}}{\text{total volume of the sample}}$$

Well-sorted sediments with round grains have a high porosity because of the large spaces between the particles. Sand that is rounded and well sorted can have porosity as high as 50 percent. But, unlike permeability, the size of the particles does not affect porosity. Figure 13-6 shows three containers of spherical particles. The container on the left holds a sample of small particles, the container in the middle holds medium particles, and the container on the right holds

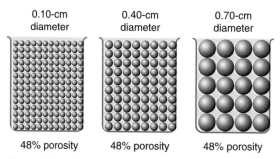

0.10-cm diameter   0.40-cm diameter   0.70-cm diameter

48% porosity   48% porosity   48% porosity

Figure 13-6 Changing the size of the particles does not change the porosity of a sample, as long as there is no change in the sorting, shape, and packing of grains. Each of the samples shown has the same porosity: 48 percent.

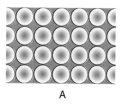 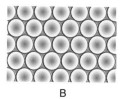

Figure 13-7 Sample A has more porosity than sample B because tighter packing in B reduces the sizes of the openings.

A   B

large particles. In each container, the particles are packed in the same way. Each sample also has the same porosity.

ACTIVITY 13-2 COMPARING THE POROSITY OF DIFFERENT MATERIALS

Obtain several different samples of sand and gravel. Determine their porosity by comparing the volume of each sample with the volume of water needed to fill each sample to its surface with water.

To remember this, consider a hollow cube that encloses the largest possible solid sphere. The solid sphere takes up about 52 percent of the volume of the cube. Therefore the percentage of open space is the remaining 48 percent. This is true of any size of cube in which the length of the side of the cube and the diameter of the enclosed sphere are equal. Although smaller particles leave smaller spaces between particles, the increase in the number of spaces balances the decrease in the sizes of the spaces.

Porosity does depend on the shape and the packing of the particles. Flat or rectangular particles can pack more closely than spherical particles. Figure 13-7 shows how tighter packing can even reduce the porosity of sediment composed of identical spherical particles.

Two other factors affect the porosity of a soil. A mixture of sediment sizes allows small particles to fit into pore spaces between larger particles. Figure 13-8 illustrates how mixing particle sizes reduces porosity. A final factor is mineral cement. When particles are held together by a substance such as calcite, clay, or silica cement, the cementing substance reduces the porosity of material.

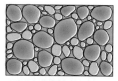

Figure 13-8 The porosity of this soil sample is reduced because small particles occupy spaces between the larger particles.

| ACTIVITY 13-3 | GROUNDWATER AND SEDIMENTS |

Materials per lab group: three transparent, plastic tubes approximately 3 to 5 cm in diameter (each tube should be fitted with a watertight bottom that can be opened quickly to allow water to flow out and a screen to hold sediment in the tubes as water flows out), supports for tubes, small, medium, and large pebbles (or plastic beads of these sizes), graduated cylinder, stop watch

Use a graduated cylinder to measure an equal volume of each size particle or bead. Place the small particles in one tube, the medium particles in another, and the large particles in the third tube. Add measured amounts of water into each tube until the water just reaches the top of the particles. For each tube, record the amount of water you added. Based on the volume of sediment and the amount of water each holds, calculate the porosity of each sample.

How long does it take for the water level to fall to the bottom of each tube after the bottom of the tube is opened? This interval is a measure of the permeability of each size of bead or particle.

How does the size of the sediments affect porosity and permeability?

 **Capillarity**

Water sticks to surfaces by means of a property called adhesion. Adhesion is related to the surface tension of water, which can be a remarkably strong force. This force draws water up into tiny spaces; this is called **capillarity**, or capillary action. Sometimes this action is able to pull water from the water table upward to where the roots of plants can reach it. Capillarity also allows trees to draw water from the soil into their leaves tens of meters above the ground. Tree trunks contain narrow passageways that draw water toward the leaves. Capillarity is also the way that a towel soaks up water or the way wax moves up through a wick to

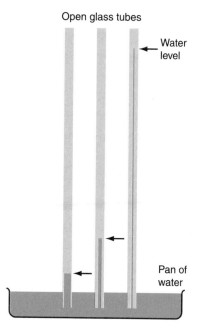

Open glass tubes

Water level

Pan of water

Figure 13-9 Water rises into narrow tubes due to capillarity. Water rises highest in the tubes with the smallest opening.

vaporize and serve as fuel for the flame of a candle. Figure 13-9 illustrates capillary action in small openings in glass tubes.

ACTIVITY 13-4 DEMONSTRATING CAPILLARITY

Place several samples of glass tubing, each with a different internal diameter, in a pan of water. (Very narrow tubing works best.) Draw a graph that shows the relationship between the internal diameter of the tubing and the height the that water rises above the level of the open water. (See Figure 13-9.)

ACTIVITY 13-5 CAPILLARITY OF SEDIMENTS

Obtain several transparent plastic or glass tubes that are about 0.5 cm in internal diameter. Cover the bottom of each tube with cotton fabric held in place by a rubber band. The fabric will allow water to pass through the opening but prevent the sediments from falling out. Fill each tube with a different size of sediment from coarse sand to silt. Place the tubes of sediment in a pan of water. Observe how high the water rises into each sample of sediment.

Capillary action occurs only in rock or sediments with very small, connected pore spaces. When openings become wider, the weight of water in the openings increases without a corresponding increase in the surface area of the openings. Adhesion cannot hold or draw water into large openings.

WHERE IS GROUNDWATER AVAILABLE?

In locations where surface water dries up, especially in the summer, groundwater may be the only reliable supply of freshwater. Unless there is a source of freshwater, an area cannot be inhabited. People need freshwater to drink and prepare food, to clean and for other household purposes, and to grow food. In most places, groundwater is more reliable and cleaner than surface water.

Most groundwater can be found within 100 meters of the surface, but the depth of solid, impermeable rock varies greatly from place to place. In some locations, solid rock without open spaces is exposed at the surface, and the impermeable rocks extends into Earth's interior. In other places, water can be brought to the surface from many kilometers underground. Digging or drilling wells can be a major expense for a property owner. The deeper the well goes, the greater the cost. Knowing where to place a well, how deep to dig, and when to stop can be difficult decisions.

About 98 percent of Earth's supply of freshwater is within the ground. Lakes, rivers, and streams are more visible than groundwater. However, there is far more freshwater stored in the ground than on the surface. The best supplies of groundwater are in underground aquifers. An **aquifer** is a zone of porous material that contains useful quantities of groundwater. Agriculture in the western United States depends heavily on well water drawn from aquifers. In some cases, water has been taken out of the aquifers so much faster than it can be replenished by rainfall and infiltration that there is danger of completely depleting the aquifer. This is sometimes called "mining" water because farmers are taking water that has been in the ground for hundreds of years. Although scientists usually consider freshwater a renewable resource, the rate of usage of some aquifers far exceeds the rate at which they receive new water.

 ## WHAT ARE SOME GROUNDWATER PROBLEMS?

Just as many streams have been contaminated by careless disposal of household, community, and industrial wastes, some aquifers are also being seriously polluted. Wherever soluble waste materials are left on or in the ground, infiltration can carry them to an aquifer. Several factors make the contamination of groundwater even more difficult to deal with than the pollution of surface waters. People are unable to see or smell buried wastes, nor can they see them flowing into an aquifer. In addition, it may take a long time for waste materials to reach the aquifer. By the time scientists know an aquifer is becoming polluted it might be impossible to prevent a very serious problem. It will also take far more time to rid an aquifer of contamination than it would to flush surface water of toxic materials.

 ### Sewage

Many people use septic systems to dispose of human waste by letting the liquid seep into the ground. Because of the nutrients in sewage, organisms that live in the ground use these nutrients, and as a result can cleanse sewage-filled water.

However, this cleansing is effective only as long as nutrients are not present in quantities too large to be consumed by the organisms that live in the ground. When communities conclude that increasing population and density of housing threaten to create a health problem, they install public sewage systems. These systems pipe sewage to a central location and treat it to speed up the rate at which organisms use and therefore remove toxic substances. Keeping waste materials out of the ground helps preserve the quality of groundwater.

Salt Water Invasion

Increasing use of groundwater in coastal locations has allowed salt water from the ocean to invade some aquifers. Because salt water is more dense than groundwater, the salt water flows in and under the freshwater. This has occurred in some places on Long Island as shown in Figure 13-10. The problem has been dealt with in several ways. Some parts of Long Island use water piped in from upstate rivers to reduce groundwater withdrawal. Returning treated wastewater to the aquifer has also been used to reduce salt water invasion.

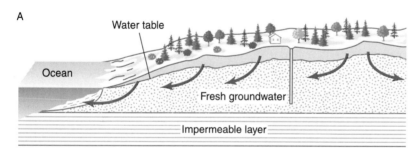

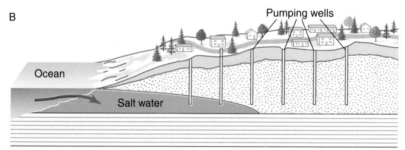

Figure 13-10 Pumping groundwater from an aquifer in a coastal region can cause salt water to flow into the aquifer and contaminate wells. Part A shows a profile before extensive pumping from the aquifer. Part B shows salt water invading the depleted aquifer.

A related problem occurs in arid parts of the United States. In these areas the soil is often salty. Irrigation of salty soil carries the salt deeper into the ground where it pollutes underground water supplies.

Sinking Ground

Taking water from the ground can cause the land to sink. Figure 13-11 shows a crack in the ground created by the use of groundwater. Usually, the sinking is a slow process that is not noticeable. However, cracks sometimes appear near the edge of the valley where deep sediments meet shallow bedrock. Sediments held up by bedrock are unable to sink with the central parts of the valley. This causes cracks to open. While the movement itself is not a danger, it can cause problems in the foundations of buildings and it can damage roads. The sinking ground level can also cause increased danger of flooding when rivers run full. Perhaps more important, this is a sign that water is being mined too quickly. In a location such as this, groundwater is a nonrenewable resource.

Figure 13-11 This crack in the ground recently appeared in Arizona. Water taken out of the ground nearby caused the land level to sink several meters.

TERMS TO KNOW

aquifer	groundwater	spring
capillarity	infiltration	transpiration
condensation	permeability	water table
convection	porosity	zone of aeration
dew point	precipitation	zone of saturation
evaporation		

CHAPTER REVIEW QUESTIONS

1. Most water vapor enters the atmosphere by the processes of

   (1) convection and radiation.
   (2) condensation and precipitation.
   (3) evaporation and transpiration.
   (4) erosion and conduction.

2. Where is most of Earth's freshwater that is in the liquid state?

   (1) in the oceans   (3) in glaciers
   (2) in the ground   (4) in clouds

3. In general, the probability of flooding decreases when there is an increase in the amount of

   (1) precipitation.   (3) runoff.
   (2) infiltration.   (4) snow melt.

4. During a rainstorm, surface runoff will probably be greatest is an area that has a

   (1) steep slope and a surface made of tiny particles.
   (2) steep slope and a surface made of large particles.
   (3) gentle slope and grass-covered surface.
   (4) gentle slope and tree-covered surface.

5. The diagrams below show the relative particle size from soils A, B, and C. Equal volumes of each soil sample were placed in separate containers.

Each container had a screen at the bottom. Water was poured through each sample to determine the infiltration rate.

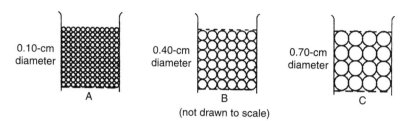

0.10-cm diameter   A

0.40-cm diameter   B

0.70-cm diameter   C

(not drawn to scale)

Which graph shows how the infiltration rates of the three soil samples would compare?

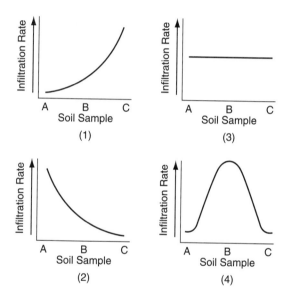

6. The water cycle runs primarily on energy from sunlight. What other factor influences every part of the water cycle?

(1) energy flow from Earth's interior   (3) porosity
(2) convection by wind   (4) gravity

7. How deep must a well be dug in order to yield a constant supply of freshwater?

(1) to the bottom of the soil
(2) to the bottom of the zone of aeration
(3) below the water table
(4) several meters into bedrock

8. Rainfall is most likely to infiltrate into a soil that is

   (1) permeable and saturated.   (3) impermeable and saturated.
   (2) permeable and unsaturated.   (4) impermeable and unsaturated.

9. Which graph best represents the general relationship between soil particle size and the permeability rate of infiltrating rainwater?

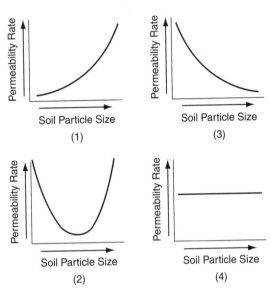

10. Which graph best shows the effect of soil permeability on the amount of runoff in an area?

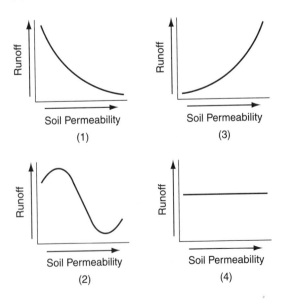

11. A student is investigating the porosity of a sample of sediment. Which of the following changes is most likely to increase the porosity of the sediment?

(1) Use a larger sample of the same sediment.
(2) Use a smaller sample of the same sediment
(3) Replace the particles of sediment with larger particles.
(4) Shake the sediment to reduce its volume, but not its mass.

Base your answers to questions 12 and 13 on the diagram below. Columns A, B, C, and D are partly filled with different sediments. Each column contains particles of a uniform size. A fine wire mesh screen covers the bottom of each column to prevent the sediment from falling out. The lower part of each column has just been placed in a beaker of water. Sediment sizes are not drawn to scale.

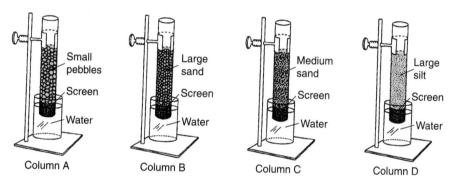

12. In which material would capillary action cause the water from the beaker to rise the farthest?

(1) small pebbles
(2) large sand
(3) medium sand
(4) large silt

13. A different experiment was performed using the same equipment. The beakers of water were removed and replaced with empty beakers. The sediments were allowed to dry. Then water was poured into each column. The permeability of the medium sand sample was shown to be

(1) less than the permeability of the silt and pebble samples.
(2) less than the permeability of the silt sample, but greater than the pebble sample.
(3) greater than the permeability of the silt sample, but less than the pebble sample.
(4) greater than the permeability of the silt and pebble samples.

14. Which is likely to be the most dependable source of freshwater for homes and agriculture in a semiarid (mostly dry) region?

    (1) rainfall   (3) deep wells
    (2) small streams   (4) dew and frost

15. Which form of waste disposal poses the greatest danger to the quality of groundwater?

    (1) burning waste in incinerators
    (2) burial underground
    (3) recycling waste material
    (4) packing wastes in waterproof containers

Open-Ended Questions

16. The diagram below shows a profile view of a location where water has been drawn from a well for a long time. The water table is shown in this diagram, but it is not labeled. The well is not used for several weeks, causing the position of the water table to change. Draw the new position of the water table after the well has been unused for several weeks.

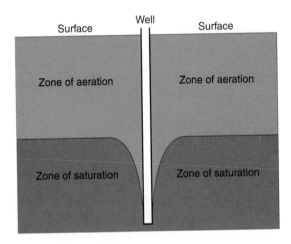

17. A student placed 4 L of gravel in a 5-L container. If 1 L of water was added to the gravel, and just reached the top surface of the gravel. What is the porosity of the gravel?

18. We generally consider water to be a renewable resource. But the rapid use of groundwater from some aquifers where precipitation is scarce is quickly

reducing the quantity of water in these aquifers. In these places, why should water be considered a nonrenewable resource?

19. When ocean water invades an aquifer near the ocean, the salt water usually flows in under the freshwater. (See Figure 13-11.) Why does ocean water flow under freshwater?

20. The residents of the city of Kingston, New York, noticed the odor and appearance of organic pollution in the Hudson River. Investigation by local authorities showed that the pollution had come from another major city in New York State. Where was the source of this pollution?

Chapter 14

Glaciers

A PUZZLING LANDSCAPE

The beauty and variety of landforms in the New World fascinated the first European settlers in North America. New York State was settled relatively quickly because it provided a route to the interior of North America along the Hudson and Mohawk rivers. In addition to natural transportation routes, as well as productive forests, and farming areas, settlers found features of the land they could not explain. Many of the rocks in the soil were varieties not found in local bedrock. Some of these exotic rock types were large boulders perched on hilltops. These boulders were far too high and too large to have been moved there by nearby streams. Between Rochester and Syracuse are numerous north-south aligned hills, most of them are blunt at the north end and trail off to the south. Even the Finger Lakes of western New York, long, narrow lakes that are also aligned north to south, could not be explained at that time by the geological processes at work in the area.

The nineteenth century expansion of settlements in New York State occurred simultaneously with a change in thinking in Europe. At that time, European geologists began to understand that prehistoric, continental glaciers played a major role in the evolution of their landscape. American geologists also found that the idea of a great ice age helped them understand puzzling aspects of North American geology. Nowhere in North America is the idea of wide-spread continental glaciers better supported than in the landscapes of New York State.

 ## WHAT IS A GLACIER?

Snow falls and accumulates on the ground everywhere in New York State. In the highest parts of the Adirondack Mountains, winter snow often lasts into early summer before it is completely melted. If the mountains were 1000 or 2000 meters higher, the reduced warmth of summer would not be able to melt winter snow. Each year more snow would accumulate and exert pressure on the underlying snow. This pressure would change the snow to ice and gravity would make the ice begin to flow downhill. This is how glaciers form. The reason that no glaciers exist today in New York State is that there are no places where the snow does not completely melt before the following winter.

Snow and ice exist as crystals. When snow falls, the flakes are usually light and feathery. After the flakes reach the ground and are buried under fresh snow, the delicate crystals gradually change to solid ice over a period of time that depends on such factors as speed of burial and temperature. Figure 14-1 on page 322 shows how the increasing density of the snow pack relates to changes in ice crystal shape.

Is ice a solid or a liquid? Ice is composed of water in the crystalline solid form. It fits the definition of a solid. Under short-term stress, ice behaves as a solid. An ice cube in an environment below freezing has a fixed shape. Hit it with a hammer and it breaks into smaller pieces. Yet, ice in a glacier flows. Glaciers do not flow because the ice is melting. They flow because solid ice responds to long-term stress by bending

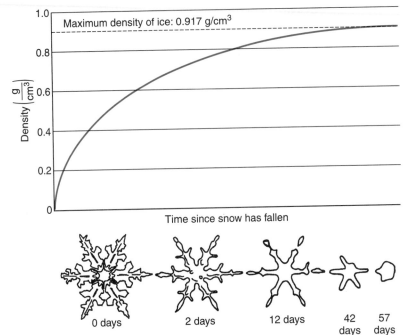

Figure 14-1 Delicate snow crystals transform into compact ice crystals. This is the first step in the formation of glaciers.

and deforming. This is similar to the behavior of solid rock within Earth's mantle. If you could strike the rock deep in Earth's mantle with a hammer, it would shatter. Forces applied over long periods of time result in fluid behavior of both rock and ice. The distinctions you have learned between solids and liquids are not as clear as you might have thought in the past.

ACTIVITY 14-1   SNOW TO ICE

This activity must be done in the winter when there are piles of snow outside. Find a location where the snow has been pushed into a deep pile. Dig out and place in several 200-mL beakers without packing it. Take the snow samples from the pile at different, measured depths. Also, gather fresh, surface snow that has not been moved. Draw a graph of snow depth versus the mass of snow in a given volume.

Valley Glaciers

Most glaciers begin where snow accumulates in the highest mountains, on coastal mountains where snowfall is abundant, or in the interiors of continents. Once permanent ice has formed, it begins to flow downhill under the influence of gravity. The type of glacier is determined by where the glacier forms. **Valley glaciers**, which begin high in mountain areas, flow from the high ice fields through valleys to lower elevations. In some locations, valley glaciers spread over flat lowlands as a piedmont glacier. Some glaciers flow into a lake or ocean where the ice breaks off, forming icebergs. Figure 14-2 is a glacier in Alaska that flows into a freshwater lake.

Other valley glaciers descend to an elevation where it is warm enough to melt the ice as quickly as it advances. If the ice in the glacier is moving forward at a rate of 1 meter per day, but the ice is also melting back 1 meter per day, the front of the glacier will not appear to move. This is known as a *dynamic equilibrium* because the rate of flow and the rate of melting are in balance. Even though the end of the glacier may be in the same place from year to year, the ice is, in fact, constantly moving downhill.

The rate at which ice advances varies from glacier to glacier. Ice in a small glacier may advance only a few millime-

Figure 14-2 The source area for the Portage Glacier in Alaska is the mountains at the upper end of the valley.

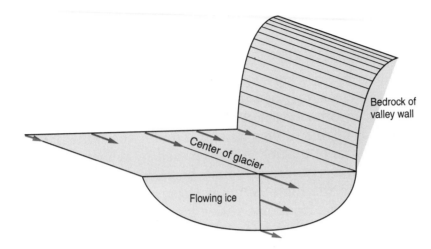

Figure 14-3 Like rivers, valley glaciers flow fastest at the center and away from the edges. Friction with the sides of the valley slows the ice near the edges and the bottom.

ters a day. Large valley glaciers commonly advance a few centimeters a day. Some glaciers experience periodic surges that cause the glacier to advance more than 30 meters per day over a period of several months. Surges are generally attributed to water accumulating under the glacier.

As gradient and volume affect the speed of water in rivers, they also influence the speed of ice flow in glaciers. Also, like the water in most rivers, ice moves fastest near the center and near the surface of a glacier. Figure 14-3 illustrates ice flow in a glacier. The long arrows indicate fast flow while the short arrows indicate slower flow.

 ## Continental Glaciers

A **continental glacier** flows outward from a zone of accumulation to cover a large part of a continent. If the process could be speeded up, it might resemble what pancake batter looks like as it is poured onto a griddle. Scratches on bedrock surfaces in the highest parts of New York's Catskill Mountains show that these mountains were completely covered by ice. The Catskills are well over 1000 meters above sea level. Furthermore, ice flowed southward over the Catskills from eastern Canada where the ice must have been even thicker. Yet this part of Canada does not have mountains as high as

New York's Catskills, so the ice in the zone of accumulation must have been very thick. Even the highest mountains in the northeastern United States were completely engulfed by the great thickness of ice covering eastern Canada.

Ice sheets that are several kilometers thick now cover most of Greenland and Antarctica. Scientists have studied both regions to gain an understanding of ice caps and how they flow outward and down to the oceans. Ice flowing into an ocean breaks away from the main body of the glacier to float away as icebergs. Because ice is less dense than water, icebergs do not sink. Icebergs from Greenland and Antarctica are large enough to pose a threat to shipping. A number of vessels, for example the *Titanic*, have sunk due to damage from collisions with icebergs. Fortunately, icebergs break up and melt before they can invade most ocean areas.

ACTIVITY 14-2 | **A MODEL OF A GLACIER**

You can create a model of a glacier with a substance called "ooblick." You can make ooblick by mixing cornstarch with water in a mass ratio of about 1 part water to 1.2 parts cornstarch. Like ice, this substance will shatter when hit hard. But, placed on a slope, ooblick will flow downhill.

 # HOW DO GLACIERS CAUSE EROSION?

When a glacier advances down a valley or over a continent, the ice pushes, carries, and drags great quantities of soil and sediment. These loose materials have little chance of remaining in place when a mass of ice hundreds or even thousands of meters thick moves over them. Ridges and knobs of bedrock are pried loose or rounded by the moving ice. Although ice is much softer than most bedrock, the rocks and sediment dragged along the bottom of a glacier scrape and scour the bedrock over which the glacier passes.

Valleys

In mountainous or hilly terrain, advancing glaciers seek the lowest passages and move through valleys first. Stream valleys often have a V shape in profile, especially in mountain areas. Streams and the sediment they carry occupy only the bottom of the valley and do not erode the sides of the valley above flood levels. The sides of a stream valley in mountainous terrain collapse under the influence of weathering and gravity, which often give them a steep but uniform slope of the V profile. When a glacier moves down a mountain valley filling it with ice, the erosive action of the glacier and its load of sediment pluck, scrape, and scour the sides of the valley changing its profile to a broader U shape. U-shaped valleys are strong evidence of glacial erosion. (See Figure 14-4.)

In the western part of New York, north-south aligned river valleys that drained the Allegheny Plateau were eroded into soft shale, siltstones, and limestones. The continental ice sheets that advanced southward into New York State also sought the lowest passages southward. The ice sheets advanced into these valleys first, making them wider and deeper.

Figure 14-4 Mountain streams originally cut Yosemite Valley in California into a narrow V shape. Later, erosion by glaciers changed the profile to a broader U shape.

Valleys west of Canandaigua Lake in the western Finger Lakes show the U shape especially well. A number of large U-shaped valleys running north-south in this part of the state were blocked at their outlets by glacial sediments to form the Finger Lakes. (See the map of New York State in the *Earth Science Reference Tables*.)

Striations and Grooves

Rocks transported along the bottom of the advancing ice leave scratches and grooves in the bedrock over which they pass. Parallel scratches called glacial **striations** can be found on hard bedrock surfaces throughout New York State, but they are best preserved in the hard metamorphic and igneous rocks of the Adirondacks and New York City regions. These scratches indicate the direction in which the glaciers moved. **Grooves** are deeper and wider cuts formed by glaciers. Grooves are often found in softer bedrock than where there are striations. Central Park in New York City has both grooves and striations in exposed bedrock. Figure 14-5 shows a bedrock surface grooved by the erosive action of rocks dragged by a glacier.

Figure 14-5 The parallel scratches, or striations, on this bedrock surface in California show the direction in which a glacier advanced.

HOW CAN WE RECOGNIZE DEPOSITION BY GLACIERS?

Sediment transported by glaciers must also be deposited. There are several differences between sediments deposited by ice and sediments deposited by water or wind. Water and wind sort sediments. (You may wish to review Chapter 11.) Moving ice transports and deposits sediment without regard to particle sizes. Therefore, sediments deposited directly by glaciers are unsorted and do not show layering. This unsorted glacial debris is sometimes called **till**.

Stream sediments are deposited where streams flow, usually in the bottom of a valley. But a glacier can move its debris anywhere that the ice covers, even to the highest parts of New York State. Glacial sediments often cover the whole land surface with an uneven blanket of till composed of mixed particle sizes.

Water and wind usually deposit the larger particles of their load relatively close to its source. On the other hand, glaciers carried their load of sediments, including boulders of granite and gneiss from Canada, hundreds of kilometers southward into New York State. In western New York State where the local bedrock is sedimentary, most often shale, siltstone, and limestone, these foreign rock types are especially noticeable. New York soils have a greater variety of minerals and they are more fertile than they would have been if they contained only local rocks. Large rocks that were transported from one area to another by glaciers are known as **erratics**. Figure 14-6 shows a large erratic located north of New York City.

Moraines

A **moraine** is a mass of till deposited by a glacier. Sometimes moraines form hills, often irregular in shape, where a glacier stopped advancing. This kind of deposit is known as a **terminal moraine**. Even though the front of the glacier

Figure 14-6 This large boulder, a perched erratic, was transported south by a glacier and deposited in till north of New York City. The smaller particles of till were washed away, leaving the large erratic balanced, or perched, on three smaller erratic boulders.

was nearly stationary, ice continued to transport sediment to the front of the glacier where it was released. The barriers of sediment that close off the Finger Lakes are moraines. Figure 14-7 illustrates the moraines that form the backbone of Long Island and extend into southern New England. Smaller mo-

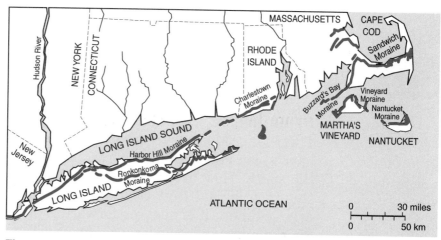

Figure 14-7 Long Island was built from sediment pushed into place by glaciers. Most rock-strewn, north shore beaches lie at the bottom of the high bluffs of the Harbor Hill Moraine. South shore beaches are composed of mostly fine sand washed from the moraines by glacial meltwater. These moraines extend into southern New England.

Figure 14-8 This kettle depression in central New York State contained a small kettle pond that froze over before the water infiltrated the ground. All that is left is a collapsed ice cover.

raines can be found throughout the state. Most of these are places where the ice front stalled and dumped its sediment load over a period of time.

Among the irregular hills in a moraine are depressions called kettles. A kettle is a small closed basin with no low-level outlet. Some kettles form when a block of ice within the till melts, leaving a closed depression. Rainwater that runs to the bottom of the kettle can escape only by evaporation or by infiltrating the ground. Some kettles fill with water to become kettle lakes. Figure 14-8 shows a kettle in central New York State.

 **Drumlins**

Drumlins are streamlined hills as much as 100 m high and 1 km long. Most of them are aligned north to south. They have steep sides, a blunt north slope, and a gentler slope to the south. How drumlins form is debated among scientists. One idea is that a glacier forms the hills by riding up and over sediment it is pushing forward. Thus drumlins show the direction in which the ice was moving. Within drumlins as in moraines, the unsorted and unlayered nature of till supports the idea that they are deposited by ice. Figure 14-9 shows a variety of features of continental glaciation found in New York State.

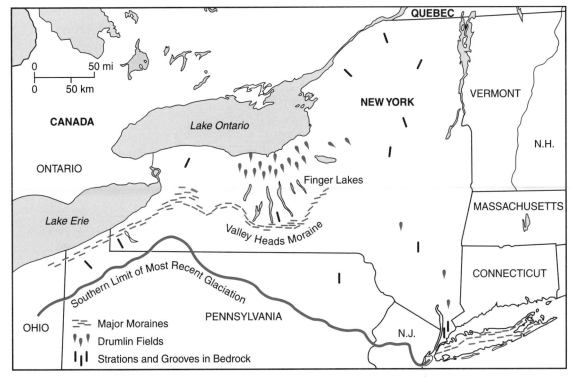

Figure 14-9 Moraines, drumlin fields, and sculpted bedrock show that the most recent advance of the continental glaciers covered nearly all of New York State.

⊕ HOW CAN WE RECOGNIZE DEPOSITION BY MELTWATER?

At the end of a glacier there are usually large quantities of water. Some flows from beneath the ice and more comes from the melting ice front. Sediments deposited by water from melting ice are known as **outwash**. The principal difference between till and outwash is that outwash deposits, like other sediments laid down by water, are sorted and layered. Outwash is generally less hilly than moraine. Kettles and kettle lakes are common in outwash plains where blocks of ice caught in the outwash later melt leaving depressions.

The difference between ice deposits and meltwater deposits can be seen very clearly on Long Island. Most of the island is made of sediments that can be traced back to glacial

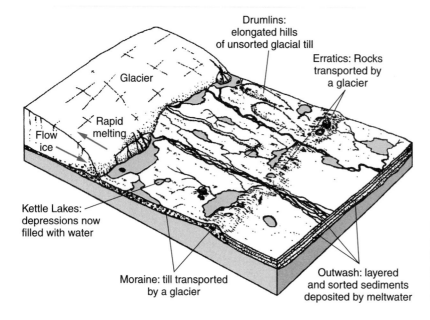

Drumlins: elongated hills of unsorted glacial till

Erratics: Rocks transported by a glacier

Glacier

Rapid melting

Flow ice

Kettle Lakes: depressions now filled with water

Moraine: till transported by a glacier

Outwash: layered and sorted sediments deposited by meltwater

Figure 14-10 These are some of the depositional features of continental glaciation that can be observed throughout New York State.

origin. The only bedrock on Long Island is in New York City at the far western end of the island. High bluffs of unsorted sediment in the Harbor Hill Moraine dominate the north shore of Long Island. The beaches of the north shore are composed of pebbles, cobbles, and even large boulders washed out of the moraine. These moraines were pushed into place by moving ice. However, most beaches on the south shore are made of sand washed out of the glaciers by meltwater. A few kilometers inland from the southern beaches are deposits of sand that show layering and sorting: strong evidence that the southern part of Long Island is made of sediment deposited by water, not by ice. Figure 14-10 shows some of the features of continental glaciation that can be observed throughout New York State.

ACTIVITY 14-3 INVENTORY OF GLACIAL FEATURES

Make a list of glacial features found near where you live. This list should include the name of the feature (moraine, large erratic, drumlins, etc.), how each formed, and its locations. If possible, add photographs of the features. A geology textbook can help you identify features of glaciation not listed in this book.

WHAT ARE ICE AGES?

Continental glaciers repeatedly covered as much as half of North America in the past. The most recent ice age ended only 10,000 to 15,000 years ago when the ice front melted back northward from Long Island and Pennsylvania. In terms of Earth's history, this is very recent. Over the past 2 to 3 million years, there have been many warm and cold periods. A drop in Earth's average surface temperature of only about 5°C is enough to cause a major ice age. The time between ice ages is variable, but it appears that there may have been 10 or more major ice advances just in the past million years. Figure 14-11

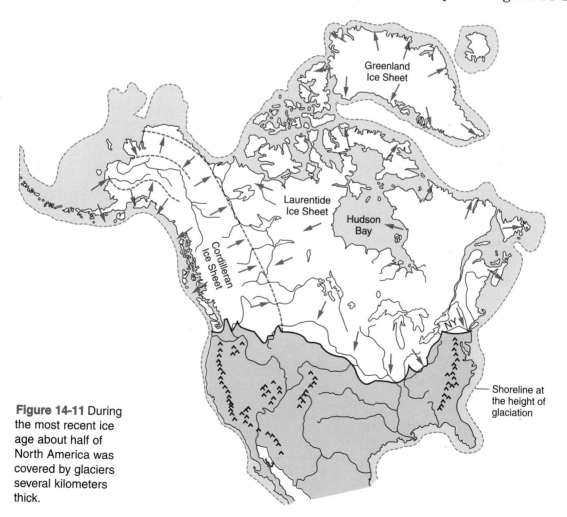

Figure 14-11 During the most recent ice age about half of North America was covered by glaciers several kilometers thick.

shows how much of North America was covered by ice during the most recent advance of continental glaciers.

Scientists have debated the cause of the ice ages since they first accepted ice ages over a century ago. Movements of the continents, changes in the tilt of Earth's axis, changes in the shape of Earth's orbit, and changes in the carbon dioxide and dust content of Earth's atmosphere have all been suggested as contributors to global climatic change. Understanding the reasons for changes in global climates is more important than simply unraveling the past. Future climatic changes of this magnitude would certainly have a major impact on human civilization.

TERMS TO KNOW

continental glacier	**grooves**	**outwash**	**till**
drumlins	**kettle**	**striations**	**valley glaciers**
erratics	**moraine**	**terminal moraine**	

CHAPTER REVIEW QUESTIONS

1. Why do large ice sheets no longer cover most of New York State?

   (1) There is no longer enough ice.
   (2) The climate is now too cold.
   (3) The climate is too warm.
   (4) There is too much freshwater.

2. The graph below shows the snow line (the elevation above which glaciers form at different latitudes in the Northern Hemisphere).

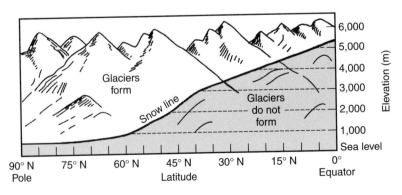

At which location would a glacier most likely form?

(1) 0° latitude at an elevation of 6000 m
(2) 15° N latitude at an elevation of 4000 m
(3) 30° N latitude at an elevation of 3000 m
(4) 45° N latitude at an elevation of 1000 m

3. What agent of erosion most likely left large boulders high above stream valleys throughout New York State?

(1) wind
(2) rivers
(3) moving ice
(4) gravity acting alone

4. The diagram below shows a glacial landscape.

Which evidence suggests that ice created this landscape?

(1) U-shaped valleys
(2) many stream valleys
(3) sorted sediments on the valley floor
(4) the landslide near the valley floor

5. The occurrence of parallel scratches on bedrock in a U-shaped valley indicate that the area has most likely been eroded by

   (1) a glacier
   (2) a stream

   (3) waves
   (4) wind

6. If a glacier moves through a mountain stream valley, the valley is likely to become

   (1) shallower and narrower.
   (2) shallower and wider.

   (3) deeper and narrower.
   (4) deeper and wider.

7. How are grooves and striations found in bedrock of granite and gneiss created by a glacier?

   (1) The moving ice wears away at the rock surface.
   (2) The ice carries rocks that abrade the rock surface.
   (3) Alternating freezing and thawing wear the rock smooth.
   (4) The weight of the ice pushes on rock to make it smooth.

Base your answers to questions 8 to 10 on the diagram below that shows landscape features formed as the most recent glacier melted and retreated from New York State.

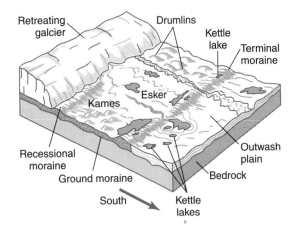

8. The moraines pictured in the diagram were deposited directly by the glacier. The sediments within these moraines are most likely

   (1) sorted by size and layered.
   (2) sorted by size and unlayered.

   (3) unsorted by size and layered.
   (4) unsorted by size and unlayered.

9. The shape of the elongated hills labeled as drumlins is most useful in determining

   (1) the age of the glacier.   (3) thickness of the glacier.
   (2) direction of glacial movement.   (4) rate of glacial movement.

10. At the stage shown in this diagram, the ice in the glacier is probably moving toward the

    (1) north.   (3) east.
    (2) south.   (4) west.

11. Many hills in western New York State are known as drumlins. These are composed of sediments transported and deposited directly by glacial ice. These sediments are likely to be

    (1) well-rounded, sand-sized particles.
    (2) in thin horizontal layers.
    (3) poorly rounded and not in layers.
    (4) found underwater, mixed with organic materials.

12. The diagram below represents the cross section of a soil deposit from a hill in central New York State. The deposition was most likely caused by

    (1) a glacier.
    (2) a windstorm.
    (3) a stream entering a lake.
    (4) wave action along a beach.

    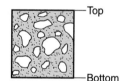

13. The diagram below shows trends in the temperature of North America during the last 200,000 years, as estimated by scientists.

    According to this graph, what is the total number of glacial periods that have occurred in North America in the last 200,000 years?

    (1) 5
    (2) 2
    (3) 3
    (4) 4

    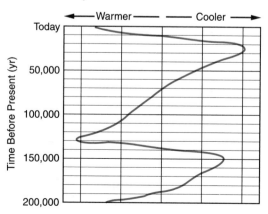

14. On a field trip 40 km east of the Finger Lakes, a student observed a boulder of gneiss on the surface bedrock. This observation best supports the inference that

    (1) surface sedimentary bedrock was weathered to form a boulder of gneiss.
    (2) surface sedimentary bedrock melted and solidified to form a boulder of gneiss.
    (3) the gneiss boulder was formed from sediments that were compacted and cemented together.
    (4) the gneiss boulder was transported from its original area of formation.

15. Which feature of glaciation common in New York State is *least* likely to help us understand the direction in which the ice was moving?

    (1) kettles and kettle lakes   (3) finger lakes
    (2) grooves and striations   (4) drumlins

Open-Ended Questions

16. Early settlers in New York were puzzled when they found that the rocks in the soil differed from the local bedrock. What major geologic event did nineteenth century geologists infer to solve this puzzle?

17. A student took the following notes at a field trip location.

    There is a good view from this windy hilltop. Rocks are visible everywhere. There are boulders, cobbles, and pebbles of many sizes and shapes mixed together. The rocks sitting on the surface are metamorphic rocks that rest on limestone bedrock. The teacher showed us parallel scratches in the bedrock. I saw almost no soil.

    (a) State the agent of erosion that deposited most of the sediments at this location
    (b) State one observation recorded by the student that supports this conclusion.

18. Ice is a natural crystalline solid of inorganic origin with a specific composition and relatively uniform properties. It is therefore a mineral. While the hardness of ice is variable depending on its temperature, the hardness of the ice at the bottom of the glaciers that covered New York State was only about 1.5 on Mohs' scale. Still, these glaciers scratched and scoured

bedrock composed mainly of quartz and feldspar, which are far harder than ice. How can glaciers scratch minerals harder than ice?

19. A student located an area of New York State where the continental ice sheet stopped and began to melt back to the north. The student attempted to find the limit of glacial erosion and deposition by looking at the sediments that cover the ground. In this location, all the ground was covered by material deposited directly by ice, or by material deposited by water that flowed out of the ice. Give one property of the ice-deposited sediments that would not be true of sediments deposited by meltwater.

20. The diagram below represents a profile view of North America at a latitude of about 50°N at the time of the most recent ice age. Draw arrows at A and at B to show the direction in which the ice is moving at each point.

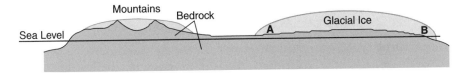

Chapter **15**

Landscapes

NEW YORK'S NATURAL WONDERS

New York is not one of the largest states of the United States, nor does it have any large national parks to rival Yosemite or the Grand Canyon. But nearly every part of the state has unique and beautiful natural wonders. No matter where you live in New York, inspiring features of geologic interest are no more than an hour or two away.

For example, Niagara Falls is a broad sweep of thundering water that carries the outflow of four of the world's largest freshwater lakes over a 60-meter plunge. Other spectacular waterfalls include Taughannock Falls north of Ithaca and Kaaterskill Falls at the eastern margin of the Catskills.

The beaches of Long Island offer miles of white sand and pounding surf. And the freshwater beaches in the Finger Lakes and the Adirondack Mountains are famous for their scenic and recreational value. The Genesee River Gorge south of Rochester has impressive cliffs hundreds of meters high. Other gorges in the southern Finger Lakes region and along the Ausable River in the Eastern Adirondacks provide equally impressive experiences with their varied erosional forms.

Figure 15-1 The Hudson River passes through the Hudson Highlands near the center of this photograph. Many people consider this part of the Hudson one of the most scenic river passages in the nation.

The high peaks of the Adirondacks and natural limestone caverns of the Catskills are scenic travel destinations. The passage of the Hudson River through the Hudson Highlands (Figure 15-1) and cliffs of the Palisades make the Hudson one of America's most scenic rivers.

 ## WHAT ARE LANDSCAPES?

A landscape is the general shape of the land surface. Landscapes include a variety of topographic features related to the processes that shaped the surface. For example, New York's glaciated landscapes include such diverse landforms as U-shaped valleys; rounded, grooved, and polished bedrock surfaces; drumlins and moraines. A landform is a single feature of a landscape. Landscapes are generally made of a variety of related landforms such as mountains, valleys, and river systems. Geological structures are major influences on landscapes. Figure 15-2 on page 342 shows how folding affects a landscape. Most land areas can be divided into regions that have similar landforms.

Figure 15-2 Folds in the rock layers have played an obvious role in shaping this Utah landscape.

| ACTIVITY 15-1 | **LOCAL LANDFORMS** |

Collect or take photographs of notable landforms in and near your community. You can use original photographs, travel brochures, or images from the Internet. For each landform, write one to four sentences about the processes that formed it.

Most geologists think of a landscape as the product of geological events over hundreds, thousands, or even millions of years. Some events are building processes such as uplift, volcanic eruptions, and glacial deposition. Other equally important events, such as weathering and erosion, wear down the land surface. Therefore, landscapes are the surface expressions of opposing geological processes.

 Plains

Most landscape regions can be classified as plains, plateaus, or mountains. Figure 15-3 shows New York State has examples of all three. **Plains** are usually relatively flat. That is, the range of elevations is small. Hill slopes are gentle and streams commonly meander over broad floodplains.

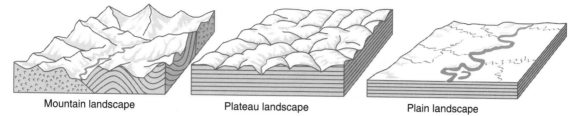

Mountain landscape   Plateau landscape   Plain landscape

Figure 15-3 Based on their elevation and underlying bedrock structure, most landscapes can be classified as plains, plateaus, or mountains. Mountain landscapes have the greatest differences in elevation, reflecting their variable rock types and complex structures. The lower relief of plateaus and plains may be due to long-term erosion, or it may be due to flat-lying, sedimentary rocks.

The bedrock underlying plains can be any kind of rock that has been in place long enough to be eroded to a low level. Shale and other layered sedimentary rocks are especially common in plains landscapes. Soils tend to be thick and stable due to the low gradients. Much of the Mississippi Valley and central part of the United States is a plains landscape that is mostly underlain by flat layers of relatively soft sedimentary rock.

 **Plateaus**

Plateau landscapes have more relief than plains but they are not as rugged as mountain landscapes. **Relief** is the difference in elevation from the highest point to the lowest point on the land surface. Some sources use the term tableland as a synonym for plateau. Indeed, some plateaus look like relatively level areas that are nearly as flat as plains but at a much higher elevation.

Parts of the Colorado Plateau of Arizona and Utah look like a plains landscape that has been pushed up more than 1000 meters. The Columbia Plateau of eastern Oregon was built up by fluid lava flows that spread over great distances. Flat-lying, sedimentary rocks underlie most plateaus, including the Colorado Plateau.

Landforms in the eastern part of the United States are generally more rounded than those in dryer areas of the West. Therefore, the Appalachian Plateau, the largest plateau in the eastern United States, is mostly a region of rolling hills.

Figure 15-4 These mountains in Alaska shows the great differences in elevation in mountain landscapes, which are often a result of differences in rock types and complex geologic structures.

Mountains

Mountain landscapes have the greatest relief. Figure 15-4 shows a rugged mountain region in Alaska. Variations in rock types, which are common in mountain areas, often include hard metamorphic and igneous rocks as well as weaker kinds of rock. The harder rocks are usually found in the highest areas, and the weaker rocks have been eroded to make deep valleys.

Tectonic uplift and deformed rock structures such as folds and faults contribute to the complexity and relief of mountain landscapes. A close look at the rocks in any major mountain area, such as the Rockies, the Alps, or the Himalayas, is likely to reveal a complex geologic history and a wide variety of rock types. Figure 15-5 shows the major landscape regions of the United States.

WHAT FACTORS INFLUENCE LANDSCAPE DEVELOPMENT?

The factors that influence landscapes are geologic and climatic. Geologic factors include crustal movement, rock type, and geologic structures. The major climatic factor in landscape development is the annual precipitation.

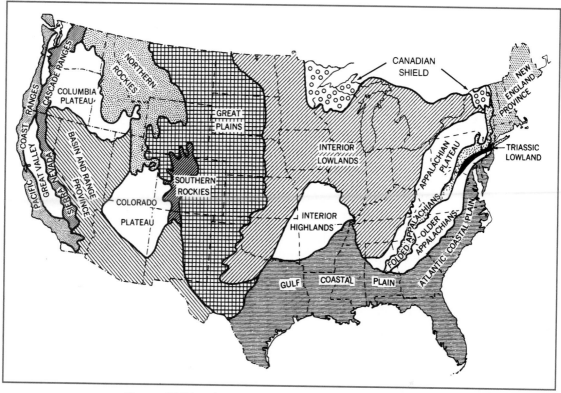

Figure 15-5 Landscape regions of the United States include plains, plateaus, and mountains.

Geologic Factors

Among geologic factors, vertical movement of Earth's crust is very important. Where Earth's crust is being pushed up, high peaks and deep valleys characterize young mountain ranges, such as the Himalayas of Asia. Rapid uplift and steep slopes also result in rapid erosion. Rivers that drain the great Himalayas carry more sediment into the oceans than rivers from any comparable area of Earth's surface.

The characteristics of rock types, especially their resistance to weathering and erosion, are very important in landscape development. The hardest rocks of any area usually are found at the highest land elevations. In places where some bedrock is more resistant to weathering and erosion than others, ridges tend to follow the hardest rock types. Valleys in these areas generally follow the softer and/or fractured rock.

A major landform called the Niagara Escarpment runs from the Niagara River south of Lake Ontario across much of western New York State. An **escarpment** is a steep slope or a cliff of resistant rock that marks the edge of a relatively flat area. The Niagara Escarpment, which separates the lowlands along Lake Ontario from higher land to the south, owes its existence to a particularly hard sedimentary rock known as the Lockport Dolostone. (Dolostone is a sedimentary rock similar to limestone but with a slightly different mineral composition.) The Helderberg escarpment southwest of Albany and the steep eastern front of the Catskill Mountains are other notable New York State escarpment landforms. Figure 15-6 is a view across the Hudson River Valley toward the Catskill Mountains in the distance.

Geologic structures also influence landscapes. The Hudson Highlands, about 30 km north of New York City, is a block of Earth's crust that has been pushed up. (This can be seen in the background in Figure 15-1 on page 341.) The Highlands are bounded by geologic faults to the north and south. In addition, the Adirondack Mountains have been pushed up into a broad dome structure. The waves of hills of the Appalachian Plateau north of Harrisburg, Pennsylvania, are the direct results of large folds in the bedrock caused by compression of the crust. Faults are zones of crushed bedrock along which rel-

Figure 15-6 Looking west across the Hudson Valley. The Catskill escarpment, which marks the edge of the Catskill Mountains, is visible in the distance.

ative movement has occurred. These lines of weakened rock influence the course of streams and enable rapid erosion. The zigzag route of the Hudson River through the Hudson Highlands is probably a result of erosion along geologic faults. Faulting also controls the direction of many northeast-flowing streams in the high peaks region of the Adirondack Mountains.

Climate Factors

Climate is the other major factor in landscape development. A humid climate favors chemical weathering, which produces rounded, less angular landforms. Most of the hill slopes of New York State are rounded and gentle because of the relatively humid climate. A moist climate also allows plants to grow and protect soil from erosion. Desert areas sometimes have a steplike profile, with flat hilltops and terraces separated by steep escarpments. Figure 15-7 is a desert area in southern Utah. Notice the cliffs and terraces that make up most of this landscape. Figure 15-8 on page 348 illustrates the contrast between the rounded shapes of a humid landscape and the angular features of many desert landscapes.

The climate of New York State does not change very much from place to place. All of New York has a moist, temperate, mid-latitude climate. Therefore the landscape differences in New York State are not the result of differences in climate.

Figure 15-7 The angular, stepped profile of this area in Utah is due partly to the dry climate. With little plant cover, soil is washed away in summer thunderstorms, leaving steep rocky slopes.

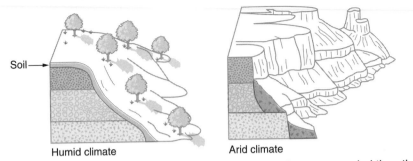

Figure 15-8 Landforms in humid climates are generally more rounded than those in desert regions. In arid climates, there are few plants to hold soil in place, so weak rocks are exposed to rapid erosion and slopes are more stepped and angular.

The variations seen in the New York landscape are caused by geological factors.

WHAT ARE THE LANDSCAPES OF NEW YORK STATE?

Figure 15-9 is the Generalized Landscape Regions map from the *Earth Science Reference Tables* that shows the landscape regions of New York State. This map can be used along with the Generalized Bedrock Geology map also in the *Reference Tables* to locate cities and other geographic features on the landscape map. For example, looking at the landscape region map and the bedrock geology map you can see that Watertown, Oswego, and Rochester are located in the Erie-Ontario Lowlands.

For such a small geographic region with a relatively uniform climate, New York has a remarkable variety of landscapes including plains, plateaus, and mountains. The boundaries between landscape regions are sometimes remarkably sharp and easy to see. For example, the eastern front of the Catskills is marked by a dramatic change in elevation from the lowlands along the Hudson River to a landscape nearly 1000 meters higher to the west. (See Figure 15-6 on page 346.) The Hudson Highlands are an uplifted block of land bounded by fault scarps both to the north and

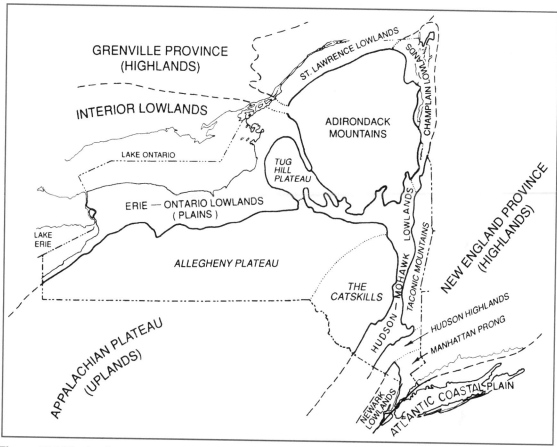

Figure 15-9 The landscape regions of New York include plains, plateaus, and mountains. In addition, there are several regions difficult to classify. Boundaries between landscape regions often follow changes in elevation and rock types.

to the south. However, the long boundary between the Allegheny Plateau and the Erie-Ontario Lowlands in western New York is more subtle, and the exact position of the transition is not easy to see in most places.

ACTIVITY 15-2 LANDSCAPE BOUNDARIES

If you live near a landscape boundary, draw that boundary on a local road map. In what ways does the land look different on the two sides of the boundary?

Long Island

A tour of New York State might start at the eastern end of Long Island where glacial moraines give this part of the island its split, fish-tail shape as seen on maps. Two moraines, which mark the most southerly advance of the recent continental glacier, dominate Long Island. These moraines form east-west lines of low hills on an otherwise low-lying and flat plains landscape largely composed of sandy outwash sediments. Wind, longshore currents, and wave action deposit and shape the sandy beaches and dunes along the south shore.

On Long Island, only the western end is hard metamorphic bedrock visible at the surface. This is within the boundaries of New York City, which also includes the bedrock islands of Manhattan and Staten Island as well as the Bronx, which is built on the bedrock mainland of North America.

The Hudson Valley

Continuing the journey upstream along the Hudson River, a traveler passes between the low region of the Newark Basin mostly hidden behind the higher cliffs named the Palisades and the rolling hills along the eastern side of the river. The Palisades themselves are composed of shallow intrusive igneous rock with a dark composition similar to basalt. Figure 15-10 shows the northern part of the Palisades.

Figure 15-10 The Palisades is a shallow igneous intrusion that forms cliffs along the western shore of the Hudson River. The cliffs extend from the New Jersey shore opposite New York City northward about 50 km into New York State.

Mostly on the east side of the Hudson River, the Taconic Mountains, the Hudson Highlands, and the Manhattan Prong are all underlain by very old metamorphic rock. This old rock gives them a complex landscape, but mostly without the steep slopes and high peaks of a true mountain landscape.

The Catskills-Allegheny Plateau

The Catskills rise above the western edge of the Hudson Valley. Although they are sometimes called the Catskill Mountains and they include some of the highest elevations in New York State, the Catskills are underlain by flat layers of sedimentary rock with rounded and even relatively flat summits. The Catskills are the higher, eastern end of the Allegheny Plateau, a large region of rolling hills that extends nearly to Lake Erie at the western end of the state. The Allegheny Plateau, an extension of the Appalachian Plateau, contains the Finger Lakes as well as other valleys deepened and U-shaped by the advancing continental glaciers.

Plains

Lakes Erie and Ontario, two of North America's Great Lakes, are bordered by a plains landscape that includes drumlins and other features of the ice ages. Local soils have a good mix of minerals due in part to the abundance of glacial till carried south from Canada. The lakes moderate the nearby climate of this agricultural region, extending the growing season. This plains landscape extends along the St. Lawrence River and around the Adirondack Mountains to join the valleys of the Hudson and Mohawk rivers.

Tug Hill Plateau and Adirondacks

Two landscape regions remain. The Tug Hill Plateau is a remote and sparsely populated highland that receives the greatest winter snowfalls in New York State.

The Adirondacks is New York's best example of a true mountain landscape. The land was pushed up onto a broad dome exposing at the surface ancient metamorphic rocks that were formed deep underground. Faulted stream valleys contribute to the mountain relief. The highest point in the state, Mt. Marcy, and some of the most scenic lakes are located in this region. Tourism is the mainstay of the economy since the area is too rugged and cold for good farming and it is too remote for commerce.

ACTIVITY 15-3   **LANDFORMS OF NEW YORK STATE**

As a class, collect photographs from individuals, from travel brochures, or from the Internet, and attach them at their appropriate place on a large wall map of New York State. Select photographs that show landforms characteristic of the various landscape regions of New York.

A similar project can be conducted collecting photographs of landform in the United States and placing them on a map of the United States.

TERMS TO KNOW

escarpment	mountain landscape	plateau
landform	plains	relief
landscape		

CHAPTER REVIEW QUESTIONS

1. Which kind of landscape is not likely to be underlain by flat lying sedimentary rocks?

   (1) lowlands   (3) plateaus
   (2) plains   (4) mountains

2. What kind of landscape region is often characterized by flat-lying layers of sedimentary rock that have been uplifted by forces within Earth without major folding and faulting, and then eroded to form deep stream valleys?

(1) plains   (3) mountains
(2) plateaus   (4) coastal lowlands

3. The table below describes three landscape regions of the United States.

Landscape	Bedrock	Elevation/Slopes	Streams
A	Faulted and folded gneiss and schist	High elevation, steep slopes	High velocity, rapids
B	Layers of sandstone and shale	Low elevation, gentle slopes	Low velocity, meanders
C	Thick, horizontal layers of basalt	Medium elevation, steep to gentle slopes	High to low velocity, Rapids and meanders

Which choice below best identifies landscapes A, B, and C?

(1) A—mountain, B—plain, C—plateau
(2) A—plain, B—plateau, C—mountain
(3) A—plateau, B—mountain, C—plain
(4) A—plain, B—mountain, C—plateau

4. In which part of the United States would you be most likely to find thin soils and steep, rocky slopes?

(1) the Great Plains of Kansas and North Dakota
(2) the coastal plains of Florida and Louisiana
(3) the hills of the Appalachian Plateau
(4) the desert plateaus of Utah and Arizona

Base your answers to questions 5–15 on the *Earth Science Reference Tables*.

5. In the 1800s the Erie Canal was built from from the Albany area through Utica and Syracuse, then just south of Rochester to Buffalo. What landscapes did the canal follow?

(1) The canal was built primarily through valleys in mountain landscapes.
(2) The canal was built mostly through plateau landscapes.
(3) The canal was built across plains landscapes whenever possible.
(4) The canal connected all the landscape regions of New York State.

6. Within New York State, the Genesee River and the Susquehanna River flow mostly through which landscape region?

   (1) the Hudson-Mohawk Lowlands   (3) the Allegheny Plateau
   (2) the Atlantic Coastal Plain   (4) the Adirondack Mountains

7. Tilted, metamorphosed bedrock is typically found in which New York State landscape region?

   (1) Taconic Mountains   (3) Tug Hill Plateau
   (2) Atlantic Coastal Plain   (4) Erie-Ontario Lowlands

8. The Catskills are actually a part of what larger landscape region?

   (1) the Adirondack Mountains   (3) the Allegheny Plateau
   (2) the New England Highlands   (4) the Manhattan Prong

9. What is the smallest distance from the southern boundary of the Adirondack Mountains landscape to the northern boundary of the Allegheny Plateau?

   (1) approximately 10 km   (3) approximately 100 km
   (2) approximately 50 km   (4) greater than 200 km

10. Which two locations are in the same New York State landscape region?

    (1) Albany and Old Forge
    (2) Massena and Mt. Marcy
    (3) Binghamton and New York City
    (4) Jamestown and Ithaca

11. Buffalo, New York, and Plattsburgh, New York, are both located in landscape regions classified as

    (1) highlands.   (3) plateaus.
    (2) lowlands.   (4) mountains.

12. It is feared that global warming could melt the polar ice caps and cause sea level to rise. Which part of New York State would most likely be flooded if this were to occur?

    (1) Allegheny Plateau   (3) Adirondack Mountains
    (2) Erie-Ontario Lowlands   (4) Atlantic Coastal Plain

13. In which New York State landscape region is surface bedrock generally composed of metamorphic rock?

    (1) Tug Hill Plateau   (3) Newark Lowlands
    (2) Adirondack Mountains   (4) the Catskills

14. What landscape region is located at 44°N latitude, 74°W longitude?

    (1) Allegheny Plateau   (3) Adirondack Mountains
    (2) Erie-Ontario Lowlands   (4) Atlantic Coastal Plain

15. Which town or city is located within the New York State landscape region with the greatest difference in elevation from the highest to the lowest land levels?

    (1) Buffalo   (3) Old Forge
    (2) Rochester   (4) Massena

Open-Ended Questions

16. How can a plains landscape change to a mountain landscape through time?

17. What kind of landscape is usually found where many types of bedrock have been folded and faulted?

18. The diagram below is a profile of a hill in a desert region. If the climate of the region were to become humid over a long time, such as the climate in New York State, what would the shape of the hill be in several thousand years? Draw a new profile in which erosion has taken place, but the maximum height of the hill is the same.

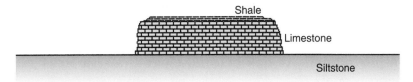

19. The Appalachian Plateau is a major landscape region of the United States. What two New York State landscape regions are parts of the Appalachian Plateau?

20. If you took an overland journey from Binghamton, New York, to Oswego, New York, what general change would you see in the topographic relief as you travel northward?

Chapter 16

Oceans and Coastal Processes

 THE BLUE PLANET

Scientists' ability to study other planets has increased greatly in the past few decades. The quality of optical telescopes has improved, providing clearer views of the planets in the solar system. Scientists have made even greater strides, placing instruments on or near other planets. In addition, astronomers have detected evidence of dozens of other planets around stars other than the sun. Through studies of other planets, it has become clear that Earth is unique among planets.

The presence of liquid water on Earth's surface is the greatest difference between Earth and other known planets. Early forms of life thrived and evolved in Earth's oceans for several reasons. The oceans protected them from harmful radiation such as the ultraviolet rays that cause sunburn. The circulating water in the oceans also transports oxygen and food to stationary organisms. Other organisms developed that could move through the oceans in search of food. So much of our planet is covered by oceans that Earth has a unique blue color when observed from space.

ACTIVITY 16-1 | **WATER ON THE PLANETS**

Prepare a report on the occurrence of water on and in other planets of our solar system. For each planet, state what astronomers have learned or what they infer about water on or in the planet.

⊕ WHAT MAKES OCEAN WATER DIFFERENT?

The explosion of a massive star left a cloud of debris in space that was drawn into various concentrated regions by gravity. The greatest concentration of mass became the sun while smaller clouds of debris were drawn together into the planets. Our planet probably began as a rocky mass without surface water. Although scientists do not know how long it took for oceans to form on Earth, evidence of surface water can be found in rocks that date back to very early in Earth's history.

The Origin of Earth's Water

There are several possible sources of the water now found in the oceans. Perhaps most of the water came from magma originating deep within Earth's molten interior. Most of the water vapor from the earliest eruptions may have remained in the atmosphere until the surface became cool enough for liquid water to accumulate. Even today, water vapor is a major gas component of erupting magma.

Some of the water could have come from outer space. Comets are composed largely of ice. They are sometimes described as dirty snowballs. Comets striking Earth probably contributed some of the water in the oceans. Even rocky meteorites, which bombarded Earth much more frequently early in its history, contained water. There are few remains of Earth's earliest rocks. Details of the formation of oceans will probably remain unknown for many years.

The Composition of Ocean Water

You may have read stories of people stranded at sea who suffered because of a lack of water. Ironically, they were surrounded by more water than they could ever need. However, ocean water contains about 3.5 percent dissolved salts. Figure 16-1 shows the average composition of ocean water.

Much of the water people drink is used to absorb and remove waste products from the body. Drinking ocean water would add unwanted salts rather than help the body get rid of them. That is why people cannot drink ocean water unless most of the salts have been removed. Other than water, the most common substance in seawater is sodium chloride, or table salt. Also present are similar magnesium, calcium,

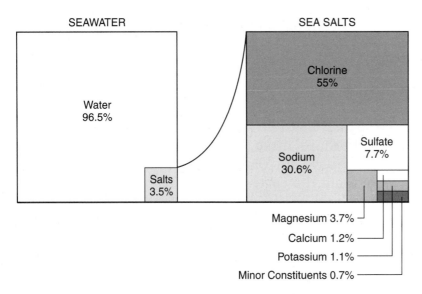

Figure 16-1 On average, about 3.5 percent of the mass of salt water is dissolved salts. The most common salt is sodium chloride, table salt.

and potassium compounds, which are also called salts by chemists.

Source of Salts

Where did all that salt come from? The water in the gases given off by volcanoes is freshwater with few dissolved salts. The kinds of salts found in ocean water are found in the land. Chemical weathering of rocks releases salts. Overland flow (runoff) and groundwater dissolve the salts in bedrock and soil, delivering about 4 billion tons of dissolved solids to the ocean each year. In spite of those salts, most of the water entering the oceans is still considered to be freshwater. Additional dissolved substances enter the oceans through deep-sea vents, which release water that has circulated through the rocks of the oceans' bottom.

ACTIVITY 16-2 | THE DENSITY OF SEAWATER

Obtain a few cups of clean ocean water, or mix your own in a ratio of 3.5 grams of table salt per liter of freshwater. Carefully pour the water into a balloon. Be sure that there is no air bubble in the balloon, and then tie the end of the balloon. Gently place the balloon in a large container of freshwater. Does it sink or float? What does this tell you about the density of ocean water? Devise a way to measure the volume and mass of the salt water in order to calculate its density.

The Hydrologic Cycle

If the oceans receive mostly freshwater, why are they salty? The oceans are part of the hydrologic cycle. The water that enters the oceans will eventually evaporate into the atmosphere. The average time a molecule of water stays in the ocean is about 4000 years. However, water cannot take along its load of dissolved solids when it evaporates; the salts are left behind.

You might think that through time the oceans would become more and more salty. However, the salinity of ocean water has been in equilibrium for millions of years. Processes that take dissolved substances out of the oceans balance the dissolved salts that enter the ocean. Some ocean organisms remove salts to make hard body parts. In addition, some salts leave the water as precipitates, forming salt deposits.

Salinity and Latitude

The balance between inflow of freshwater and evaporation of salt water depends on latitude. Figure 16-2 shows that at about 25° north and south of the equator there are regions where the oceans are a little more salty than average. This is because the climate at these latitudes is generally dry. Consequently, there is relatively little rainfall and more evaporation of ocean water in these regions. Therefore, the oceans are a little saltier in these desert latitudes.

Near the equator, precipitation is plentiful and rivers such as the Amazon dilute the salt water of the oceans. Ocean

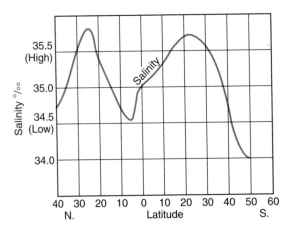

Figure 16-2 In the desert latitudes about 25° north and south of the equator where evaporation of ocean water is most rapid, the salt concentration of the oceans is highest.

water salinity is also lower at high latitudes where temperatures are cool and evaporation is low.

HOW CAN WE INVESTIGATE THE OCEANS?

Until the middle of the last century, finding the depth of the oceans was a tedious process. Rolling out miles of steel cable to reach the ocean bottom took a long time. However, since the Second World War, scientists have been able to measure the depth of the oceans by bouncing sound waves off the seafloor. Figure 16-3 shows the distribution of land elevations

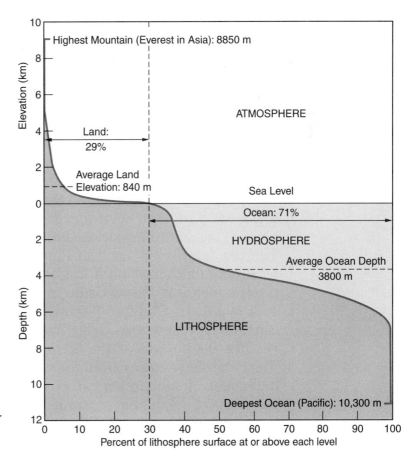

Figure 16-3 This curve shows the surface of the lithosphere at various elevations over Earth's surface. Notice the large portion of the surface just above sea level and about 4 km below sea level. This is not a map, but a variation of a kind of graph called a histogram.

and ocean depths over Earth. From this figure it is clear that the depth of the oceans ranges between sea level at the shore to a maximum depth of more than 10 km.

 ## The Shallow Ocean

Scientists know the most about the shallowest parts of the oceans. Here they can observe the ocean bottom most easily. This is the portion of ocean bottom where light can penetrate, and bottom life is abundant. Divers can descend to several hundred meters, but deeper exploration by humans must be done in special diving chambers known as submersible vehicles. The greatest danger to humans in deep-ocean exploration is the extreme pressure caused by the weight of overlying water. For this reason, most exploration of the deep oceans is now done with remote-controlled diving devices. However, much of the ocean remains unexplored.

 ## The Deep Oceans

Investigations of the oceans have revealed that igneous rocks of mafic composition, such as basalt and gabbro, usually underlie the sediments covering the ocean floor. These rocks are darker in color and more dense than granite and rocks of similar composition that are found in the continents. The two "platforms" seen in Figure 16-3, one just above sea level and another about 4 km below the ocean's surface, are a consequence of this division of Earth's crust into two basic rock types. In Chapter 9, you learned that plate motions constantly renew the ocean bottoms. Upwelling material reaches the surface at the ocean ridges creating new crust. The crust moves away from the ocean ridges toward trenches and zones of subduction carrying the continents with it. At the zones of subduction, oceanic crust is drawn back into the interior while continental rocks are deformed as they resist subduction.

⊕ HOW DOES THE WATER IN THE OCEAN CIRCULATE?

The water of the oceans is constantly moving. The primary cause of deep currents is differences in density. Dense water sinks to the bottom and forces water that is less dense to the surface. Near Earth's poles water is cooled, becomes more dense, and sinks to the bottom. Water reaches its greatest density at a temperature of 4°C. Over all the planet, deep-ocean water is near freezing.

Surface temperatures vary considerably with latitude: warmer near the equator and colder near the poles. The sinking of cold water at the poles must be balanced by upwelling that brings deep water back to the surface. Cold water can hold more oxygen and support more marine life than warm water. For this reason, upwelling, cold currents commonly bring nutrients to the surface in some of the world's best fishing grounds.

The Coriolis Effect

The circulation of surface water follows wind circulation. Both are affected by Earth's rotation. Winds and ocean currents generally curve as they travel long distances over Earth's surface. This curvature is called the **Coriolis effect**. Actually, the winds and ocean currents are going as straight as they can, but Earth's rotation makes them appear to curve to the right in the Northern Hemisphere and to the left in the Southern Hemisphere.

Figure 16-4 on page 364 is a map of the world showing the most common surface current directions. Notice that most of the currents in the North Atlantic follow a circular path curving constantly to the right in a great clockwise circle. The currents in the northern part of the Pacific Ocean also follow this clockwise (to the right) pattern. Currents in the South Atlantic and southern parts of the Pacific Ocean curve to the left in a counterclockwise pattern.

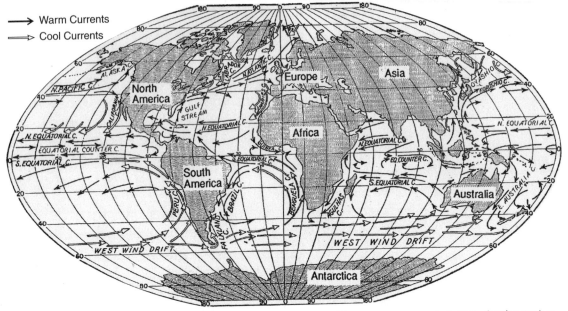

Figure 16-4 Surface currents in the world's oceans generally follow circular routes curving to the right, in clockwise patterns, in the Northern Hemisphere and circling to the left, in counterclockwise motion, in the Southern Hemisphere.

| ACTIVITY 16-3 | **OBSERVING GYRES** |

A gyre is a large, curving pattern of circulation in the ocean. Use the surface ocean current map in the *Earth Science Reference Tables*, or Figure 16-4, to locate gyres in the Northern Hemisphere as well as in the Southern Hemisphere. For each gyre, list the ocean or part of an ocean it occupies, the names of the surface currents that form the gyre, and the direction in which it circulates (clockwise or counterclockwise). What is the most common direction of circulation in each hemisphere?

 Currents and Climate

Ocean currents influence the climates of coastal locations. The temperature of ocean water does not change as quickly as the temperature of rock and soil. Therefore, coastal locations usually have a smaller range of temperature than do inland locations. Cold and warm currents also affect coastal temperatures. For example, people who live along the coast of California are not as likely to swim in the ocean as people who live along the Gulf of Mexico or the Atlantic coastline of the United States. Cold ocean currents along the California coast keep the water temperature too chilly for most swimmers, even in the summer. The cool ocean water also prevents summer temperatures from getting too hot along the coast of southern California.

The climate along the southern coast of Alaska is moderated by relatively warm ocean water. Summer and winter temperatures are relatively mild in this part of Alaska. Palm trees grow in some areas along the west coast of Great Britain where the warm currents of the Gulf Stream and the North Atlantic Current moderate winter temperatures. These coastal locations have winters far less severe than central European cities that are far from the ocean. By looking at the arrows on Figure 16-4 you can tell where warm or cool ocean currents affect coastal areas. The black arrows show warm currents and the white arrows show cool currents.

 **El Niño**

In recent decades, scientists have become more aware of how changes in ocean currents can affect the climate over large areas. Most of the time, cold ocean currents and nutrient-rich water are found off the western coast of South America. Good fishing in this region provides food and employment in oceanside villages. However, in some years the upwelling of cold water is replaced by warm water, which reduces fish production. This usually happens about the time of the Christmas holidays. Local inhabitants call it **El Niño**, a Spanish term for the Christ Child, although it is an unwelcome "Christmas present." However, this event affects more than the local fishing industry. A strong El Niño can cause increased winter rain and flooding along the coast of California and drought in the western Pacific. The relationship between ocean currents and regional climatic changes is giving scientists new methods to predict weather and prepare for its consequences.

 ## WHAT ARE THE TIDES?

Many people who live along ocean coastlines are aware of the periodic rise and fall of the oceans. The twice-daily cycle of change in sea level is known as the **tides**. Currents associated with tides can affect fishing and the ability of boats to navigate in some places. If a storm strikes a coastal area at high tide, wave and water damage is likely to be greater than from a storm that comes ashore at low tide.

Tidal range is the difference between the lowest water level and the highest water level. Most locations have two high tides and two low tides every day, but some places have only a single daily cycle. In some locations, the change in sea level is too small to be noticeable. However, in the Bay of Fundy, along the eastern coast of Canada, the tidal range can be as much as 15 meters. This is about the height of a four-story building. Figure 16-5 shows a beach along the coast of Mexico at high tide. Figure 16-6 shows the same beach at low tide about 6 hours later.

Figure 16-5 High tide at Puerto Peñasco, Mexico. The gravitational pull of the moon and the sun cause sea level to change in a cycle known as the tides.

Figure 16-6 Low tide at Puerto Peñasco, Mexico. Most locations experience two cycles of high and low tides in a period of just over 24 hours.

ACTIVITY 16-4   **EXTREMES OF TIDAL RANGES**

Prepare a report that identifies places around the world that have an unusually large or small range of ocean tides. Plot these locations on a world map. Why do some locations have higher tides than others, and how do these extreme tides affect the local area?

 **Gravity**

Gravity is the force of attraction between any two or more objects. The force of gravity holds Earth in its path around the sun. It also keeps the moon in orbit around Earth.

People do not feel the gravitational attraction between their body and most of the objects around them because the objects are too small. People certainly do feel the attraction between their body and Earth. That force is weight. It is a strong force because Earth is so massive and because we are so close to it. If you climb a tall mountain, you move a little farther from the center of Earth. This decreases your weight, although the change is too small to observe without careful measurement. If you could move far enough above Earth into space you would actually notice a decrease in your weight. Astronauts in orbit around Earth feel completely weightless as the result of their distance above Earth and their orbital motion.

The tides are caused by gravity of the moon and the sun. Although the moon is much smaller than the sun, it is much closer to Earth. Therefore, the moon has a greater gravitation effect on Earth than does the sun.

Moon and Tides

Consider how the moon's gravity affects the solid Earth and the oceans. The moon pulls most strongly on the part of Earth closest to it. When the moon is directly over the ocean, this part of the ocean experiences a high tide. The moon pulls less strongly on the solid Earth, but it does pull Earth away from the water on the side of Earth away from the moon. So there is also a high tide on the side of Earth away from the moon. That is why most locations have two high tides each 24-hour day. Figure 16-7 shows how the moon pulls more strongly on the side of Earth closer to the moon.

Sun and Tides

The sun also influences ocean tides. When Earth, sun, and moon are in a line with one another, the highest, or **spring tides**, occur. At spring tides, the sun and moon do not need to

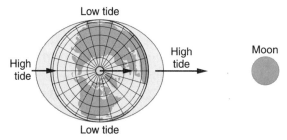

Figure 16-7 Tides are caused mainly by gravity of the moon. The moon pulls water on the near side of Earth away from the solid Earth. It also pulls Earth away from water are the far side. This is why most locations have two high tides every 24 hours. (The length of the arrows represents the strength of the moon's gravity.) Distances are not to scale.

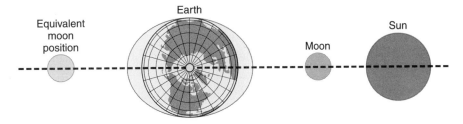

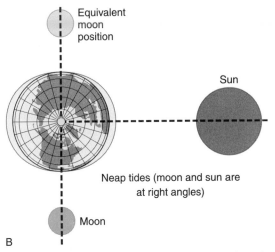

A   Spring Tides (Earth, moon, and sun are in a line)

B

Figure 16-8 A and B: The tidal range at any location depends on the angle between Earth, sun, and moon. When they are in a line (A), the greatest range, or spring tides, occur. The smallest tidal range happens when the sun and moon are at right angles to Earth (B). (Neither distances nor size of the sun are to scale.)

be on the same side of Earth. (See Figure 16-8A.) The range of the tides is the lowest when the sun and moon are at right angles to Earth, and **neap tides** occur. Figure 16-8B shows the configuration of Earth, sun, and moon at neap tides.

The moon orbits Earth every 27 days. As the moon orbits Earth, you see a different portion of its lighted surface—that is why the more of the moon is visible on some nights than on others. Therefore, the period of the tides is not exactly 12 or 24 hours. Each day the moon seems to fall behind the sun by about an hour. A full cycle of the tides is about 12.5 hours, or 25 hours in places that only experience one cycle per day.

ACTIVITY 16-5   GRAPHING THE TIDES

Graph the height and time of tides. (Graph time on the horizontal axis and water height on the vertical axis.) If you live on the coast, you may be able collect your own data, or you can use data from a local newspaper. If you live inland, you can use data from the Internet.

HOW DO COASTLINES CHANGE?

In earlier chapters, you read about erosion caused by glaciers, wind, runing water, and gravity acting alone. It is now time to consider coastal erosion and the movement of sediments along coastlines. When you think of visiting an ocean beach you may picture a broad strip of sand where you can play, rest, get a suntan, or enjoy the water. You may not realize that the beach is a dynamic part of a system that transports sediment. The sediment of which the beach is composed is on a journey that transports weathered rock from the land into the ocean. A beach is one of Earth's most active environments of deposition and erosion.

There are two primary sources of sediment for beaches. Waves, particularly in storms, erode the coast and cause the shoreline to migrate toward the land. Rock and sediment fall or are washed onto the beach. Streams and rivers sweep other material into the ocean. Beaches are zones of transport where sediments move along the shore by wave action and currents. Figure 16-9 shows a wide beach composed of sediment eroded from the cliff behind the beach.

Waves

The energy of most waves comes from wind. The greater the wind's velocity and the greater the distance it blows over open water, the larger the waves it creates. Because winds

Figure 16-9 The sand on this beach came from erosion of the cliffs behind the beach. There are no rivers nearby to supply sediment from inland areas.

can blow for greater distances over the ocean than over a lake, ocean waves are usually larger than waves on lakes. Friction between moving air and the surface of the water sets up waves that move forward in the direction of the wind.

The waves you observe can be deceiving. It may look as if the water is moving forward with the waves. However, energy not water is transferred by waves. Figure 16-10 on page 372 shows that as the energy of the wave moves forward, surface water moves in circles. Deep water is not affected by waves. When the wave enters shallow water near shore, the crest moves faster than the bottom of the wave and a breaker forms. As the wave breaks, it gives up its energy along the shore. This energy can do three things along the beach:

1. By causing abrasion, wave energy breaks up sand and rock in the surf zone. (You may recall from Chapter 11 that abrasion is the wearing away of sediments caused by collisions.)

2. Wave energy can erode the beach, including sediments and rock behind the beach.

3. Wave energy transports sand and sediment parallel to the shore.

 **Longshore Transport**

Most beaches have a region called the surf zone. The **surf zone** extends from where the waves' base touches the bottom (a depth of about half the distance between wave crests) to

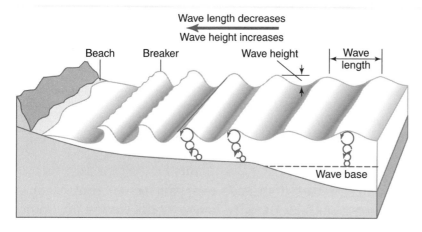

Figure 16-10 Ocean waves are driven by winds. In deep water, waves make the surface water move in circles as they carry their energy forward. Waves break in shallow water, giving up their energy to abrade and transport beach sediments.

the upper limit the waves reach on the beach. The surf zone along most beaches is like a river. Waves cause sand to wash onto the beach with the breakers and then wash back into the water with the return flow. If the waves approach the beach at a right angle, head on, this could be the whole story. However, most waves approach at a different angle. The result is a zigzag motion that carries sand (or whatever sediment the beach is made of) downwind along the beach as shown in Figure 16-11. The resulting motion of the water along the shore

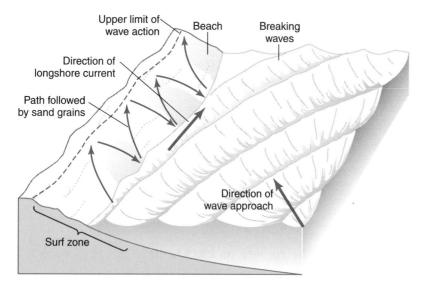

Figure 16-11 As waves wash onto a beach, beach sediment moves forward and back in the surf zone. Beach sediment is also carried parallel to the shore in the downwind direction. The result is longshore transport.

is called a longshore current, and the motion of the sediment is known as **longshore transport**.

Depositional Features

A variety of coastal features are related to wave erosion and longshore transport. Sometimes the advance and retreat of the waves deposits sand that forms low ridges along the shore. They are called **sandbars**. If you have ever waded in the ocean along a sandy beach and encountered a shallow area separated from the shore by deeper water, you have found an underwater sandbar.

A spit is a sandbar that forms a continuation of a beach into deep water. Spits sometimes grow across bays, forming a baymouth bar. Similar offshore features that rise above sea level are **barrier islands**. A shallow bay called a lagoon separates barrier islands from the mainland.

The maps of New York State in the *Earth Science Reference Tables* (Figures 15-9 and 18-8) show the series of islands that separate the south shore of Long Island from the Atlantic Ocean. Jones Beach and Fire Island are a part of this series of barrier islands. These features are common on gently sloping coastlines with abundant sand.

BEACHES   Figure 16-12 A, B, and C on page 374 illustrates a sequence of events in a shore area with a sandy beach. Part A shows a shoreline in balance. Beach sand originates from sediment carried by the river on the right and eroded from the bluffs along the shore. Waves from the southeast bend as they enter shallow water near the shore, and a longshore current carries sand westward. The sand spit growing across the bay makes it clear that the principal direction of sand carried by longshore transport is toward the west. Part B shows a breakwater built parallel to the shore to protect boats from large waves. A groin/pier has been built from the shore out into the ocean. The structures are very new in Part B and no changes in the beach are visible. Part C shows how the beach changes in response to these two obstructions. The beach gets

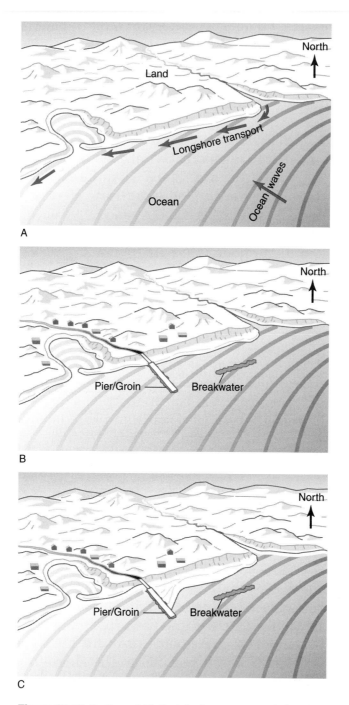

Figure 16-12 (A, B, and C) Part A shows a natural shoreline in dynamic balance. The construction of a breakwater and a groin/pier are shown in Part B. Part C shows how these structures cause the beach and shoreline to change.

wider behind the offshore breakwater as sand builds outward from the beach. This is because wave energy has been reduced behind the breakwater and deposition increases. Westward transport deposits sand on the upwind side of the solid pier. However, the beach shrinks on the downwind side where the flow of sand has been stopped. Even the sand spit is reduced because sand movement was stopped by the groin/pier. In general, when a groin or solid pier is constructed into the ocean in a region of longshore transport, the beach becomes wider on the upwind side and narrower in the downwind side.

ACTIVITY 16-6   COASTLINES AND HUMAN INTERVENTION

While this activity may be more meaningful if you live near a coastline, you may be able to find useful library references or information on the Internet. Prepare a brief report about how a real shore area changed when one or more structures were built into the water. If possible, use maps and images as a part of your report.

HOW SHOULD WE MANAGE ACTIVE SHORELINES?

Humans affect shorelines in many ways. People increase shoreline erosion by trampling on protective vegetation, especially in sand dunes. To protect the beach or unstable features, people build breakwaters, groins, and jetties. This is common in areas where shoreline erosion threatens buildings or other property. Dunes and hills are flattened to make building sites and parking areas. It is important to understand that shore areas are delicate and dynamic features. A growing number of citizens are recognizing that the best way to manage changing coastal regions is to restrict development and allow natural processes to continue without human interference.

Social Issues

Coastal regions are popular home and vacation sites. Recreational opportunities, including swimming, boating, and fishing, make beachfront property highly valued. However, there is discussion of whether there should be private ownership of beaches. In addition, people debate the wisdom of building on unstable areas near shorelines.

This is especially true around New York City and near other urban areas. Should the beaches of Long Island be playgrounds for the wealthy, or should they be available to everyone? Should anyone be allowed to walk along ocean beaches? Should people be allowed to construct homes in unstable, sandy areas and low areas that are subject to storms and flooding? Do roads and buildings seriously affect the natural resources of oceanfront property? Beachside communities constantly deal with these issues. There are probably no solutions acceptable to everyone. Governments simply try to balance the factors and select the best policies from a wide range of controversial solutions.

ACTIVITY 16-7   ZONING FOR COASTAL PRESERVATION

In a cooperative group, develop a set of policies to guide both public and private development of ocean coastal areas. Prepare a document that could be given to coastal communities to help them develop zoning regulations for their ocean front areas.

TERMS TO KNOW

barrier island	**El Niño**	**neap tides**	**spring tides**
Coriolis effect	**longshore transport**	**sandbar**	**surf zone**

CHAPTER REVIEW QUESTIONS

1. A student took home several gallons of unpolluted ocean water and boiled away all the water in a clean metal pot. Which statement below best describes the appearance of the bottom of the pan?

   (1) The pan was as clean as it was before the water was boiled away.
   (2) A film of calcite was left in the bottom of the pan.
   (3) A substance resembling table salt was left in the bottom of the pan.
   (4) The pan contained a transparent film of quartz along the bottom.

2. Swimmers notice that it is easier to float in ocean water than it is to float in freshwater. Why is it easier to float in salt water?

   (1) Salt water is more dense than freshwater.
   (2) The ocean has larger waves than lakes and rivers.
   (3) Ocean water is usually deeper than freshwater.
   (4) Ocean water has more dissolved gases than freshwater.

3. Which is the best source of nearly pure water?

   (1) ocean water from near the south shore of Long Island
   (2) ocean water from 100 km off the south of Long Island
   (3) water from a shallow desert lake that has no outlet
   (4) water from the Hudson River north of Albany

4. A research ship in the middle of the Pacific Ocean took measurements of ocean water both at the surface and near the ocean bottom 6 km below the surface. In what way is most water from deep in the oceans different from the water they observed near the surface?

   (1) The water near the bottom of the ocean is warmer than surface water.
   (2) Water near the bottom of the ocean is more dense.
   (3) Surface water is salty but bottom water is freshwater.
   (4) Water near the bottom receives more light than water near the surface.

5. Which location has a coastal climate that is generally made warmer by the influence of a nearby ocean current?

   (1) Southern California
   (2) Peru in South America
   (3) Brazil in South America
   (4) Northwestern Africa near the Canary Islands

6. The Canaries Current along the west coast of Africa and the Peru Current along the west coast of South America are both

   (1) warm currents that flow away from the equator
   (2) warm currents that flow toward the equator
   (3) cool currents that flow away from the equator
   (4) cool currents that flow toward the equator

7. Warm water from tropical oceans is carried to northern Europe by the Gulf Stream and the

   (1) Alaska Current
   (2) Canaries Current
   (3) North Atlantic Current
   (4) Brazil Current

8. Which surface ocean current transports warm water to higher latitudes?

   (1) Labrador Current
   (2) Falkland Current
   (3) Koroshio Current
   (4) Peru Current

9. What is the usual period of time from one high tide the next high tide for most oceanfront locations?

   (1) about 1 hour   (3) about 1 week
   (2) about 12 hours   (4) about 2 weeks

10. A student recorded these times of three successive high tides at a single location:

    9:12 A.M.
    9:38 P.M.
    10:04 A.M.

    What is the approximate time of the next high tide?

    (1) 10:12 P.M.   (3) 10:38 P.M.
    (2) 10:30 P.M.   (4) 11:04 P.M.

have been able to piece together a remarkable account of Earth's history.

Uniformitarianism

Geological features sometimes show the results of catastrophic events of the past such as violent volcanic eruptions and continental glaciers. The record of these dramatic events has led some people to assume incorrectly that these events were more common in the geological past than they are today. As another example, when some people see fossils of marine organisms in mountains thousands of meters above sea level, they may not understand that slow uplift over millions of years elevated the land.

Geologists have responded to these misconceptions by adopting uniformitarianism as one of their guiding principles. **Uniformitarianism** is the concept that the geological processes that took place in the past are generally similar to those that occur now. This does not mean that all geological changes are slow and steady. For example, erosion caused by most streams is very slow except when the streams are in flood. Those limited flood events do most of the erosional work of streams and rivers. Most volcanic eruptions are sudden events that are not very frequent.

However, the present is not really different from most of the geological past. Scientists can understand most of the features in the geological record by observing similar events in the present. Uniformitarianism is sometimes summerized by the expression, the present is the key to the past. This means that our best way to understand the geological events that shaped our planet is to look at the geological processes at work today.

Superposition

Layers of different rock are the results of a series of events. In most places, the lowest layers of sedimentary rocks are the oldest. After all, the lowest layers must be in place before

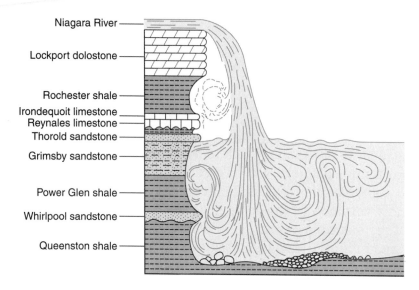

Niagara River

Lockport dolostone

Rochester shale
Irondequoit limestone
Reynales limestone
Thorold sandstone
Grimsby sandstone

Power Glen shale

Whirlpool sandstone

Queenston shale

Figure 17-1 The profile of Niagara Falls shows the oldest rocks are at the bottom and the layers become younger toward the top. This illustrates the law of superposition.

younger sediments can be deposited on top of them. **Superposition** is the concept that unless rock layers have been disturbed, each layer is older than the layer above it and younger than the one below it. Therefore, when geologists assign relative ages to rocks, the oldest rocks are on the bottom and the layers become progressively younger toward the top.

Figure 17-1 is a profile of the rock layers at Niagara Falls. Each rock layer is named for the location where it can be observed and studied. For example, the hard cap rock at the top that is responsible for the falls is the Lockport Dolostone. If you want to observe this layer in the field, there is a bedrock exposure near Lockport, New York. The first layer deposited at Niagara Falls is the layer on the bottom, the Queenston Shale. After this layer was deposited, sand was deposited that became the Whirlpool Sandstone. One by one, each layer was deposited before the layer above it.

Superposition does not always apply. In some places, folding has overturned the rocks or faulting has pushed older layers on top of younger layers. These exceptions to the law of superposition will be explained in the next section. However, if you are looking at rock layers or a diagram of rock layers, unless you see evidence for overturning or faulting you should assume that the law of superposition can be applied. The older layers are those on the bottom. Exceptions are unusual.

Original Horizontality

Sedimentary rock is usually the result of deposition of sediments in layers. In some places, the layers are tilted rather than horizontal. The principle of **original horizontality** states that no matter the present angle or orientation of sedimentary layers, it is almost certain that the layers were originally horizontal and were tilted after deposition. The cause of the tilting could be folding or an uneven regional uplift.

Figure 17-2 shows a sequence of events in the creation of a bedrock outcrop. An **outcrop** is a place where bedrock is exposed at the surface. In diagram 1, sandy sediments are washed into a curved basin. Most sedimentary rocks are formed from deposits in water. The low spots in the basin are the first to fill with sediment. Therefore, no matter what the shape of the basin, the layers of sediment are flat and hori-

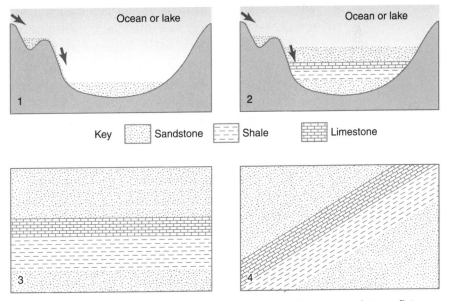

Figure 17-2 When sediments are deposited in a basin, they accumulate as flat, horizontal layers as shown in diagrams 1 and 2. This occurs no matter what the original shape of the basin. Diagrams 3 and 4 represent a later time and a bedrock outcrop that is from the center of diagram 2. When geologists see tilted sedimentary layers as in diagram 4, they usually infer that the layers were deposited flat and horizontal.

zontal. Diagram 2 shows a sequence of layers that are horizontal in spite of the curvature of the basin. Diagram 3 represents an outcrop of rock layers from the depositional basin. In diagram 4, the layers have been pushed up on one side (or down on the other) to create tilted layers in the outcrop. The point is that when you see tilted layers like those in diagram 4, you can infer that they started as flat, horizontal layers of sediment that later were tilted.

 **Inclusions**

Geologists sometimes find that bedrock contains and surrounds pieces of a different kind of rock. An **intrusion** is a body of rock that was injected into the surrounding rock as hot, molten magma. Therefore, all intrusions are igneous rocks. An **inclusion** is a fragment of one kind of rock that is enclosed within another rock. This can happen in several ways. An intrusion of basaltic magma following a zone of weakness through sedimentary rocks can pick up fragments of the sedimentary rock. When the magma solidifies it contains and surrounds pieces of the rock it invaded. Alternatively, consider a region of granite in which the granite weathers to boulders of granite sitting on granite bedrock. If the region then collects layers of sediment, the sediments will surround the granite boulders. The result could be sedimentary rocks that contain granite boulders. Figure 17-3 shows an inclusion of sandstone rock surrounded by granite.

Whenever a body of rock contains inclusions of another rock, the inclusions are older than the surrounding rock. Some of the oldest masses of rocks found on Earth contain inclusions. This tells us that another rock unit existed before these very old rocks. Unfortunately, it may not be possible to determine how much older the inclusions are than the main body of rock, but the sequence can be inferred. Particles of sediment can be thought of as inclusions in sedimentary rock. Most sedimentary particles are the weathered remains of an older rock. Therefore, grains of sand in sandstone or

Figure 17-3 It is clear that this fragment of sandstone fell when the granite was still magma. Therefore, the inclusion of sandstone must be older than the granite that surrounds it.

pebbles in conglomerate are actually older than the rock they now occupy.

 Cross-Cutting Relationships

When a fault is seen in an outcrop, the fault must be younger than the rocks it cuts through. Some rocks also contain igneous intrusions. Cross-cutting means that something goes through previously existing rock. The cross-cutting feature, usually a fault or an intrusion, is always younger than the rock in which it is found. After all, the rocks must be there before they are faulted and before magma cuts through them.

Figure 17-4 on page 390 shows changes in the bedrock through time at a single location. This series of diagrams illustrates cross-cutting relationships. The layers shown in diagram 1 are offset by the fault in diagram 2. Therefore, the layers are older than the fault. Both rock layers and the fault cut by basaltic magma in diagram 3 are older than the magma. The fault shown in diagram 4 cuts through the sedimentary layers, the original fault, and a branch of the basalt intrusion. Therefore, all three of these features are older than the second fault.

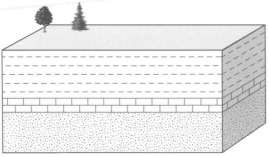

1. Deposition and rock formation

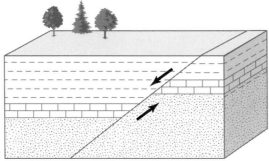

2. Faulting

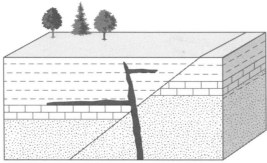

3. Intrusion

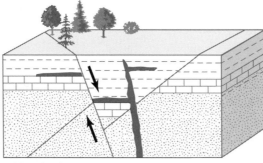

4. More faulting

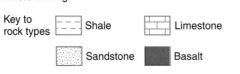

Key to rock types: Shale, Limestone, Sandstone, Basalt

Figure 17-4 The cross-cutting relationships in this series of diagrams show that faults and intrusions must be younger than the rocks in which they are found. The layers in diagram 1 must exist before they can be faulted as shown in diagram 2. The layers and the fault are crossed by the intrusion in 3, so both must be older than the intrusion. The second fault in diagram 4 is younger than everything it cuts through: sedimentary layers, the intrusion, and the original fault.

Contact Metamorphism

Metamorphism includes all the changes that occur when a rock is subjected to extreme heat and/or pressure. In Chapter 5, you learned that there are two kinds of metamorphism. Deep burial of a very large mass of rock can subject the rock to conditions of extreme heat and pressure that exist far underground. This is regional metamorphism. Contact metamorphism occurs in narrow zones adjacent to molten magma.

Magma can be confined underground or it may erupt onto the surface. Intrusion is an internal process. As hot magma squeezes into cracks and zones of weakness, the cooling magma passes its heat energy to the nearby rock. This creates a baked zone of contact metamorphism adjacent to the intrusion. Because the magma is surrounded by preexisting rock, the zone of contact metamorphism extends from the intrusion in all directions.

Extrusion occurs at the surface; it is external. While rock below an extrusion is altered by contact metamorphism, there is no rock on top to be baked. Sometimes geologists find a layer of igneous rock that has not altered the rock layer immediately above it. This is evidence that the layer above was deposited after the magma cooled.

Figure 17-5 on page 392 shows layers of sedimentary rock that are intruded by magma that also flows onto the surface as an extrusion. Later, a new layer of sedimentary rock was deposited on top of the extrusion. From the final diagram, it is clear that the top layer is younger than the magma. It is also evident that the magma was extruded at the surface because the layer above the extrusion does not show contact metamorphism.

Unconformities

No location shows a continuous record of geological events throughout Earth's history. If an area was above sea level for part of its past, it is likely that sediments or rock were removed by weathering and erosion. Thus erosion causes gaps in the geologic record. When new layers are deposited on the

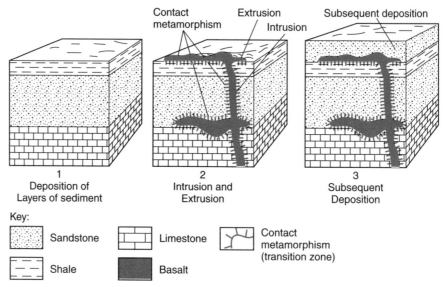

Figure 17-5 Contact metamorphism—the first diagram shows sedimentary rocks that will be intruded by magma in diagram 2. The portion of the magma that comes to the surface is known as an extrusion. In diagram 3, new deposition has left a sandstone layer on top of the basalt. By looking at diagram 3 alone, you can tell that the basalt magma was extruded and that the top sandstone layer is younger than the extrusion because there is no contact metamorphism on top of the basalt.

erosion surface, the buried erosion surface is known as an **unconformity**. Figure 17-6 shows the formation of an unconformity. Deposition, which most often occurs in water, creates sediments that become sedimentary rock. The region is uplifted and the layers are exposed to weathering and erosion. After erosion has occurred over a period of time, subsidence (sinking of the land) and flooding leads to the deposition of a new layer on top of the erosion surface.

Unconformities are common throughout New York State. New York's sedimentary rocks show a continuous record of the early development of fish and land animals. For a long time, a shallow sea covered New York. Then, about 350 million years ago, the land was pushed up above sea level. This was probably the result of a collision between ancient North America and Africa. The disappearance of the sea not only stopped the deposition, but it also led to the erosion of some layers of previously deposited sediments. Wherever new layers were deposited on top of the erosion surface, a buried erosion surface, or unconformity, was created. Figure 17-7 on page 394 shows an unconformity in Utah.

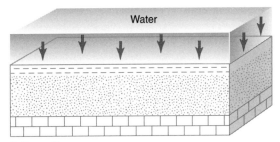

1. Deposition

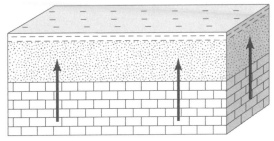

2. Uplift

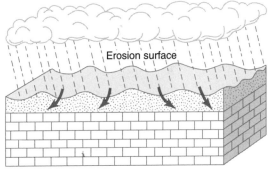

3. Erosion

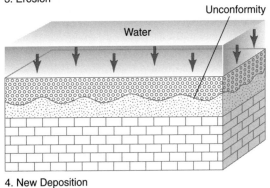

4. New Deposition

Figure 17-6 Formation of an unconformity—layers of sedimentary rock are created by deposition and compression. Uplift exposes the layers to weathering and erosion, which wear down the land to a new level. Deposition on top of the erosion surface creates an unconformity: a gap in the geologic record.

Key to Rock Types

- ---- Shale
- Sandstone
- Limestone
- Conglomerate

Figure 17-7 The cutoff layers along this cliff make it clear that the top layer was deposited over an erosion surface creating this unconformity in Utah.

| ACTIVITY 17-2 | **LOCAL ROCK FEATURES** |

Identify as many of the above rock features (inclusions, cross-cutting relationships, contact metamorphism, and unconformities) as you can in local bedrock outcrops. For each feature you identify, state the location where it may be observed locally. If possible, take photographs of these features.

 # HOW CAN WE INTERPRET GEOLOGIC PROFILES?

The profile of an outcrop shows a series of layers that can help a geologist interpret a history of geological events. A geologist would probably make a drawing of the outcrop to help identify evidence of specific processes. However, as a student of Earth science, you are more likely to be supplied with diagrams of outcrops and asked to list the events, in chronological order, that created the profiles.

If a sedimentary rock is present, deposition (followed by compaction and/or cementing) must have taken place. Erosion is indicated by an uneven boundary between layers or layers that are cut off. Igneous rocks must be the result of melting, intrusion, extrusion, and solidification, also known as crystallization. If you are not familiar tilting, folding, faulting, and metamorphism, ask your teacher to explain them.

Most profiles begin with deposition or solidification. (The rock needs to be formed before it is changed.) If the bottom layer is metamorphic rock, deposition or solidification is followed by metamorphism.

Keys to Rock Types

Do not expect to see the same symbols on every map or profile. However, a key should make the meaning of the symbols clear. A key should accompany each profile. Pages 6 and 7 of the *Earth Science Reference Tables* have charts to help you to identify igneous, sedimentary, and metamorphic rocks.

The charts for sedimentary and metamorphic rocks contain symbols that are sometimes used on maps and diagrams of geologic profiles to indicate various kinds of rock. Many of these symbols represent characteristic textures of the rocks they symbolize. For example, sandstone is composed of gritty particles, which is usually represented by a dotted pattern. Shale tends to break into thin layers, which is usually shown by a pattern of short horizontal line segments. A combination of line segments and dots usually represent siltstone, a sedimentary rock of intermediate-size particles. Limestone is shown by a symbol that looks like stacked bricks because limestone often breaks along joints and bedding surfaces into large, somewhat rectangular blocks.

ACTIVITY 17-3   SYMBOLS AND ROCKS

Copy map symbols from the *Earth Science Reference Tables,* or profile diagrams and/or from geologic maps. Next to each symbol tell how the symbol resembles the texture of the rock it represents. Adding photographs or samples of the various kinds of rocks can enrich this assignment.

Interpreting Profiles

The profiles in Figure 17-8 illustrate how a geologic history can be inferred for each profile. The law of superposition tells us that unless there is evidence of overturning or an igneous intrusion, the youngest layers are on the bottom.

Profile A starts with deposition of sediments to form conglomerate, sandstone, and shale, in that order. These layers are faulted, followed by deposition (of limestone) and erosion.

Profile B has igneous rock at the bottom, so it starts with crystallization of magma. This must have happened before the deposition of sediments because no metamorphism is shown in the siltstone layer. The law of original horizontality tells you that shale, sandstone, and conglomerate above were deposited and then folded. Later, the top layers were eroded.

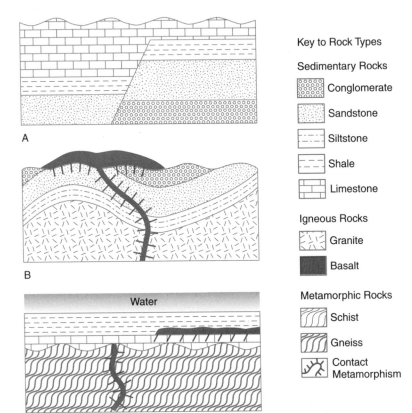

Figure 17-8 Geologic profiles contain clues to the sequence of events that created them. The series of events responsible for each of these three profiles is explained in the text.

Intrusion and extrusion at the surface are the final events, unless you wish to point out that erosion at the surface is an ongoing process.

Profile C has a metamorphic rock at the bottom. This rock could have two possible origins. Therefore, this history begins either with deposition and sedimentary rock formation followed by metamorphism or by solidification of an igneous rock followed by metamorphism. Gneiss can begin as a sedimentary or an igneous rock. Intrusion of basalt occurred next. Erosion (note the unconformity) is followed by deposition of limestone. The second basalt body must have been an extrusion because there is contact metamorphism below it, but not above. Deposition of shale sediments is probably the last step in this profile because the water above (probably a lake or an ocean) would support renewed deposition.

Do you know the square root of 4? Most people think only of 2, but negative 2 is also correct. This question has two equally valid answers. In the same way, geologic profiles are often subject to different interpretations. As long as the sequence of events supports the features shown in the profile, and all the features are explained, a sequence is considered correct. Multiple interpretations are common in science. This is especially true in an environment as complex as the geosciences.

HOW DO GEOLOGISTS ESTABLISH ABSOLUTE TIME?

In the first part of this chapter, you learned about techniques to determine a sequence of events in relative time. However, scientists want to know more. They want to know how old Earth is, how long the events of mountain building and erosion have been taking place, and how long Earth has been inhabited by living creatures. Recent discoveries about the structure and stability of atoms, along with the technology to count atoms, have finally made it possible to answer these questions.

Radioactive Isotopes

Not all atoms are stable. The atoms of some elements spontaneously break down into atoms of different elements. These unstable elements are called **radioactive**. In the process of breaking down, they radiate energy and subatomic particles. Radioactivity was discovered when a scientist working with uranium left the substance on a photographic plate. When the plate was developed he realized that the film was exposed even though it had never been exposed to light. This discovery led to further investigations and the recognition of a group of substances that are now known as radioactive elements.

STRUCTURE OF ATOMS You learned earlier that all matter is composed of extremely small components called atoms. Atoms, in turn, are made primarily of protons, neutrons, and electrons. Protons and neutrons are found in the core of the atom: the atomic nucleus. The nucleus makes up 99.9 percent of the mass of the atom. Each element, such as oxygen, hydrogen, or iron, has a unique number of protons. That is, all atoms of oxygen have 8 protons; hydrogen atoms have just one proton, and iron atoms have 26 protons. Only 92 different elements occur naturally on Earth. Each element has a different number of protons in the nucleus of its atoms. To balance the positive electrical charge of the protons, neutral atoms have the same number of electrons in orbit around the nucleus. Figure 17-9 is a representation of an atom of carbon.

Unlike protons, the number of neutrons in an element is not fixed. It appears that the number of neutrons in the nucleus affects the stability of atoms. Carbon-12, which contains six neutrons its nucleus, is stable. However, carbon-14, whose nucleus contains eight neutrons, is radioactive and unstable. Atoms of the same element that contain different numbers of neutrons are known as **isotope**s.

Uranium exists in two common isotopes, both of which are radioactive. Uranium-238 is the more stable isotope.

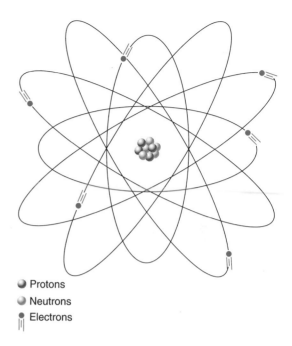

Figure 17-9 Carbon atoms always contain six protons. The most common form of carbon also has six neutrons in the nucleus. If the atom is neutral, six electrons orbit the nucleus. This diagram is not to scale. If it were, the atomic nucleus would be much smaller than shown here. Any diagram of an atom is a model because the particles in an atom are too small to be visible at any magnification.

● Protons
● Neutrons
║ Electrons

Uranium-235 is used in nuclear power plants and atomic weapons.

RADIOACTIVE DECAY When a radioactive isotope, or **radioisotope**, breaks down, it often changes to a stable isotope of a different element. For example, carbon-14 changes to nitrogen-14. Uranium-238 transforms through a series of more than a dozen steps until it reaches a stable decay product: lead-206. Potassium-40 can change to either argon-40 or calcium-40. The stable end material of radioactive decay is known as the **decay product**.

Radioactivity can be used to find absolute ages because the rate at which these nuclear changes take place is predictable. No matter how much carbon-14 you start with, after 5700 years just half of it will remain carbon-14 and half of it will have changed to nitrogen-14. The time it takes for half of the atoms in a sample of radioactive isotope to decay is called its **half-life period**. Each radioactive isotope has its own measurable and consistent half-life period. Some radioisotopes have a half-life period of a fraction of a second. For others, the half-life period is billions of years.

For example, no matter how much pure uranium-238 you start with, after a period of 4.5 billion years half of it will have changed to lead-206. Table 17-1 is from the *Earth Science Reference Tables*. It shows the half-life periods of four substances commonly used by geologists to determine the absolute age of rocks or fossils. It is important to note that chemical combination or environmental factors such as heat or pressure do not affect the decay rate of these isotopes. That fact makes measurements of radioisotopes such a powerful tool in determining absolute time.

Table 17-1. Radioactive Decay Data

Radioactive Isotope	Disintegration	Half-life (years)
Carbon-14	$C^{14} \rightarrow N^{14}$	5.7×10^3
Potassium-40	$K^{40} \rightarrow Ar^{40}$ or $K^{40} \rightarrow Ca^{40}$	1.3×10^9
Uranium-238	$U^{238} \rightarrow Pb^{206}$	4.5×10^9
Rubidium-87	$Rb^{87} \rightarrow Sr^{87}$	4.9×10^{10}

ACTIVITY 17-4 A MODEL OF RADIOACTIVE DECAY

Place about 25 pennies or similar objects with two different sides in a flat-bottomed container about the size of a shoe box. All the pennies should be heads up (or tails up) at the start. Close the container and shake it for a few seconds. Open the lid and remove the pennies that have flipped over. Count the pennies left in the box, record your data, and repeat the procedure until no pennies remain. Graph your data with the number or percent of pennies remaining versus the number of periods of shaking. (Each period of shaking represents one half-life period.) How is this procedure similar to radioactive decay? How is unlike radioactive decay?

The nuclear decay of an individual atom is a random event. The atoms of a radioactive substance have highly erratic release mechanisms. There is no way to tell when an individual atom will disintegrate. However, the atoms of each radioisotope have a characteristic average decay time. For the millions of atoms in even a very small sample of a radioactive substance, the decay rate, that is, the half-life period, is predictable and unchanging.

Working with Half-Life Periods

Scientists can tell the age of a sample of a radioactive isotope according to how much of it remains unchanged. Figure 17-10 shows a model of radioactive decay in which each segment of the circle represents one half-life period. (That period of time depends on the radioisotope.) Therefore, every segment of this circle, no matter how small, represents the same amount of time. However, as time goes on, the portion of the sample that remains the original radioisotope becomes smaller, and the amount that can decay also becomes smaller. In theory, this circle can be divided into smaller and smaller segments

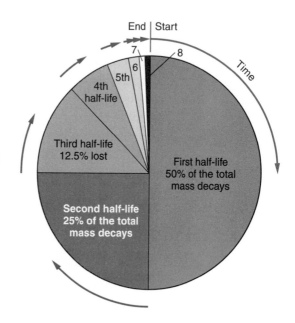

Figure 17-10 A model of radioactive decay—each arrow around the outside of this circle represents one half-life period. In that time, half of the original radioactive isotope changes to its decay product. Through time the amount of decay product increases, and the amount of original radioisotope decreases in a predictable way.

without limit. The later segments at the top would eventually become too small to be visible, although each would represent the same amount of time. In fact, a sample of a radioactive substance does have a limited number of atoms. But that number is so large that, in practice, the number of atoms does not become an issue.

The way scientists determine the age of some radioactive samples is to compare the amount of the original radioisotope with the amount of its decay product. This comparison is known as the **decay-product ratio**. For example, if half the atoms in a sample have changed, the age of the sample is one half-life period. For carbon-14 this would be 5700 years. (See Table 17-1.)

One half-life period for uranium-238 is 4.5 billion years. Suppose that in a sample, $\frac{3}{4}$ of the radioisotope has changed to the decay product. In this case, it took one half-life for half of the radioisotope to change and an additional half-life for half the remaining radioisotope to change. That is, the sample is 9 billion years old, two half-life periods.

Perhaps the best way to solve problems like this is to think through each half-life period until you get to the needed decay-product ratio. For example, suppose you want to know the number of half-life periods that have passed for a sample that is $\frac{7}{8}$ decay product. It must therefore contain $\frac{1}{8}$ of the original radioisotope. After one half-life period, half of the original radioisotope would remain. After a second half-life period, $\frac{1}{4}$ would remain. After three half-life periods, $\frac{1}{8}$ would remain unchanged. Keep in mind that with the passage of each successive half-life period, one half of the remaining mass of radioisotope decays. Figure 17-11 is a graphical model of radioactive decay in which the radioactive element decreases as the decay product increases through time.

Selecting the Best Radioactive Isotope

A carpenter has many tools. Selecting the right tool for a job is an important first step in any project. Radiometric dating (determining absolute age with radioactive isotopes) involves

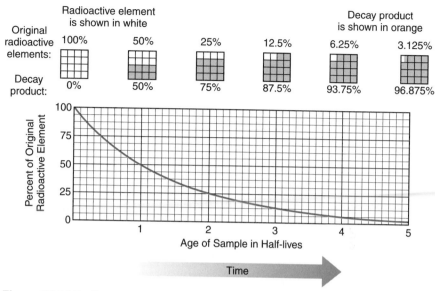

Figure 17-11 Radioactive decay—as time progresses, the amount of decay slows because less and less of the original radioisotope remains. However, the half-life period does not change. The line on the graph shows the radioisotope remaining. The black shading in the boxes above the graph shows the increase in the amount of the decay product through time.

similar choices. Different radioisotopes are best suited to different applications. The two critical issues are the presence of any particular radioisotope in the rocks being dated, and the estimated age of the rocks.

CARBON-14   Carbon-14 is often used to find the absolute age of organic material that is less than about 50,000 years old. Carbon-14 decays so quickly that there is little left in older samples. This makes measurements difficult and precision poor beyond that age. Organic materials always contain carbon. Plants absorb carbon from the atmosphere. The air contains the more common isotope, carbon-12, and a smaller amount of carbon-14. The ratio of these two forms of carbon in the atmosphere is thought to be in equilibrium and relatively stable through time.

In living plants, carbon-14 decays to nitrogen-14. However, it is continually replenished. When plants die, they no longer absorb carbon, and the carbon-14 continues to decay.

Scientists use the ratio of carbon-12 to carbon-14 to establish with great precision when the plant died.

Radiocarbon dates have been verified by checking them with dates obtained from counting tree rings. The oldest trees alive today are more than 5000 years old. Radiocarbon dating has been especially useful in work on prehistoric human habitations and events during or since the last ice age.

OTHER ISOTOPES   The remaining three isotopes in Table 17-1 have longer half-life periods. This means that they are more useful for dating older rocks, from 100,000 years old and back to the origin of Earth. These three radioisotopes are found in selected minerals. Each can be measured with great precision. The age of an igneous rock is generally the age of its minerals. In some places, these isotopes have been used to find the age of layers of volcanic ash. Therefore, ash beds are often useful horizons of absolute time that can be found over a broad geographic area.

TERMS TO KNOW

absolute age	intrusion	radioisotope
decay product	isotopes	relative age
decay-product ratio	original horizontality	superposition
geologist	outcrop	unconformity
half-life period	radioactive	uniformitarianism
inclusion		

CHAPTER REVIEW QUESTIONS

1. Which geological events destroy a part of the rock record?

   (1) uplift and erosion over a long period of time
   (2) continuous deposition of sediments in a deep ocean basin
   (3) folding of rock layers deep under Earth's surface
   (4) a long period of volcanic activity

Base your answers to questions 2–4 on the cross sections below that show widely separated outcrops at locations X, Y, and Z.

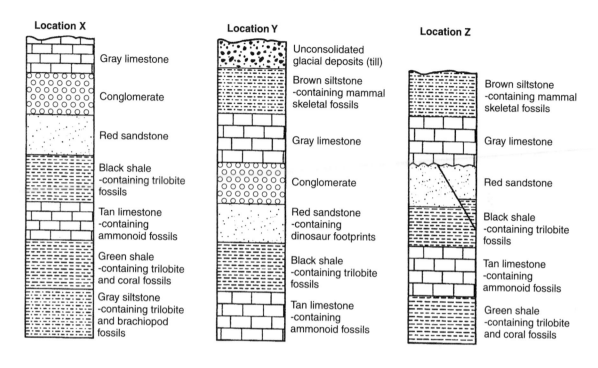

2. Which rock layer is probably oldest?

    (1) gray siltstone   (3) tan limestone

    (2) green shale   (4) brown siltstone

3. An unconformity can be observed at location Z. Which rock layer was most probably removed by erosion during the time represented by the unconformity?

    (1) conglomerate   (3) black shale

    (2) gray siltstone   (4) brown siltstone

4. The fossils in the rock formations at location X indicate that this area was often covered by

    (1) tropical rain forests   (3) desert sand

    (2) glacial ice   (4) seawater

The geologic cross section below has not been overturned. Use this diagram to answer questions 5 and 6.

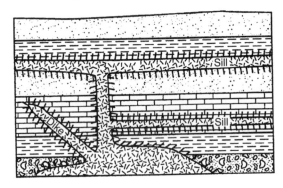

5. Which rock type is oldest?

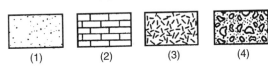

    (1)   (2)   (3)   (4)

6. What feature is represented by the symbol along the edges of the dike and sill in the figure below?

   (1) contact metamorphic rock   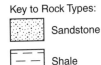
   (2) an unconformity
   (3) a glacial moraine
   (4) index fossils

7. Which event is most likely to occur as a direct result of an igneous intrusion?

   (1) deposition   (3) faulting
   (2) metamorphism   (4) folding

8. The diagram below represents an outcrop of rocks.

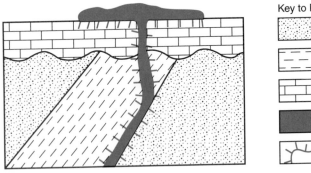

Key to Rock Types:

   Sandstone

   Shale

   Limestone

   Basalt

   Contact Metamorphism

Which sequence of events below is best shown by this profile?

(1) deposition, extrusion, tilting, new deposition, erosion
(2) deposition, extrusion, tilting, new deposition, erosion
(3) deposition, tilting, erosion, new deposition, extrusion
(4) extrusion, tilting, erosion, deposition, metamorphism

9. Which graph most accurately indicates the relative age of the rocks along line AB in the geologic cross section below if no overturning has occurred?

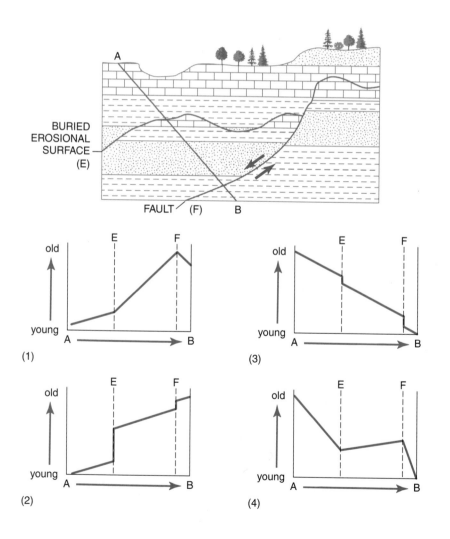

10. The absolute age of an igneous rock can best be determined by

    (1) comparing the amount of decayed and undecayed radioactive isotopes in the rock.

    (2) comparing the sizes of the crystals found in the upper and lower parts of the rock.

    (3) examining the rock's relative position in a rock outcrop.

    (4) examining the environment in which the rock is found.

The diagrams below represent two different geologic cross sections in which an igneous formation is found in sedimentary rock layers. The layers have not been overturned.

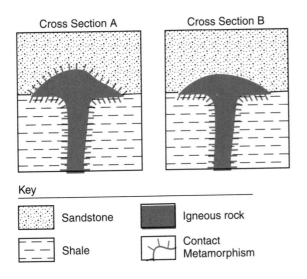

11. Which statement best describes the relative age of each igneous formation compared to the overlying sandstone bedrock?

    (1) In A, the igneous rock is younger than the sandstone, and in B, the igneous rock is older than the sandstone.

    (2) In A, the igneous rock is older than the sandstone, and in B, the igneous rock is younger than the sandstone.

    (3) In both A and B, the igneous rock is younger than the sandstone.

    (4) In both A and B, the igneous rock is older than the sandstone.

12. The graph below represents data from an experiment conducted in a geology laboratory with natural radioactive materials.

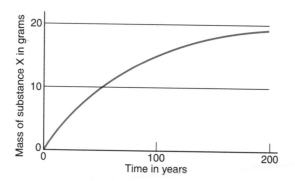

The substance represented by the line on this graph can best be described as

(1) unweathered rock

(2) an unconformity

(3) an unstable substance

(4) a nuclear decay product

13. What is the approximate half-life of the radioactive material used in this experiment?

(1) 5 years

(2) 10 years

(3) 50 years

(4) 100 years

14. The diagram below represents the present number of decayed and undecayed atoms in a sample that was originally 100% radioactive material.

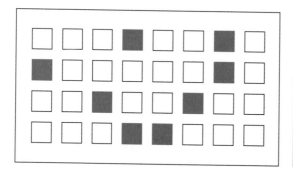

If the half-life period of the radioactive material is 1000 years, what is the age of the sample represented by the diagram?

(1) 1000 years

(2) 2000 years

(3) 3000 years

(4) 4000 years

15. Carbon-14 would most likely be useful in determining the age of

    (1) an igneous intrusion 1 billion years old
    (2) an igneous intrusion 10,000 years old
    (3) a buried dinosaur bone 100 million years old
    (4) remains of a campfire 10,000 years old

Open-Ended Questions

16. The diagram below represents a cliff of exposed bedrock that was investigated by an Earth science class.

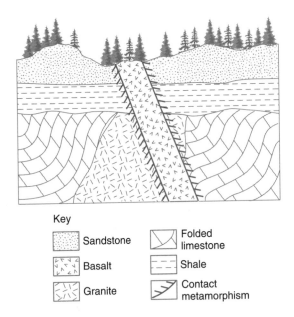

Key

Sandstone   Folded limestone

Basalt   Shale

Granite   Contact metamorphism

After the students examined the cliff, they made three correct inferences about the geologic history of the bedrock.

Inference 1: The shale layer is older than the basalt intrusion.
Inference 2: The shale layer is older than the sandstone layer.
Inference 3: An unconformity exists directly under the shale layer.

Explain how *each* inference is supported by evidence shown in the diagram.

Use the diagram below for questions 17 and 18. It shows a cross section of a portion of Earth's crust. Overturning of rock layers has not occurred.

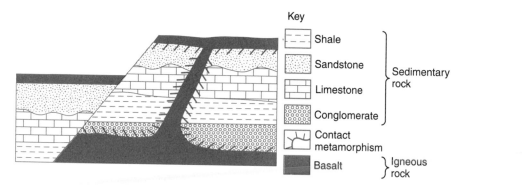

Key

Shale
Sandstone   } Sedimentary rock
Limestone
Conglomerate
Contact metamorphism
Basalt   } Igneous rock

17. State one form of evidence that indicates basalt is the youngest rock unit in the cross section.

18. How can you tell that crustal movement has taken place?

19. The table below shows the results of a student's demonstration modeling radioactive decay. To begin, the students placed 50 pennies heads up in a container. Each penny represented one radioactive atom. The student placed a top on the box and shook the box. Each penny that flipped over to the tails up side was replaced with a bean that represented the stable decay product. The student continued the process until all of the pennies had been replaced by beans.

Shake Number	Number of Radioactive Atoms (pennies)	Number of Stable Decay Atoms (beans)
0	50	0
1	25	25
2	14	36
3	7	43
4	5	45
5	2	48
6	1	49
7	0	50

Graph the data shown on the table by following the steps below.

(a) Set up your graph paper with "Number of Atoms" on the vertical axis and "Shake Number" on the horizontal axis. Use numbers on the two axes so that your data extends at least half way along each scale.

(b) Mark with a dot the number of radioactive atoms (pennies) after each shake.

(c) Connect the dots with a solid line.

20. If each shake represents an additional 100 years, what is the half-life of the radioactive material represented in this model?

Chapter 18

Fossils and Geologic Time

DINOSAURS

The remains of dinosaurs had been noted in literature as early as the 1600s. However, not until 1824 did English scientists recognize that these fossils were the remains of ancient reptiles. In 1841, the term *dinosaur*, meaning "terrible lizard," was first applied to this group of extinct animals.

An historic event in the early development of public interest in dinosaurs occurred in the northeastern part of the United States in the mid-nineteenth century. The first dinosaur skeleton that could be displayed as a whole animal was discovered in New Jersey. It was unearthed in 1858 in Haddonfield, across the Delaware River from Philadelphia. The skeleton was put on public display 10 years later in the Philadelphia Academy of Natural Sciences. Museum officials hoped that the three-story-high monster would spark public interest in the natural sciences. It succeeded well beyond their expectations. Museum attendance tripled, and for 15 years this was the only mounted dinosaur skeleton in the world.

The financial success of the display led other museums to seek dinosaur skeletons, as they became a focal point for public interest in natural history museums. More extensive dinosaur discoveries in the American West soon eclipsed the find in New Jersey. Nevertheless, the New Jersey dinosaur started a public fascination with these creatures that has continued for well over a century and shows no sign of ending. Many people, even professional scientists, recall that an interest in dinosaurs was their introduction to the fascinating world of science.

 WHAT ARE FOSSILS?

Fossils are any remains or evidence of prehistoric life found in the natural environment. Historical records are generally considered to extend about 10,000 years into the past. Therefore, fossils must be older than this. In fact, the earliest fossils are traces of primitive life-forms that existed on Earth very early in its history, billions of years ago. Fossils provide critical information about the development of life on Earth and about past environments. The study of fossils is called **paleontology** and scientists who study fossils are known as paleontologists. Fossils can be grouped into two broad categories: body fossils and trace fossils.

 Body Fossils

Among the most complete fossils are ice age elephants, known as mammoths. These creatures were recovered from permanently frozen ground where they have been preserved since the last ice age. Mammoth fossils sometimes include soft parts such as hair and flesh.

Whole-body preservation is not limited to large mammals. Insects that land on the sticky parts of pine trees may become encased in resin that flows over their body. Amber is the hardened form of this tree resin. Complete bodies of insects have been preserved in amber that is millions of years old. However, fossil remains of soft tissue are relatively rare. Usually the soft parts of an organism are eaten or decay before they can be preserved.

Bones, teeth, and shells are far more likely to become fossilized than soft tissue. For example, a homeowner enlarging a pond in the Hudson Valley of New York discovered the bones of another ice age elephant, a mastodon, in 1999. At the invitation of the landowner, scientists and students from several universities excavated a nearly complete skeleton of this relative of modern elephants. This is just one of three very recent finds of mastodon skeletons in New York State. Like mammoths, mastodons became extinct about 10,000 years ago.

How did the scientists know what animal left these bones? The bones of every animal are different. For this reason, an experienced paleontologist can often identify fossil animals from a single bone or tooth.

Replacement is another form of fossil preservation. It occurs when minerals in groundwater gradually take the place of organic substances. Petrified wood is a common example. In the process of replacement, minerals take the shape and may even show the internal structures of the original living organism. However, these fossils are composed of mineral material such as quartz rather than the wood or other organic substance that made up the original organism.

A mold is created when fine sediments surround a fossil organism or part of an organism, preserving its shape. Leaf impressions are molds. Other sediments that fill in the hole left by the organism are known as casts. In fact, filling a hollow mold makes a cast. Fossils of seashells found in rocks that extend from the lower Hudson valley through western New York State are often molds and casts.

 Trace Fossils

Traces fossils are signs that living organisms were present. However, they do not include or represent the body parts of an organism. For example, the only dinosaur fossils found in New York State are footprints found in the lower Hudson Valley. Worm burrows are common in some sedimentary rocks in western New York State. Coprolites can be described as "the only material a fossil animal intended to leave behind." They

Figure 18-1 Modern footprints in sand—wherever footprints are preserved they give clues to events of the past.

are, in fact, fossilized dung, or feces. Figure 18-1 shows modern footprints in a sand dune.

ACTIVITY 18-1 **NEARBY FOSSIL BEDS**

Where are the nearest rocks that contain fossils? Local residents, park naturalists, or university professors may be able to help you locate fossils nearby. You may even be able to visit this location if it is on public land or if you have the permission of the landowner. Road cuts sometimes reveal fossils although there may be restrictions about stopping along busy highways. Be sure to investigate and secure any needed permission before you visit a fossil bed or collect specimens. As an alternative, some museums and universities have collections of fossils that you can visit.

ACTIVITY 18-2 **INTERPRETING FOSSIL FOOTPRINTS**

The pattern of footprints below was found in bedrock millions of years old. (See Figure 18-2.) What do you think happened to produce this particular pattern of footprints? Propose a sequence of events that could account for the details you see in the footprints. Can you think of alternate explanations for this pattern? How can you determine which explanation is correct?

Figure 18-2 Footprint Puzzle

Fossils tell us more than just what life-forms inhabited Earth. They also provide clues to ancient environments. When paleontologists find fossil fish or the remains of other organisms that resemble modern life-forms that live in oceans or lakes, they infer that they are looking at an ancient underwater environment. Although most sedimentary rocks are the result of material deposited in water, land derived sedimentary rocks are also common. Fossil wood, such as the petrified logs pictured in Figure 18-3 on page 418, helps scientists identify terrestrial sediments.

Figure 18-3 These petrified logs in Utah help identify the rocks in which they are found as the remains of a land environment.

 # HOW DID LIFE BEGIN ON EARTH?

Scientists are not sure how life began on Earth. It is clear that chemical reactions in the environment of early Earth could have produced amino acids. Living organisms are made of these basic compounds. Still, how amino acids become organized to make even the simplest life-forms is not understood. Experiments to produce living material from nonliving processes have not been conclusive. However, millions of years of chemical changes on the primitive Earth might have produced results that cannot be duplicated in short-term laboratory experiments.

Life could have started in places not usually considered good places for living organisms. Some of the most primitive forms of life have recently been discovered in nearly solid rock many kilometers below the surface. Life-forms have been found near hot water vents in the deep ocean bottom where sunlight cannot penetrate. Perhaps conditions in one of these places gave rise to the first life-forms.

Rock fragments from Mars have been found in Antarctica. These rocks provide another clue to the ability of life to exist in extreme environments. It appears that the impact of objects from space, such as meteorites, blasted rock fragments from the surface of Mars. Some of these rocks fell to Earth. Microscopic examination of these rocks has revealed tiny objects that many scientists believe to be evidence of life. Could very simple forms of life travel from planet to planet as spores in rock fragments like these? While this is an interesting possibility, scientists do not know for sure. Further investigations are needed to help scientists decide how life on Earth began.

Very few rocks survive from the earliest part of the planet's history. Furthermore, metamorphism has drastically changed most of the oldest sedimentary and igneous rocks. Some scientists think that they have found evidence of carbon compounds that must have been created by living organisms nearly 4 billion years ago. This evidence is controversial. Stronger evidence is provided by patterns of life structures. The oldest recognizable fossils are probably clumps of primitive bacteria called stromatolites. These fossils date from about 3.5 billion years ago. Similar clusters of bacteria are alive today in Western Australia.

 ## WHAT IS ORGANIC EVOLUTION?

The oldest fossils are remains of tiny single-celled organisms. These life-forms were so primitive that their cells did not contain a nucleus. They reproduced by splitting into two or more new cells. The first critical advance in the development of more complex life-forms is found in rocks about 1.4 billion years old. Cells with a nucleus appeared. These cells could reproduce sexually and inherit characteristics from two parent cells.

Colonies of single-celled organisms developed into simple multicellular organisms, such as jellyfish and worms, less than a billion years ago. The first organisms with shells and internal hard parts appeared about 545 million years ago. These organisms eventually gave rise to the complex life-forms that exist today. The gradual change in living organisms from generation to generation is known as **evolution**. Although it is called a theory, organic evolution is one of the most fundamental ideas of science.

The Work of Darwin

Organic evolution became accepted by most scientists after Charles Darwin published *On the Origin of Species by Means of Natural Selection"* in 1859. A **species** can be defined as a group of organisms so similar they can breed to produce fertile offspring. Humans are a good example of a biological species. Darwin based his book on scientific observations and extensive travel. He spent many years organizing his ideas about the development of life before he published his book. In fact, Alfred Russell Wallace independently arrived at similar conclusions at about the same time.

Four Principles of Evolution

Organic evolution is based on four principles. First, variations exist among individuals within a species. Besides differences in size and shape, individual organisms have different abilities to find food, resist disease, and reproduce effectively. Second, organisms usually produce more offspring than the environment can support. Think of the thousands of seeds many trees produce or the number of eggs one salmon can lay. If all these offspring survived, the world would be overrun with just trees and salmon. The third principle is the effect of competition among individuals of a given generation. Organisms die for a variety of reasons, but those best suited to their environment are most likely to survive. Fourth, the individuals best suited to the environment will live long enough to reproduce and pass their traits to following generations. This is what Darwin meant by natural selection.

ACTIVITY 18-3 VARIATIONS WITHIN A SPECIES

Many characteristics of individual members of a species center on some average value. Length or height is a good example. Record the height in centimeters of all the individuals in your science class. Group the heights in 10-cm intervals (150 cm–159 cm, 160 cm–169 cm, etc.). Count the number of individuals whose height falls within each 10-cm interval. Graph your data as a histogram with the intervals of height on the horizontal axis and the number of individuals within that interval on the vertical axis. Does the data show a pattern of distribution around a typical value? What other characteristics that vary from individual to individual might show a similar distribution?

Evolution and DNA

Darwin did not understand how organisms pass their desirable characteristics to succeeding generations. Chromosomes, genes, and DNA were discovered and studied during the twentieth century. These chemical-messengers pass organic "building plans" from parent to offspring and guide the growth of individuals. As this information passes from generation to generation, errors, or mutations occur. Most mutations are harmful and are not passed on to offspring because they make the individual less able to survive and reproduce. However, the rare helpful changes in genetic plans that make an individual better adapted to the environment are passed to succeeding generations.

Many changes in the genetic makeup of species are slow and subtle. However, recent discoveries have shown that other important changes occur relatively rapidly in small interbreeding populations. The accumulation of genetic changes eventually produces a group of individuals that can no longer interbreed with other groups. This is how a new species originates.

Extinction

Competition is not restricted to individuals of a single species. Different species also compete to survive. When one species is better adapted to live in a particular environment than another, the better-adapted group may take food, shelter, or other natural resources needed by the less well-adapted species. Predation, or hunting, of one species by another can also eliminate whole species. The American bison, or buffalo, once roamed the prairies of North American in great numbers. Hunting, especially by European settlers with rifles, nearly eliminated the American bison. A much smaller number of bison survive today on farms and game reserves. When all the individuals of a particular species die, **extinction** has taken place. A number of species including the marsupial Tasmanian tiger and the passenger pigeon have become extinct in the past few centuries. Most of the life-forms found as fossils have, in fact, become extinct.

ACTIVITY 18-4 | AN EXTINCT SPECIES

Choose a species that is in danger of extinction or has become extinct in recent years. Report on why that species is threatened or became extinct.

The geologic record includes times when large numbers of species became extinct. Recent discoveries have shown that at least one of these mass extinctions occurred about the time that a giant meteorite or other object from space collided with Earth. The impact of this object threw great amounts of dust into the atmosphere causing global changes in climate. The dust-laden atmosphere prevented much of the sunlight from reaching Earth's surface. This killed plants and the animals that depended on them. Most dinosaurs became extinct at that time.

Other scientists dispute the importance of a meteor impact, suggesting similar atmospheric changes result from volcanic eruptions. A third suggestion accounts for global changes in climate related to plate tectonic motions and the formation of a supercontinent known today as Pangaea.

 # HOW HAS GEOLOGIC TIME BEEN DIVIDED?

In the late eighteenth and early nineteenth centuries, engineers and geologists working in Great Britain and Europe noticed that the fossils contained in rock formations could often be used to identify them. They also noticed that layers containing certain other fossils were consistently located above or below these layers. In fact, they began to identify whole groups of fossils according to where they were found. For example, any rocks with the fossils found in Devon, in the south of England, were called Devonian. These rocks were always located above those with fossils found in certain parts of Wales and below the rocks of the coal mining areas near Bristol. These observations were used to establish a geologic time scale of relative ages as shown in Figure 18-4 on page 424.

Some fossils are more useful than others in establishing the age of rocks. These are called index fossils. The best **index fossils** are easy to recognize, and are found over a large geographic area, but they existed for a brief period of time.

Trilobites are a group of animals with external skeletons that lived on the bottom of the oceans for nearly 300 million years. This group of animals existed for a long time. Different species of trilobites evolved and became extinct. Each species looks distinctly different from the others. Furthermore, trilobites left good fossils all over the world. Therefore, geologists find trilobites to be one of the most useful groups of index fossils. Figure 18-5 on page 425 illustrates some of the variety of trilobites used as index fossils.

It is possible that humans also will be good index fossils at some time in the distant future. Humans have distinct hard parts, and we often bury our dead. We have left signs of

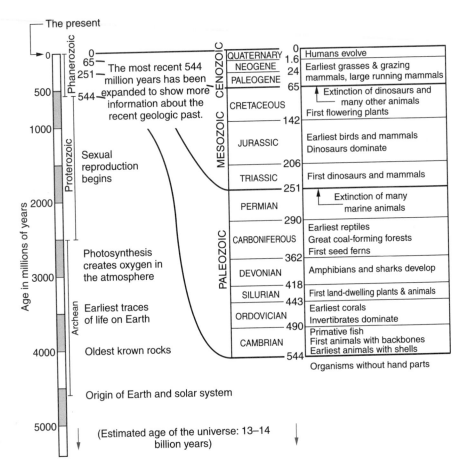

Figure 18-4 The geologic time scale, originally a scale of relative time based on fossils, has become an absolute scale through radiometric dating.

habitation on every continent. Humans have also experienced rapid evolution in geologic time.

Precambrian Time: The Dawn of Life

Scientific evidence indicates that Earth and the solar system formed at the same time. The processes of plate tectonics coupled with metamorphism, weathering, and erosion has destroyed Earth's original crust. Our planet is too dynamic to preserve its oldest surface materials. Nevertheless, scientists have found rocks from the original formation of the solar system outside Earth. The moon has been a good place to look for these rocks because the moon has no atmosphere. The

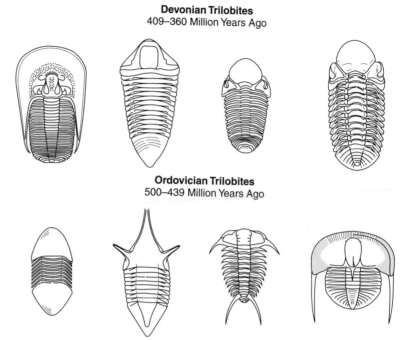

Devonian Trilobites
409–360 Million Years Ago

Ordovician Trilobites
500–439 Million Years Ago

Figure 18-5 The wide geographic range of trilobites and the relatively short periods of time that each species existed make them excellent index fossils. Most adult specimens range in size from 25 to 100 mm (1 to 4 inches).

only destruction of moon rocks is caused by the impact of meteorites. The oldest rock samples returned from the moon by Apollo astronauts have been dated at 4.6 billion years.

Meteorites are another source of primitive rocks because they are the materials from which the solar system formed. Meteorites also yield an age of 4.6 billion years. Earth's formation marks the beginning of geologic history and the start of Precambrian time. The oldest rocks from Earth discovered so far are just under 4 billion years old. However, some mineral inclusions have been dated at 4.4 billions years.

Early in the development of the geologic time scale, fossil collectors could not recognize fossils in rocks older than Cambrian time. Some of them concluded that older rocks (Precambrian) do not contain fossils. However, scientists now know that the apparent lack of fossils in these oldest rocks is deceiving. Fossils are there. Organisms alive at the time did not have hard parts, like shells and skeletons, that would form obvious fossils. Although the Precambrian time makes up about 88 percent of Earth's history, events of the Precambrian are not as well known as more recent events. This is because later rocks have covered most of the rocks from this time, or the rocks were recycled by weathering and erosion, or altered by metamorphism.

Most of the earliest forms of life inhabited the oceans, which provided them with nutrients and protected them from harmful solar radiation. Carbon dioxide and nitrogen probably dominated Earth's early atmosphere, making it similar in composition to the present atmosphere of Venus and Mars. Photosynthesis eventually began in single-celled Precambrian organisms. They used carbon dioxide to store energy, releasing oxygen as a waste product. The addition of oxygen to the atmosphere had two very important results. Oxygen in the form of ozone absorbs harmful radiation, such as ultraviolet rays, from outside Earth. Oxygen also is necessary for air-breathing animals. Both of these factors made possible the later development of life outside the oceans.

Paleozoic Era: The Origin of Complex Life-forms

The name *Paleozoic* can be translated from its word origin as "the time of early life." This era began a little more than half a billion years ago. The abundance of fossils that marks the beginning of the Paleozoic was a result of rapid evolution. Hard parts, such as skeletons and shells, allowed some organisms to move rapidly in search of food and others to protect themselves from becoming food. Many of the specialized features of today's most complex life-forms, including our sense organs, can be traced back to structures that appeared early in the Paleozoic Era.

The first fish as well as the first plants and animals to inhabit the land appeared in the Paleozoic along with the earliest insects and amphibians. Amphibians lay their eggs in water; but as adults, many move onto the land. Reptiles lay eggs with protective shells, and live and reproduce entirely on land. Trilobites appeared in the oceans early in the Paleozoic. They evolved many variations in shape before they became extinct at the end of the Paleozoic.

The end of the Paleozoic Era is marked by the extinction of as much as 95 percent of living species. The cause of this ex-

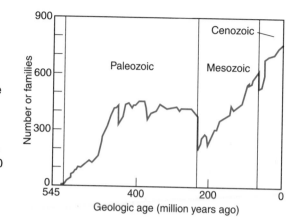

Figure 18-6 Changes in the number of families of marine animals indicate mass extinctions that mark the ends of two geologic eras. The event at the end of the Paleozoic era wiped out about 50 percent of animal families, and a large number of species.

tinction is not known, although some dramatic change in the world's climates is suspected. This might have been related to the formation of the supercontinent Pangaea, the impact of a large meteorite, or great volcanic eruptions. Regardless of the cause, this catastrophic event led to the appearance of new life-forms in the next era. Figure 18-6 shows when major mass extinctions occurred in the geological past.

Mesozoic Era: The Age of Dinosaurs

The word *Mesozoic* can be translated as "middle life." This era began about 250 million years ago following the mass extinction at the end of the Paleozoic Era. Some forms of fish, insects, and reptiles survived the Paleozoic extinction. Mammals appeared in the Mesozoic, but they were small creatures. Dinosaurs dominated life on land. The evolution of dinosaurs produced a wide range of animals that could inhabit nearly every terrestrial environment. The first birds probably evolved from flying dinosaurs in the Mesozoic.

When people or things are called dinosaurs, it may imply that they are not very intelligent or successful. However, dinosaurs existed for nearly 200 million years. Some of them probably achieved remarkable intelligence before they became extinct at the end of the Mesozoic. Recently discovered evidence has linked that extinction to the impact of meteorites and climatic change resulting from dust blown into the atmosphere. However, like the even more catastrophic extinction at the end of the Paleozoic, the cause of this mass extinction is still under debate.

Cenozoic Era: The Age of Mammals

The word *Cenozoic* means "recent life." The Cenozoic Era began 65 million years ago and continues to the present. With the extinction of dinosaurs, mammals evolved as the dominant group of vertebrate animals. Mammals inhabit nearly every terrestrial environment. Whales, dolphins, and seals inhabit the seas, and bats take to the air. The first humans evolved in the late Cenozoic. The oldest human fossils were found in Africa and are about 2 to 4 million years old.

In New York State, the most dramatic geological events of the Cenozoic Era were probably the repeated advances of continental glaciers. Glaciers nearly covered the state until about 15,000 years ago. The formation of the glaciers resulted in a drop in sea level that exposed a land connection between Asia and North America. This allowed humans to migrate from Asia into North America. Unlike earlier eras, the Cenozoic is the only era with no specific end. The age of mammals is expected to continue into the foreseeable future.

ACTIVITY 18-5 GEOLOGIC TIME LINE

Use a strip of paper about 5 meters long to make a time line of the following events. Develop a scale that indicates how much time is represented by each unit of distance. According to your scale of time, plot each event where it belongs on the strip.

Origin of Earth and solar system

Oldest known rocks on Earth

Oldest microfossils

Fossil organisms become abundant

Earliest traces of terrestrial (land) organisms

First birds

Extinction of dinosaurs

Earliest humans in Africa

End of Wisconsin Glacial Episode

Columbus sails to the Americas

When the events have been plotted, divide your time line into Precambrian, Paleozoic, Mesozoic, and Cenozoic eras. You may wish to make each era a different color.

GEOLOGIC HISTORY OF NEW YORK STATE

Figure 18-7 on pages 430–431 is taken from pages 8 and 9 of the *Earth Science Reference Tables*. It is a chart of Earth's geologic history. Geologic events and fossils of New York State are highlighted. You should be familiar with the information in this chart and how to use it. Look at the figure and note the following.

- The left page is primarily used to show the major divisions geologic time and the absolute age of these divisions in millions of years before the present. For example, Precambrian time ended at the beginning of the Paleozoic Era, 544 million years ago.

- The column labeled "Life on Earth" identifies important changes in the evolution of life. This information expands on the information given in the gray area on the left side of the page.

- In the next column, thick, black, broken vertical lines indicate the age of bedrock in New York State. The open parts of the thick line represent intervals for which there is no bedrock in New York. Notice that there are whole geological periods missing. These incomplete parts of the line represent major unconformities in New York bedrock.

GEOLOGIC HISTORY OF NEW YORK STATE

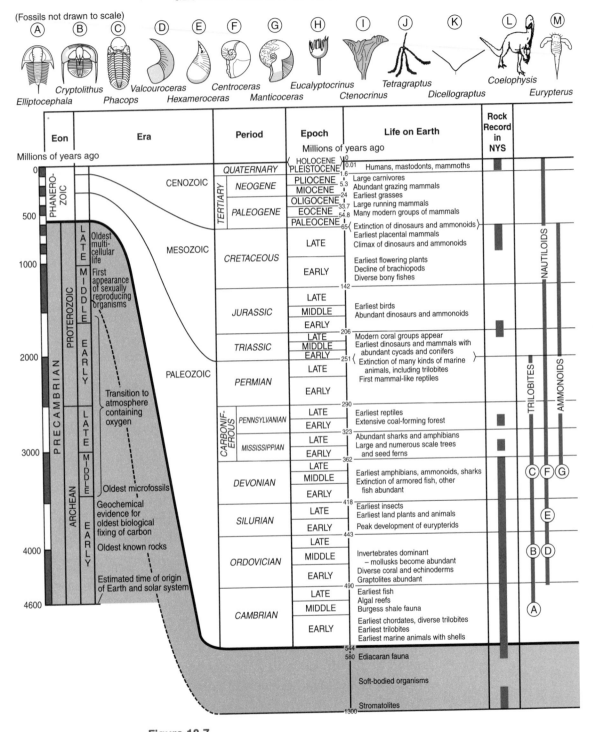

(Fossils not drawn to scale)

A Elliptocephala
B Cryptolithus / Phacops
C Valcouroceras
D Hexameroceras
E Centroceras
F Manticoceras
G Eucalyptocrinus
H Ctenocrinus
I Tetragraptus
J Dicellograptus
K Coelophysis
L Eurypterus

Figure 18-7

430

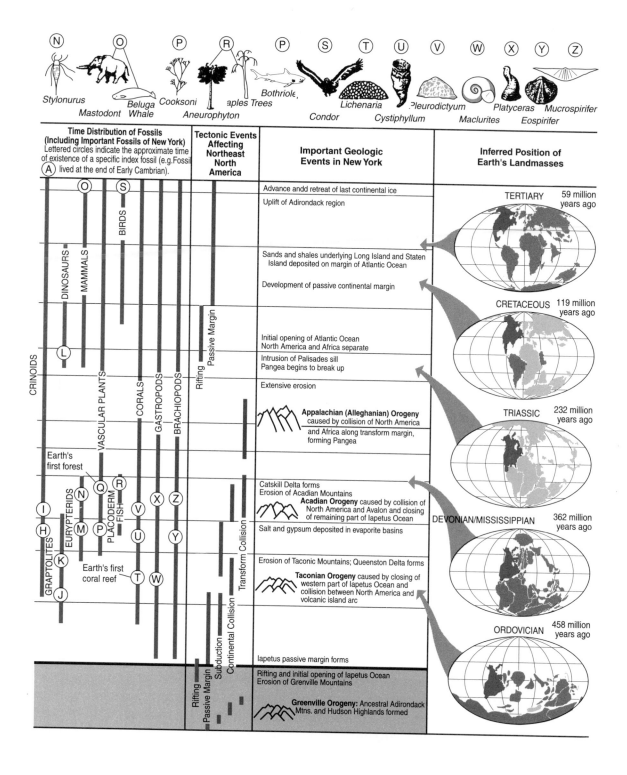

The Geologic History of New York State

| Time Distribution of Fossils (Including Important Fossils of New York) Lettered circles indicate the approximate time of existence of a specific index fossil (e.g. Fossil Ⓐ lived at the end of Early Cambrian). | Tectonic Events Affecting Northeast North America | Important Geologic Events in New York | Inferred Position of Earth's Landmasses |

Fossil name labels (top):
Ⓝ Stylonurus
Ⓞ Mastodont, Beluga Whale
Ⓟ Cooksoni
Ⓡ Aneurophyton, Maples Trees, Bothriole
Ⓟ
Ⓢ Condor
Ⓣ Lichenaria
Ⓤ Cystiphyllum
Ⓥ Pleurodictyum
Ⓦ Maclurites
Ⓧ Platyceras
Ⓨ Eospirifer
Ⓩ Mucrospirifer

Important Geologic Events in New York:
- Advance and retreat of last continental ice
- Uplift of Adirondack region
- Sands and shales underlying Long Island and Staten Island deposited on margin of Atlantic Ocean
- Development of passive continental margin
- Initial opening of Atlantic Ocean North America and Africa separate
- Intrusion of Palisades sill Pangea begins to break up
- Extensive erosion
- **Appalachian (Alleghanian) Orogeny** caused by collision of North America and Africa along transform margin, forming Pangea
- Catskill Delta forms Erosion of Acadian Mountains
- **Acadian Orogeny** caused by collision of North America and Avalon and closing of remaining part of Iapetus Ocean
- Salt and gypsum deposited in evaporite basins
- Erosion of Taconic Mountains; Queenston Delta forms
- **Taconian Orogeny** caused by closing of western part of Iapetus Ocean and collision between North America and volcanic island arc
- Iapetus passive margin forms
- Rifting and initial opening of Iapetus Ocean Erosion of Grenville Mountains
- **Greenville Orogeny:** Ancestral Adirondack Mtns. and Hudson Highlands formed

Inferred Position of Earth's Landmasses:
- TERTIARY 59 million years ago
- CRETACEOUS 119 million years ago
- TRIASSIC 232 million years ago
- DEVONIAN/MISSISSIPPIAN 362 million years ago
- ORDOVICIAN 458 million years ago

Fossil/organism time bars labels:
CRINOIDS, DINOSAURS, MAMMALS, BIRDS, VASCULAR PLANTS, CORALS, GASTROPODS, BRACHIOPODS, GRAPTOLITES, EURYPTERIDS, PLACODERM FISH

Earth's first forest
Earth's first coral reef

Tectonic events labels: Rifting, Passive Margin, Transform Collision, Continental Collision, Subduction, Passive Margin, Rifting

- In the column labeled "Time Distribution of Fossils," thinner vertical lines indicate when certain fossil organisms lived. Notice the creatures labeled A-Z on at the top of the page. These letters appear on the vertical lines to show when these organisms were alive. For example, trilobites A, B, and C lived at different times during the Paleozoic Era.

- The maps at the right side of the page show plate tectonic motions of the continents. Specific positions of North America are indicated. The maps also show how North America was positioned with respect to other continents at five specific times in the past.

New York Bedrock

Figure 18-8 is a geologic map from the *Earth Science Reference Tables*. New York's oldest rocks are Precambrian. These rocks are exposed in the Adirondacks of northern New York State and the Hudson Highlands between New Jersey and Connecticut. Paleozoic sedimentary rocks extend westward from these regions to lakes Erie and Ontario. Long Island has the youngest "bedrock." It is composed of geologically recent sediments of glacial origin. Note that the absolute age of these bedrock regions can be determined by using this map along with the geologic history chart in Figure 18-7, also found at the center of the *Earth Science Reference Tables*.

HOW DO GEOLOGISTS CORRELATE ROCK LAYERS?

Geologists often compare rock layers in two or more locations to determine whether they are the same layers, or their relative age. Matching bedrock layers by rock type or by age is called **correlation**. The principle of superposition (explained in Chapter 17) and the use of index fossils help geol-

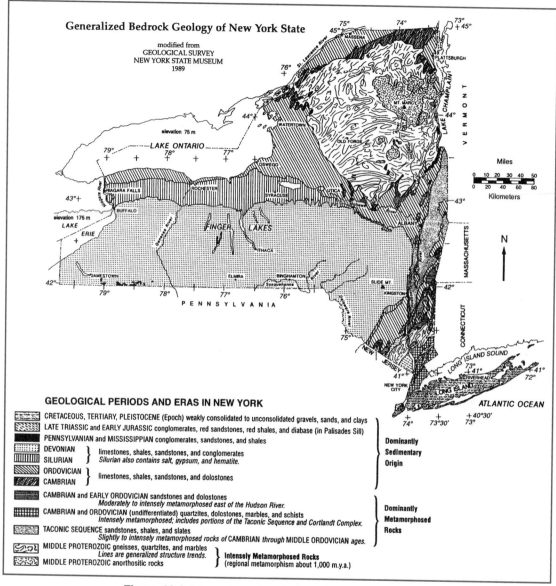

Generalized Bedrock Geology of New York State

modified from
GEOLOGICAL SURVEY
NEW YORK STATE MUSEUM
1989

GEOLOGICAL PERIODS AND ERAS IN NEW YORK

CRETACEOUS, TERTIARY, PLEISTOCENE (Epoch) weakly consolidated to unconsolidated gravels, sands, and clays

LATE TRIASSIC and EARLY JURASSIC conglomerates, red sandstones, red shales, and diabase (in Palisades Sill)

PENNSYLVANIAN and MISSISSIPPIAN conglomerates, sandstones, and shales

DEVONIAN } limestones, shales, sandstones, and conglomerates
SILURIAN } Silurian also contains salt, gypsum, and hematite.

ORDOVICIAN } limestones, shales, sandstones, and dolostones
CAMBRIAN }

} Dominantly Sedimentary Origin

CAMBRIAN and EARLY ORDOVICIAN sandstones and dolostones
  Moderately to intensely metamorphosed east of the Hudson River.

CAMBRIAN and ORDOVICIAN (undifferentiated) quartzites, dolostones, marbles, and schists
  Intensely metamorphosed; includes portions of the Taconic Sequence and Cortlandt Complex.

TACONIC SEQUENCE sandstones, shales, and slates
  Slightly to intensely metamorphosed rocks of CAMBRIAN through MIDDLE ORDOVICIAN ages.

} Dominantly Metamorphosed Rocks

MIDDLE PROTEROZOIC gneisses, quartzites, and marbles
  Lines are generalized structure trends.

MIDDLE PROTEROZOIC anorthositic rocks

} **Intensely Metamorphosed Rocks**
(regional metamorphism about 1,000 m.y.a.)

Figure 18-8 Paleozoic sedimentary rocks underlie the largest part of New York State, beginning in the western part of the state and ending near the Hudson River. The Adirondack region and Hudson Highlands are composed mostly of metamorphic rocks. Long Island is primarily composed of very recent sediments.

ogists make decisions about these issues. Correlation is an important part of drawing geologic maps such as Figure 18-8. Geologic maps can be used to establish the geologic history of an area.

Correlation is also useful in locating natural resources such as petroleum (crude oil), mineral ores, and groundwater. These resources are often found by drilling to a particular level in the rock sequence. Study of rock fragments from the drill hole and knowledge of the local strata can help geologists determine whether the drill is above or below the level at which the resource would most likely be found.

The first step in correlating rock layers in different locations is to find outcrops of bedrock. In most places, soil and sediment cover the rocks. An *outcrop* is a place where the solid lithosphere is exposed at Earth's surface. In some places, bedrock is exposed by erosion such as on the steep bank along a stream or river. Fresh outcrops are also common at road cuts where making the road straight or level required the crew to drill and blast solid rock. Sometimes builders expose bedrock when they level building lots to construct homes or other structures.

 ## Comparing Fossil Types

Correlation is usually accomplished by comparing rock types or fossils in different locations. The fact that two places have the same kind of rock does not mean they are the same age. According to the principle of uniformitarianism, the processes and events that make different kinds of rock have occurred throughout geologic history. Furthermore, at any given time, different kinds of rock are made in different environments all over the world. However, if geologists observe the same sequence of rock types in two nearby locations, there is a good chance that they are observing rocks of the same age in both locations. The probability that the layers correlate is even greater if they can follow the rock units from one outcrop to the next. Geologists try to do this when they create geologic maps.

Fossils are very useful in correlation, especially if a geologist can use index fossils that are widespread in geographic

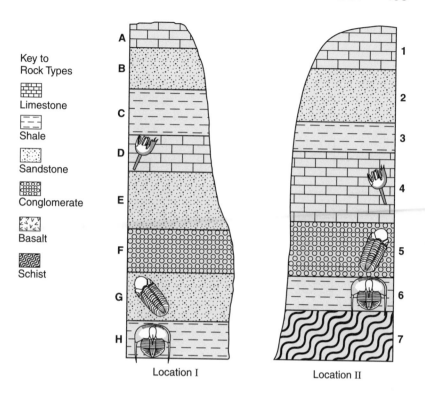

Key to
Rock Types

Limestone

Shale

Sandstone

Conglomerate

Basalt

Schist

A B C D E F G H

Location I

1 2 3 4 5 6 7

Location II

Figure 18-9 Correlation.

range, but restricted in time by rapid evolution. Most rocks, even sedimentary rocks, do not contain abundant fossils. A geologist must look hard to find fossils that can be used to correlate rocks from one location to another.

Figure 18-9 shows how rock types and fossils can be used in correlation. If locations I and II are not too far apart, the following inferences can be made. Layers A, B, and C are probably the same age as layers 1, 2, and 3. The rock type and the sequence are the same. Layers D and 4 not only are the same rock type and occur in the same sequence, but they contain the same fossil. Fossils indicate that layers G and H correlate with layers 5 and 6, although G and 5 are not the same kind of rock. (Fossil correlation is usually more dependable than correlation by rock types.) It appears that layers E and F are missing at location II. E and F may not have been depositated at location II or they may have been eroded This would indicate an uncormity between layers 4 and 5 at location II. Furthermore, it is likely that the oldest rock layer in both locations is the schist at the bottom of location II.

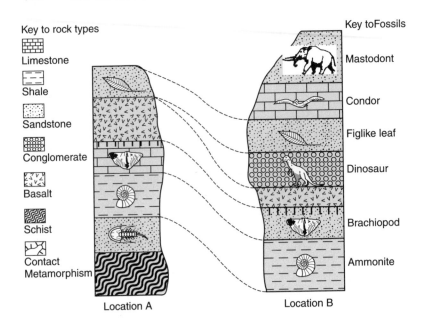

Figure 18-10 Application of correlation principles.

In Figure 18-10, the dashed lines help to illustrate the following. (1) The mastodont and condor, as well as the dinosaur layers at B are missing from location A. (2) The brachiopod fossils at A and B show that the limestone and sandstone that contain them are similar in age. (3) The two bottom layers at location A seem to be older than any exposed rock at B.

Figure 18-11 is an exercise in correlation. In using these diagrams, assume that if a fossil species is not shown in its ocean or land environment, it did not exist at that time. How can the following be concluded from this diagram? (1) One outcrop is a marine environment while the other is terrestrial. (2) Layer A can be only Cambrian or Ordovician in age. (3) Outcrop I (only) is overturned. (4) Layers D and F must both be Cretaceous in age.

Catastrophic Events

Catastrophic events can create very distinct time markers in rock layers. A large volcanic eruption can eject great quantities of ash into the atmosphere. When Mount St. Helens exploded in 1980, a cubic kilometer of ash and rock debris was

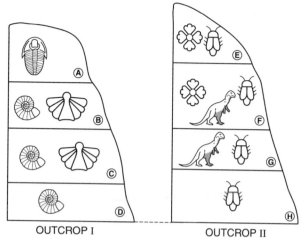

I. Fossls in two distant outcrops

	OCEAN ORGANISMS			Land organisms		
GEOLOGIC TIME PERIODS WHEN ORGANISMS LIVED (time expressed in millions of years before the present)	Trilobites	Ammonites	Spirifers	Insects	Dinosaurs	Flowering Plants
24 Paleogene 66				🪲		🌸
Cretaceous 144		🐚		🪲	🦕	🌸
Jurassic 190		🐚	🐚	🪲	🦕	
Triassic 245		🐚	🐚	🪲	🦕	
Permian 286	🐛	🐚	🐚	🪲		
Carboniferous 360	🐛	🐚	🐚	🪲		
Devonian 408	🐛	🐚	🐚	🪲		
Silurian 438	🐛		🐚			
Ordovician 505	🐛					
Cambrian 540	🐛					

Figure 18-11 An exercise in correlation.

II. Correlation chart

thrown as high as 18,000 m into the atmosphere. While this is a major event in modern American history, Tambora, a volcano in Indonesia, ejected an estimated 30 cubic kilometers in 1815. Most of the debris from these explosions settled near the volcano. However, fine volcanic ash carried into the strong winds of the stratosphere can be deposited over a large area in a matter or days or weeks, a mere instant of geological time. If such a layer of ash is found in the rock strata over a broad area, it represents a very precise time marker that can be found in almost any environment of deposition. These precise time horizons can be very useful in the regional correlation of rock layers.

Scientists also know that the impact of large asteroids, meteorites, or comets results in similar time markers. The impact event that marks the end of the Mesozoic Era was caused by an object estimated to be 10 km across. It created a crater more than 100 km in diameter on the Yucatan Peninsula of Mexico. An unknown quantity of rock from the crater and the remains of the flying object formed a dense cloud of dust. The dust was so widespread it blocked the sun and led to dramatic climatic changes over the whole planet. This event also deposited a distinctive layer of ash that can be found in sedimentary rocks on all the continents and in places on the floor of the ocean. This layer of ash very precisely marks the top boundary of rocks of Mesozoic age.

TERMS TO KNOW

correlation	extinction	paleontology
evolution	index fossil	species

CHAPTER REVIEW QUESTIONS

1. On what basis was the geologic time scale divided into eras and other divisions?

   (1) Earth's history was divided into ages of time of equal length in millions of years.

(2) The length of each age depends on the half-life of a radioactive substance.

(3) The evolutionary development of life-forms was used to divide geologic time.

(4) Geologic time was based on motions of the moon and Earth.

2. What group of fossil organisms was alive during the Paleozoic Era, but did not live in any other time period?

(1) dinosaurs   (3) trilobites
(2) brachiopods   (4) corals

3. Nautiloids are a group of marine animals that have been alive for almost 500 million years. The form of nautiloid that paleontologists call *Centroceras* is known to have been abundant how long ago?

(1) 500 million years ago   (3) 300 million years ago
(2) 400 million years ago   (4) 200 million years ago

4. The diagram below is a portion of a geologic time line. Letters A, B, C, and D represent the time intervals between labeled events.

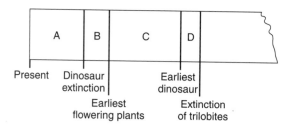

Fossil evidence indicates that the earliest birds developed during which time period?

(1) A   (3) C
(2) B   (4) D

5. During which era did the initial opening of the present-day Atlantic Ocean most likely occur?

(1) Cenozoic   (3) Paleozoic
(2) Mesozoic   (4) Late Proterozoic

6. The diagram below shows the abundance of organisms called crinoids, blastoids, and echinoids throughout different geologic periods. The number of species living at any given time is represented by the width of the blackened areas.

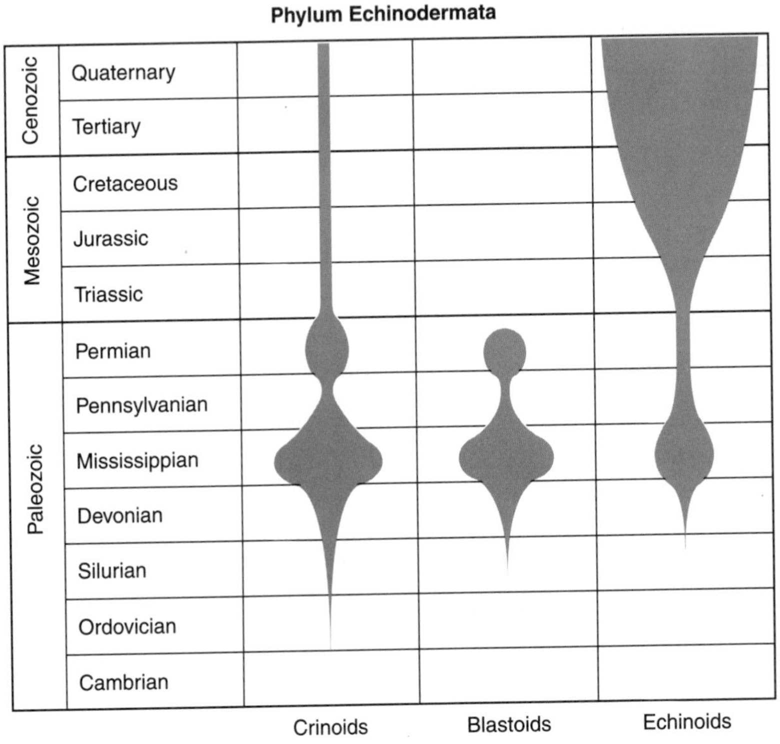

Phylum Echinodermata

| | Crinoids | Blastoids | Echinoids |

Which statement about crinoids, blastoids, and echinoids is best supported by the diagram?

(1) All three are now extinct.
(2) They came into existence during the same geologic period.
(3) They all existed during the Devonian.
(4) All three have steadily increased in number of species since they first appeared.

7. Historians have discovered that the great pyramids near Cairo, Egypt, were built nearly 5000 years ago. During which geologic period were they constructed?

(1) Quaternary
(2) Neogene
(3) Paleogene
(4) Cretaceous

8. What is the geologic age of most of the bedrock that lies directly above Precambrian bedrock in New York State?

   (1) Paleozoic

   (2) Cenozoic

   (3) Mesozoic

   (4) Archean

9. Rocks near Hamilton, New York, contain fossils of Devonian corals and crinoids. What does this suggest about the local bedrock in that location?

   (1) It had a terrestrial environment about 400 million years ago.

   (2) It had a terrestrial environment about 300 million years ago.

   (3) It had a marine environment about 400 million years ago.

   (4) It had a marine environment about 300 million years ago.

10. During which geologic epoch did the most recent continental glaciers retreat from New York State?

    (1) Pleistocene

    (2) Eocene

    (3) Late Pennsylvanian

    (4) Early Mississippian

11. Bedrock at Massena, New York, is most like to correlate in age with bedrock at which other New York State location?

    (1) Niagara Falls

    (2) Elmira

    (3) Syracuse

    (4) Albany

12. Surface bedrock of the Allegheny Plateau is most likely to contain fossils of the earliest

    (1) grasses.

    (2) flowering plants.

    (3) dinosaurs.

    (4) amphibians.

13. Carbon-14 is most useful for determining the age of the bones of which animal?

    (1) trilobites that lived during the Paleozoic Era

    (2) mastodonts that lived during the Cenozoic Era

    (3) *Coelophysis* that lived during the Mesozoic Era

    (4) soft-bodied organisms that lived during the Precambrian

14. What is the most useful technique a geologist can use to correlate the age of rock layers in widely separated outcrops of sedimentary rock?

    (1) compare the thickness of the rock layers
    (2) compare the fossils found in the rock layers
    (3) compare the grain sizes of the rock layers
    (4) compare the densities of minerals in the rock layers

15. The diagrams below are drawings of index fossils. Which pair can be found in Ordovician bedrock?

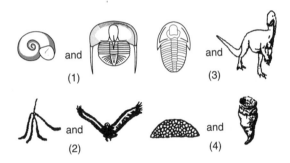

Open-Ended Questions

Base your answers to questions 16–18 on the following information and your knowledge of Earth science.

Scientific evidence indicates that the earliest mammals probably evolved approximately 225 million years ago from an ancient reptile group called the therapsids. For millions of years afterward, early mammals and therapsids coexisted until the therapsids apparently became extinct 65 million years ago. However, geologists have recently found a fossil they believe to be a therapsid that is only 60 million years old. They found the fossil, which they have named *Chronoperates paradoxus* (paradoxical time wanderer) near Calgary in Canada. This find suggests that for 105 million years after the apparent extinction of the therapsids, a few of the reptiles continued to live in a narrow geographic range in Canada.

16. According to fossil evidence, during which geologic period did the earliest mammals appear on Earth?

17. Explain briefly two reasons why *Chronoperates paradoxus* would *not* be a good index fossil.

18. State one method geologists could have used to determine that *Chronoperates paradoxus* lived 60 million years ago

19. Name one species of trilobites that might be found at Watertown, New York.

20. New York State has a natural record of most of its geologic history since the end of the Precambrian. Why are geologists unable to determine any New York State events from the Permian Period of the Paleozoic Era?

Chapter 19

Weather and Heating of the Atmosphere

 WEATHER

We are all affected by the weather. Anything we do outdoors depends on the weather. People dress for the weather and plan their lives around it. Summer weather conditions can make the difference between a pleasant family outing and a terrifying encounter with rain and lightning. Snow and ice can make driving nearly impossible. Fog has been responsible for major traffic accidents that can involve dozens of vehicles. The airline industry is especially sensitive to weather conditions. Bad weather can cause massive delays and sometimes loss of life in accidents that are weather related.

Weather also has indirect effects. Winter storms can close schools and businesses, bringing the local economy to a standstill. A cold winter can drive up the demand and the price of heating oil, causing economic difficulties for families. A summer heat wave can strain electric power supplies, even leading to a total loss of power for large areas. The price of food is often affected by weather conditions. A freeze in Brazil can have a major effect on the price of coffee. Floods, drought, and abnormal temperatures affect the prices of nearly all agricultural products.

Over the years, people have become more successful at protecting themselves from the direct effects of weather extremes. Homes and other buildings have total climate control systems. Meteorologists have developed skills and technology to predict weather with ever increasing accuracy. Nevertheless, nature can still bring us to our knees. Hurricane Andrew slammed into south Florida in 1992 with sustained winds of 145 miles per hour. It included wind gusts of 175 miles per hour and tornadoes of unknown wind speed. In some parts of the storm's path no buildings escaped damage. Property losses in excess of $25 billion made this the most expensive natural disaster in the history of the United States.

WHAT ARE THE ELEMENTS OF WEATHER?

Weather is the short-term condition of Earth's atmosphere at a specific time and place. Weather is usually described in terms of the present conditions or the conditions over a few days. The principal difference between weather and climate is the amount of time under consideration. **Climate** is the average weather conditions of a specific region over a long time, including the range of weather conditions. Scientists who study weather are known as *meteorologists*. The science of Earth's atmosphere and how it changes is called **meteorology**. There are seven variables that meteorologists measure to describe the weather: temperature, humidity, air pressure, wind speed and direction, cloud cover, and precipitation.

Temperature

Most people's first concern about weather is the temperature. This gives us important information about how we need to dress to stay comfortable and maintain our health. Feeling

Figure 19-1 New York State has a mid-latitude climate in which temperatures change considerably with the seasons.

hot or cold does more than give information about our body and our surroundings. It also causes us to take action that ensures our health and survival. If we are too hot or too cold we change our clothing or we move to a more comfortable environment. Figure 19-1 is a winter scene in New York State.

Scientists define **temperature** as a measure of the average kinetic energy of the molecules in a substance. You learned in earlier chapters that matter is made of extremely small units called atoms. Most atoms join in fixed ratios to form molecules, the basic units of most elements such as oxygen and nitrogen, as well as chemical compounds such as water. Above an extremely cold temperature known as absolute zero ($-273°C$), all atoms and molecules are in motion. This motion is a form of kinetic energy known as heat. In solids and liquids, most of the molecular motion is confined to vibrational motion, but the molecules in a gas have more freedom of movement.

Matter is always exchanging kinetic energy with its surroundings. If the matter is relatively cool, it will absorb more energy than it gives off and its temperature increases. If the substance is at a higher temperature than its surroundings, it will give off more energy than it absorbs and become cooler. Matter in temperature equilibrium (balance) with its environment continues to exchange energy, but the energy lost equals the energy gained.

MEASURING TEMPERATURE Temperature is measured with an instrument called a **thermometer**. Common thermometers, such as the one in Figure 19-2, operate on the principle that matter usually expands when it absorbs heat energy. Many

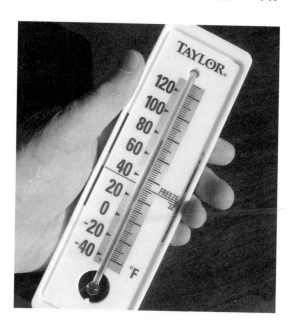

Figure 19-2 A thermometer contains colored alcohol in a glass bulb at its bottom. The alcohol expands or contracts in the tube, registering changes in temperature.

thermometers contain a liquid such as alcohol in a small reservoir at the bottom of a narrow tube. When the liquid absorbs energy from its surroundings, it expands upward into the tube. This indicates an increase in temperature. When the liquid cools, it contracts and registers a lower temperature. A coloring agent is added to alcohol to make the liquid more visible. (Some thermometers use mercury, a liquid metal, instead of alcohol. Mercury is silver in color, so it requires no coloring agent. However, mercury is a poisonous substance that can pose health and environmental risks if the thermometer breaks.) A graduated scale printed on the thermometer tube or on its base allows a person to read temperatures with considerable accuracy.

ACTIVITY 19-1 MAKING A THERMOMETER

You can construct a thermometer using a flask, a narrow glass tube, and a one-hole stopper. You can use water as the liquid that expands and contracts. A ruler along the tube or attached to the tube can be used as a scale.

You can observe an interesting and unusual property of water if you start with water very close to the freezing temperature in the flask (thermometer bulb). As your thermometer warms to near room temperature, record the changing water level in the tube. What unusual property of water does this procedure show?

When you use a thermometer to measure temperature, you should keep the thermometer in its measurement position for a minute or so. Heat energy must flow into or out of the thermometer until the instrument and its environment have reached equilibrium. Only at that time will the thermometer correctly indicate the temperature of the substance in which the thermometer bulb is placed.

TEMPERATURE SCALES Figure 19-3 from the *Earth Science Reference Tables* shows three commonly used temperature scales. The Fahrenheit scale is generally used in the United States. On this scale, ice melts, or water freezes at 32°F and boils at 212°F. Since the difference between those values is 180°, each Fahrenheit degree is $\frac{1}{180}$ of the difference.

The Celsius scale is used in nearly every other country and in scientific investigations. On the Celsius scale, reference points are determined by designating 0°C and 100°C as the freezing and boiling points of water. Since the difference between these values is 100°, each Celsius degree is $\frac{1}{100}$ of the difference, Therefore, a change of 1°C is a larger change than 1°F.

The third scale of temperature is the Kelvin scale. Although a change of 1 K is the same as a change of 1°C, the Kelvin scale starts at absolute zero, the lowest possible temperature of matter. Molecules have no kinetic energy at this temperature. Like the speed of light (300,000,000 meters per second), absolute zero is one of the few natural limits known to scientists. Therefore, there are no negative temperatures on the Kelvin scale.

Figure 19-3 can be used to convert temperatures among the Fahrenheit, Celsius, and Kelvin scales. To convert from one scale to the other, simply look horizontally to the corresponding value on the desired scale.

Temperature

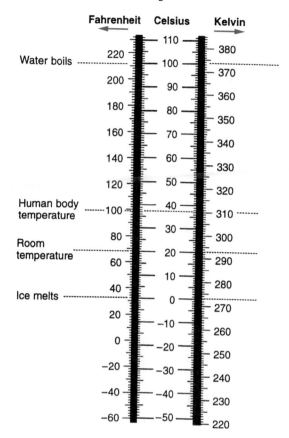

Figure 19-3 Although the Fahrenheit scale is familiar to most residents of the United States, most other nations and most scientists prefer to use the Celsius scale. The advantage of the Kelvin scale is that Kelvin starts at the absolute lowest temperature of matter.

SAMPLE PROBLEM

Problem 1   What Celsius temperature is equal to 0° Fahrenheit?

Solution   On the temperature conversion chart in the *Earth Science Reference Tables*, 0° Fahrenheit is next to its equivalent of −18° Celsius.

Problem 2   What is the normal human body temperature on the Celsius scale?

Solution   On the temperature conversion chart in the *Earth Science Reference Tables*, human body temperature is labeled at about 99°F or in the range of 37°C to 38°C.

Practice Problem

Which of the following temperatures is the coldest: −20°F, −20°C, or 220 K?

 Humidity

The water vapor content of the air is called **humidity**. The amount of water vapor air can hold depends on its temperature. Warm air can hold far more moisture than cold air. Figure 19-4 shows the relationship between air temperature and the ability of the atmosphere to hold water vapor. Although we do not usually feel humidity as readily as we feel temperature, the air may seem "clammy." In humid weather, sweat does not evaporate readily. This makes it more difficult to keep a comfortable body temperature.

Meteorologists usually express humidity as **relative humidity**. This is a comparison of the actual water vapor content of air with the maximum amount of water vapor air can hold at a given temperature. For example, if the relative humidity is 50%, the air is holding half as much moisture as it could hold at the present temperature. At 100% relative humidity, the air is holding as much moisture as it can at that temperature, and the air is said to be **saturated**. Although relative humidity does not tell the amount of water vapor that a given volume of air holds, most people find relative humidity more useful than humidity expressed in other ways. Later in this chapter and in

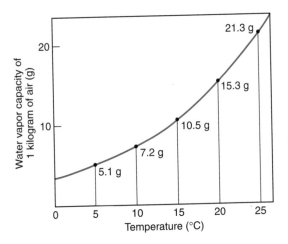

Figure 19-4 When air temperature increases, its ability to hold water vapor also increases. This graph does not show the water vapor content of air. It shows the ability of air to hold water vapor.

the next chapter, you will learn more about the importance of atmospheric humidity. Water vapor is the most important reservoir of energy in the atmosphere, and it plays a critical role in weather changes and violent weather.

There are a variety of instruments to measure humidity. Some instruments use materials that expand or contract when they are exposed to moisture. In Chapter 20 you will learn about how humidity is determined with a psychrometer and the role of humidity in cloud formation.

 ## Air Pressure

You may be familiar with the concept of pressure as it relates to your daily life. For example, bicycle and automobile tires require a specific pressure in order to work properly. If you swim underwater, you may feel extra pressure when you dive deep below the water's surface. *Pressure* is generally defined as force per unit area.

Although you may think air is light, Earth's atmosphere is so large that it has a total mass of 5,000 million tons. Gravity acting on the atmosphere exerts a weight of approximately 14.7 pounds on each square inch of Earth's surface. (See Figure 19-5 on page 452.) We seldom notice changes in air pressure because it varies within a small range near Earth's surface.

 ## Measuring Air Pressure

Air pressure is measured with an instrument called a **barometer**. There are two kinds of barometers, as shown in Figure 19-6 on page 452. Air pressure pushing on the surface of a dish of mercury can push a column of mercury into a tube to a height of nearly 30 inches. (Mercury is a dense metal that is a liquid at normal temperatures.) If the air pressure is high, the mercury is pushed a little higher in the tube. If it is low, the column of mercury does not rise as high. For this reason, atmospheric pressure is sometimes measured in inches of mercury.

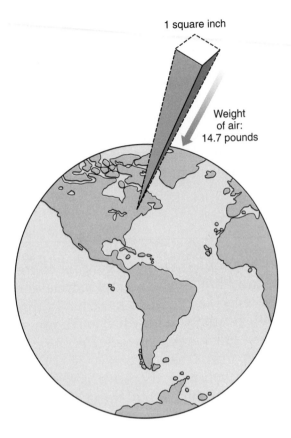

Figure 19-5 Air pressure is caused by the weight of the atmosphere. Above each square inch of Earth's surface at sea level is a column of air that weighs 14.7 pounds.

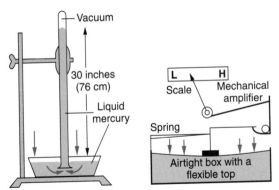

Figure 19-6 In a simple mercury barometer, the force of atmospheric pressure balances the weight of a column of mercury. As air pressure changes, so does the height of the mercury in the tube. For this reason air pressure is sometimes measured in inches of mercury. An aneroid barometer has an airtight box that expands and contracts in response to changes in air pressure. That motion is linked to a pointer on a scale.

The second kind of barometer, an aneroid barometer, contains a flexible airtight box. The box expands or contracts depending on changes in air pressure. This motion is transferred to an indicator dial that is calibrated in units of atmospheric pressure. Figure 19-7 is a scale to convert units of pressure from millibars to inches of mercury. For example, normal atmospheric pressure is about 1013.2 millibars, which is equal to 29.92 inches of mercury.

Air pressure is something we seldom notice. Why do meteorologists usually include it in their weather reports? In Chapter 22, you will learn how differences in air pressure cause winds. Atmospheric pressure (barometric pressure) can also be used to predict weather changes. Fair weather is often associated with rising barometric pressure. Falling air pressure can mean that a storm is approaching.

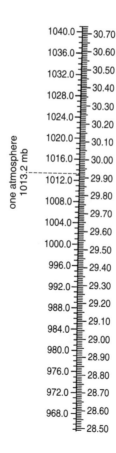

Figure 19-7 This scale can be used to convert readings between millibars and inches of mercury.

ACTIVITY 19-2 MAKING A BAROMETER

You can make a simple barometer using a wide-mouth jar, a balloon, and a long toothpick or splint. Cut the neck off the balloon. Stretch the balloon over the mouth of the jar, forming an airtight seal. Air in the jar will expand or contract in response to changes in atmospheric pressure. This will cause the balloon to bend up or down. The toothpick, or a similar splint of wood, can be used as an indicator, Glue the toothpick or splint to the balloon near the center of the mouth of the jar, allowing it to extend over the edge of the opening. Compare the movement of the toothpick/splint with the changes in barometric pressure listed in the local weather report.

 ## Wind

Wind is the flow of air along, or parallel to, Earth's surface. Winds help to transfer energy in the atmosphere by mixing warm tropical air with cold polar air. In this way, winds help to moderate temperatures on our planet. Usually, the wind is gentle. However, in major storms winds may be in excess of 160 km/hour (100 miles/hour) and do great damage.

MEASURING WIND Wind is a vector quantity. This means that, like ocean currents, winds have a speed and a direction. Therefore, two quantities are needed to describe winds. For example, the wind might be from the southwest at 8 km/hour. Wind speed is usually measured with an anemometer, which has a cup assembly that catches the wind, causing the assembly to rotate. The speed at which it rotates is usually transferred to an indicator on a calibrated dial. Anemometers are often combined with a wind vane, which points into the wind, indicating the wind direction. Wind is always labeled by the direction it comes from, not the direction it is moving toward. Figure 19-8 shows a combination anemometer and wind vane.

Figure 19-8 Wind speed is measured with an anemometer. The speed of the wind is measured by how fast the cup assembly rotates. Wind direction is determined with a wind vane, such as the one attached to this anemometer. The wind vane points to the direction the wind comes from.

ACTIVITY 19-3 | MEASURING WIND

Devise and apply a procedure that uses a scrap of tissue paper, a stopwatch, and a meterstick to measure the speed of the wind. You can find the wind direction with a magnetic compass. Remember that the wind direction is the direction from which the wind is coming.

Take measurements every five minutes. How much does the speed of the wind change from measurement to measurement?

ACTIVITY 19-4 | MAKING A WIND GAUGE AND WIND VANE

You can make a simple wind gauge by wrapping one end of a piece of paper or thin cardboard around a pencil or coat hanger wire, letting the other end fall free. Wind will move the free end of the paper at an angle. If a protractor is attached, the angle can be related to the wind speed. To make the device work better in strong breezes, a paper clip or a coin can be added as a weight as shown in Figure 19-9.

You can make a wind vane from a cardboard cutout mounted on a vertical pivot as shown. The larger surface catching the wind should be at the tail end of the arrow.

Cloud Cover

Clouds affect us by determining the amount of sunlight that reaches Earth's surface. The amount of cloud cover and types of clouds can help predict the likelihood of precipitation. Cloud cover is often expressed as the percent of the sky that is covered. If clouds cover one-quarter of the sky, there is 25 percent cloud cover. Very low clouds that reach ground level are known as **fog**. Fog can cause problems with visibility and therefore affects transportation. Cloud formation and its role in weather development will be discussed in Chapter 20.

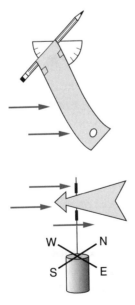

Figure 19-9 Wind gauge and wind vane.

Precipitation

Clouds are made of tiny water droplets or ice crystals that are too small to fall through the air. In this way, they are like sedimentary particles suspended in water. If these droplets or crystals grow large enough, they can fall from the air as **precipitation**. A raindrop can be a million times the size of a cloud droplet. Figure 19-10 shows a comparison of cloud and raindrops.

Most clouds are high enough in the atmosphere that the temperature is below freezing. Even when cloud droplets are below the freezing, they can remain in the liquid state. However, if it is cold enough, separate snow crystals form and grow much faster than cloud droplets. The snow crystals also must become large enough to fall as precipitation. In warm weather, the snow melts as it falls into warmer air, changing to rain before it reaches the ground.

OTHER FORMS OF PRECIPITATION While rain and snow are the most common forms of precipitation, there are several others. *Drizzle* is made up of very small raindrops that fall more

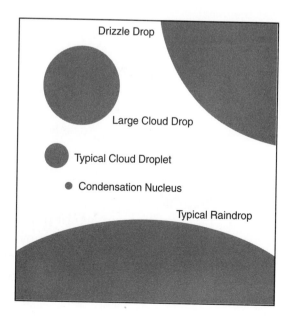

Figure 19-10 Raindrops are much larger than cloud droplets. That is why they fall as precipitation. These droplets are actually about 50 times smaller than they are shown in this diagram.

Drizzle Drop

Large Cloud Drop

Typical Cloud Droplet

Condensation Nucleus

Typical Raindrop

slowly than normal rain. *Sleet* is a partly frozen mixture of rain and snow that occurs when the temperature is close to freezing. *Freezing rain* is precipitation that falls as a liquid, but freezes, forming a coating of ice on exposed surfaces.

Hail is precipitation in the form of ice that usually occurs in violent thunderstorms. Hailstones usually begin as snowflakes that start to melt and then gather a coating of moisture as they fall. Rapid updrafts in the storm repeatedly blow the precipitation back up into colder air where the coating of moisture freezes. Hailstones can be water-coated and refrozen many times before they become heavy enough to fall through the updrafts and reach the ground. Although hailstones are usually less than a centimeter in diameter, hail the size of baseballs has fallen during especially severe thunderstorms in the American Midwest.

Rain is measured with a rain gauge. A cylindrical container with straight sides that is open at the top can be used as a simple rain gauge. The amount of rain is measured as a depth over the surface of the land. It can be reported in inches, centimeters, or any other convenient units of length. Snow can be measured as a depth of accumulation, or it can be melted to find the equivalent quantity of rain.

ACTIVITY 19-5 | EXTREMES OF WEATHER

Use the Internet and/or printed resources to research the world's record (high and low) temperature, air pressure, wind, and precipitation. Plot the locations where these records occurred on a world map.

ACTIVITY 19-6 | RECORDING WEATHER VARIABLES

If you make or purchase basic weather instruments (thermometer and barometer or hygrometer), you can record daily weather conditions. You may want to graph your observations through time. What patterns do you see in weather changes? How do the variables relate to one another? In Chapter 22, you will learn how meteorologists use weather maps and computer analysis to make predictions of future weather. However, you can use your readings to make your own predictions. Future measurements will enable you to check your own predictions.

Placement of Weather Instruments

When weather observers record atmospheric conditions, measurements must be taken carefully. For example, a temperature taken in direct sunlight will not be accurate because the liquid in the thermometer is heated by sunlight. For this reason, official thermometers and psychrometers are placed in a weather shelter such as the one shown in Figure 19-11. Precipitation should be recorded where buildings neither block nor add to the readings. Snow depth can be particularly difficult to measure, especially when wind moves snow from open areas, forming drifts where the wind is blocked. Wind should be measured in an open, flat area where it is not affected by buildings or other structures.

Figure 19-11 Official measurements of temperature and humidity are usually taken in a weather shelter. The shelter protects the instruments from direct sunlight; the openings in the shelter allow air to circulate.

ACTIVITY 19-7	VISIT A WEATHER STATION

Most professional weather stations welcome visitors by appointment. The meteorologists at the weather station can provide interesting information about local weather and how it is measured. Report on what a professional meteorologists does on the job.

HOW DOES THE SUN WARM EARTH?

In Chapter 8, you learned that changes in the solid Earth are driven by heat energy. That energy comes from Earth's formation billions of years ago and from radioactive decay of elements underground. Changes in the atmosphere, or weather, also require energy.

The Sun's Energy

Sunlight is the primary source of atmospheric energy. For many years, scientists tried to understand how the sun could create so much energy and never appear to change. The same discoveries that led to radiometric dating of rocks and fossils also helped scientists understand solar energy.

The most abundant substance in the sun is the element hydrogen. Under the extreme conditions of heat and pressure within the sun, hydrogen nuclei join in a process known as **nuclear fusion**. When four hydrogen nuclei transform into a helium nucleus, a large amount of energy is released. The amount of energy is so great that humans have been trying to use this process to generate electrical power. Unfortunately, it is very difficult to sustain the extremes of temperature and pressure needed to maintain fusion in laboratories or in power plants. (Nuclear fission, in which large nuclei such as uranium break apart into smaller nuclei, has been used to provide energy for nearly half a century.)

The energy created by fusion moves to the surface of the sun where it radiates into space as sunlight. **Radiation** is the transfer of heat energy in the form of electromagnetic waves. It is the only way that energy can travel through the vacuum of space. Therefore, radiation is the only way heat can travel from the sun to Earth. Radiation is very fast. The energy travels at the speed of light. In fact, it takes only 8 minutes for solar energy to reach Earth. Although Earth intercepts only a tiny portion of the energy given off by the sun, this energy powers hurricanes, tornadoes, and all other dynamics of the atmosphere.

 Electromagnetic Waves

Electromagnetic energy includes a wide range of radiant energy. This energy travels as waves similar to the waves on a lake or an ocean. You may have been surprised to learn in Chapter 16 that waves on water do not carry the water with them. Like electromagnetic waves, waves on water carry energy. However, we are not able to see electromagnetic waves in the same way that we observe waves on water. Still, they have similar measurable properties.

One property common to all energy waves is frequency. **Frequency** is a measure of how many waves pass in a given period of time. Visible light has a frequency of about 600 trillion cycles per second. A second property is amplitude. In waves on water, amplitude is the height of the wave. For electromagnetic waves, amplitude is a measure of their strength or brightness. The third measure is wavelength. This is the distance from the top of one wave to the top of the following wave. Figure 19-12 on page 462 illustrates these properties for waves on water.

Figure 19-13 on page 462, from the *Earth Science Reference Tables*, shows the range of wavelengths of electromagnetic energy. Only a narrow band of this energy is visible as light. Although we cannot see other wavelengths of electromagnetic energy, they are very important to us. Gamma rays and X rays are the forms of electromagnetic energy with the short-

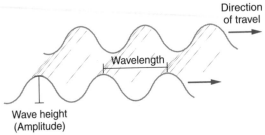

Figure 19-12 All waves carry energy. This diagram of waves on water illustrates two characteristics common to all waves: wavelength and wave height (or amplitude). Frequency is the number of waves that pass in a given length of time.

est wavelengths. These highly penetrating radiations have applications in industry and medicine.

Ultraviolet radiation is longer in wavelength than gamma rays and X rays, but the wavelength is still too short for the energy to be visible. Most of the sun's energy reaches Earth's surface as visible light, which warms our planet and enables us to see. On the other side of the wavelength scale is invisible long-wave radiation. This includes infrared and microwaves, which can heat matter. Radio waves are used in communications such as radio and television.

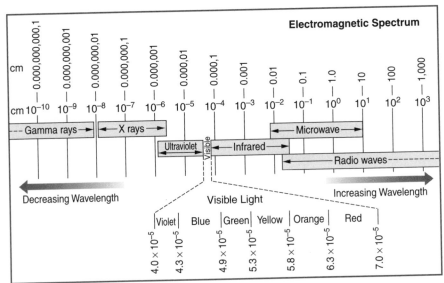

Figure 19-13 A narrow band of electromagnetic energy is visible to us as light. However, other wavelengths of radiant energy are important in natural processes and scientific applications.

Ozone and the Sun's Energy

Earth's atmosphere is not transparent to all forms of electro-magnetic energy. High-energy gamma rays and X rays are stopped by Earth's blanket of air. Although some ultraviolet rays reach Earth's surface, a form of oxygen known as ozone shields us from most of them. Normal oxygen has two atoms of oxygen in each molecule (O_2). When oxygen drifts into the upper atmosphere, the energy of ultraviolet radiation converts oxygen to a three-atom molecule called ozone (O_3). This process absorbs ultraviolet energy.

Oxygen was not present in Earth's early atmosphere. The beginning of photosynthesis in the mid-Precambrian added oxygen to the atmosphere. Oxygen then formed a layer of ozone in the upper atmosphere. Until the creation of the ozone layer, life inhabited only the oceans where it was protected from harmful ultraviolet radiation. The protection provided by ozone enabled the evolution of terrestrial forms of life, including humans. Even with the protection of Earth's ozone layer, too much exposure to ultraviolet radiation can lead to sunburn. This painful condition can cause aging of the skin and skin cancer. Figure 19-14 shows the range of energy wavelengths in sunlight that penetrates to Earth's surface.

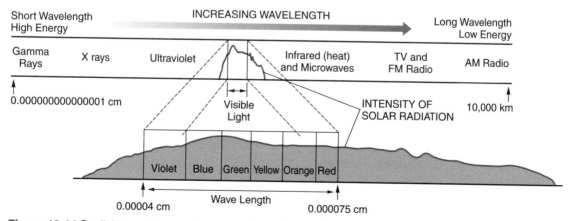

Figure 19-14 Sunlight contains a wide range of wavelengths. Most of the short-wave radiation is absorbed by Earth's atmosphere. This curve shows wavelengths that reach Earth's surface. Insolation is about 10% ultraviolet, 40% visible light, and 50% infrared. The visible wavelengths are the only part humans can see.

In the late twentieth century, scientists studying the upper atmosphere discovered that the concentration of ozone in the upper atmosphere was decreasing rapidly. Further investigation revealed that a group of chemicals used in spray cans and air conditioners called CFCs (chlorofluorocarbons) was drifting into the upper atmosphere. CFCs cause ozone (O_3) to convert rapidly to oxygen (O_2), which does not protect us from ultraviolet radiation. Fortunately, compounds that do not destroy ozone are being substituted for CFCs in the industrial nations. This is one technological threat to our environment that appears to be yielding to international efforts and advancing technology.

Insolation and Earth

Solar energy that reaches Earth is known as **insolation**. This word can be thought of as a shortened form of the phrase *in*coming *sol*ar radi*ation*. As sunlight travels deeper into the atmosphere, its speed decreases. This decrease causes the light to bend toward the ground. The bending of light as it moves from one substance to another of different density is known as **refraction**.

ACTIVITY 19-8 | OBSERVING REFRACTION

Place a long pencil or any straight object of similar size into a transparent container filled with water. If you look along the pencil, you can observe how it appears to bend at the surface of the water. Viewed from the side, the pencil may appear in two straight segments. Both observations are caused by the refraction of light rays.

Solar energy that reaches the ground can be reflected or absorbed. Earth absorbs about half the insolation that reaches the ground, which raises the ground's temperature. Radiation that is not absorbed is reflected. **Reflection** is the process by which light bounces off a surface or a material. (Some objects such as clouds have no distinct surface, but they

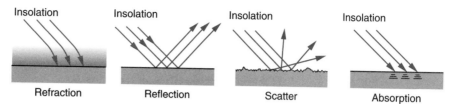

Figure 19-15 Sunlight reaching Earth is affected in four ways. It is refracted (bent) as it travels through the atmosphere. It is reflected by smooth surfaces or scattered by rough surfaces. Dark-colored surfaces absorb most of the light that strikes them.

reflect light.) Figure 19-15 shows how Earth's atmosphere affects insolation.

Whether light is reflected or absorbed depends on the properties of the surface it strikes. Dark surfaces absorb more insolation than light-colored surfaces. After all, dark surfaces appear dark because relatively little light bounces off them to reach your eyes. In addition, rough surfaces absorb even more light. When light reflects off a rough surface, it is reflected in many different directions as shown in Figure 19-16. This is called **scatter**. This diagram also shows that some of the reflected light strikes the surface more than once. The result is more opportunities for the surface to absorb energy.

What portion of the energy that Earth receives from the sun is absorbed and what part is lost by reflection? Figure 19-17 on page 466 shows that about half of the insolation energy reaching Earth is absorbed by the ground. Nearly one-third of it is reflected back into space. The rest is absorbed within the atmosphere.

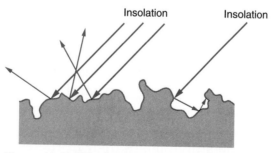

Figure 19-16 This diagram is a magnification of the rough surface in Figure 19-15 Parallel rays of light are reflected in different directions by an irregular surface. In addition, multiple reflections of the same light ray allow rough surfaces to absorb more energy than smooth surfaces.

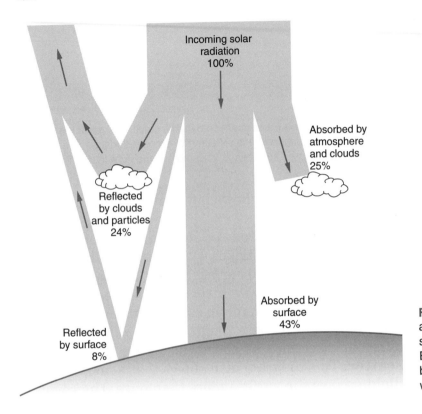

Incoming solar radiation 100%

Absorbed by atmosphere and clouds 25%

Reflected by clouds and particles 24%

Absorbed by surface 43%

Reflected by surface 8%

Figure 19-17 The ground absorbs nearly half of the solar energy that reaches Earth. The rest is reflected back into space or absorbed within the atmosphere.

 Angle of Insolation

The strength of sunlight depends on the sun's position in the sky. The angle between Earth's surface and incoming rays of sunlight is known as the **angle of insolation**. The angle of insolation is also the angle of the sun above the horizon. If the sun is directly overhead, the angle of insolation is 90°. At sunrise and sunset, when the sun appears along the horizon, the angle of insolation is near 0°. The angle of the sun changes with time of day and with the seasons. When the sun appears highest in the sky, as it is around 12 noon and early in the summer, the strength of sunlight is greatest. There are several reasons for this.

When the sun appears high in the sky, light rays spread over a smaller area as they reach Earth's surface. A given amount of energy confined to a smaller area results in sunlight that is more intense. Figure 19-18 shows that light is

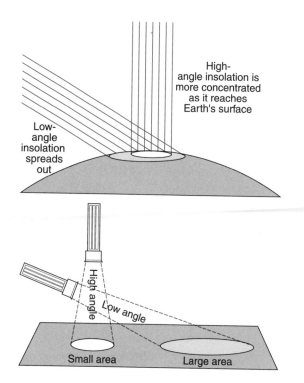

Figure 19-18 When the sun is directly overhead, its rays are concentrated in a small area. When the sun is low in the sky, the same energy covers more area, so the sunlight is not as strong.

stronger when the source shines straight down on a surface. For any location on Earth's surface, sunlight is most intense at noon on the first day of summer. You should note that for any location in New York State, although the angle of insolation is highest at noon on the first day of summer, even at that time the sun is not directly overhead. Only within tropical latitudes can the angle of insolation be 90°.

The second reason that sunlight is stronger when the sun appears high in the sky is related to absorption of energy by the atmosphere. Earth's atmosphere is not completely transparent, even to visible light. The lower the sun appears in the sky, the more atmosphere light must pass through to reach Earth's surface. Red light penetrates the atmosphere better than blue light. That is why the sun appears red when it is rising and setting, but not when it is high in the sky.

Furthermore, reflection increases when the sun appears low in the sky. You may have noticed that if you look into a still lake when the sun nearly overhead, you can see deep into the water. If the sun appears low in the sky, reflected light

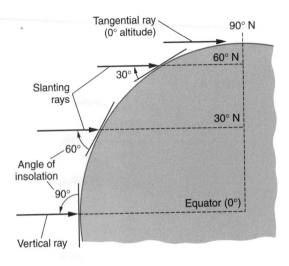

Figure 19-19 The angle of the noon sun in the sky depends on latitude. As you move out of the tropics, the noon sun appears lower in the sky. The closer the sun appears to the horizon, the weaker sunlight becomes. This diagram shows Earth on the first day or spring or the first day of autumn.

bounces off the lake's surface and prevents you from seeing under the surface. Most surfaces absorb more light when the angle of insolation is greater.

In addition to the time of day and the season, the intensity of insolation at Earth's surface also depends on latitude. In the tropics, the noon sun always appears high in the sky, so sunlight is relatively strong. However, near the poles the sun never appears far above the horizon. Polar regions receive less solar energy because the sun's rays spread over a larger area and the light has to pass through more atmosphere. Figure 19-19 shows how the angle of the sun depends on latitude. The angles on this diagram represent noon on the first day of spring or the first day of autumn. Throughout the year, the tropical latitudes have the greatest angle of insolation.

 ## Duration of Insolation

The longer you leave a container under an open water tap, the more water flows into the container. The same principle holds true for solar energy. In 1 year, every place on Earth has a total of 6 months of daylight and 6 months when the sun is below the horizon. Tropical regions have 12 hours of daylight every day throughout the year. For this reason, tropical climates are sometimes called the seasonless climates. Neither

LAB 19-1: Angles of Insolation

Materials: clip lamp with metal shade, three thermometers, protractor, black construction paper, tape, timer or clock

Problem: How does temperature depend on the angle of insolation?

Procedure:

1. Use the construction paper to make three square sleeves about 2 cm on a side. If tape is used to hold the sleeves together, leave one side of the sleeve exposed as black paper.

2. Cover the bulb of each thermometer with a sleeve and place the thermometers about 20 cm from the light source as shown in Figure 19-20. The surface of one sleeve should be angled at 90°, another at 60°, and the third at 30° to a line from the light-bulb to the thermometer bulb.

3. Record the starting temperature on each thermometer and turn on the

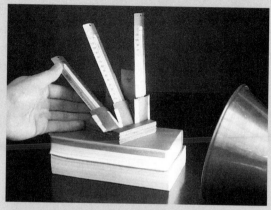

Figure 19-20 Angles of insolation.

lamp. Record the temperature on each thermometer once a minute for 15 minutes.

4. Plot your data as three lines on the same graph. (Put time on the horizontal axis.)

What does the graph indicate about the relationship between the angle of insolation and temperature?

the hours of daylight nor the temperature changes very much throughout the year.

At higher latitudes, such as New York State, days are longer in summer and shorter in the winter. This is true both north and south of the equator. Therefore, these areas receive more solar energy in the summer because the sun appears higher in the sky and also because the sun is visible in the sky for a longer period of time. These changes are the cause of the seasons.

Duration refers to a length of time. Therefore, the **duration of insolation** is the amount of time the sun is visible in the sky, or the number of hours from sunrise to sunset. The greatest range of day length occurs near the poles. The poles experience 6 months of daylight followed by 6 months when the sun is below the horizon. Although the poles experience the longest duration of insolation, the angle of insolation is always low. Therefore, although summer sunlight is constant, it is too weak to produce the warm temperatures that occur in the tropics.

Earth's atmosphere actually extends the amount of time that the sun is visible in the sky. Suppose you lived on a very small island in the middle of the ocean. At sunrise, you could observe when the center of the sun's disk rises to the level of the ocean in the east. You could then measure with great accuracy how long it took the sun to cross the sky and set when its center was level with the western horizon. Performing this procedure at the equinox (the first day of spring or the first day of autumn) you might expect it to take exactly 12 hours. However, it is actually a little longer than 12 hours. Refraction of sunlight in the atmosphere allows the sun to be visible after it has dipped below the horizon.

HOW DOES SOLAR ENERGY CIRCULATE OVER EARTH?

Heat energy can move from one place to another in three ways. You already learned that the only way solar energy can reach Earth is by radiation through space. Although radiation is very fast (300 million meters per second) it can occur only through empty space or through transparent materials. Once solar energy reaches Earth, it is distributed over the planet's surface by conduction and convection.

 Conduction

Earlier in this chapter you learned that when molecules absorb energy and their temperature rises, the molecules move more rapidly. When molecules come into contact, this energy

LAB 19-2: Conduction

Materials: two calorimeters with perforated tops, metal bar bent at each end, two Celsius thermometers, timer or clock, hot and cold tap water, graph paper

Problem: To determine the direction of heat flow

Procedure:

1. Fill one calorimeter $\frac{3}{4}$ full with hot tap water. Fill the other calorimeter $\frac{3}{4}$ full with cold water.

2. Insert a thermometer and one end of the metal bar through each cover and into the calorimeters as shown in Figure 19-21.

3. Record the initial water temperatures and repeat these measurements every minute for 20 minutes.

4. Graph the data as two lines on a single graph. (Label temperature on the

Figure 19-21

vertical axis and time on the horizontal axis.)

What was the direction of heat flow? What changes could be made in this set-up to increase the rate of energy flow? (Please explain as many ways as you can to alter the apparatus in order to maximize conduction.)

is passed along from molecule to molecule. The transfer of energy by collisions of molecules is known as **conduction**. Conduction occurs most efficiently in solids. Heat energy is conducted within Earth as well as between the solid Earth and the atmosphere.

Some substances such as metals are good conductors. We use metal containers such as frying pans to cook our food because heat is conducted readily through the metal from the heat source to the food. On the other hand, you do not want to burn yourself when you hold the frying pan. So, the handle of a frying pan is made from a poor conductor, an insulator. Wood and plastic are good insulators.

Conduction can be an effective way for heat to flow between two materials if they are in contact. The atmosphere exchanges some heat energy with Earth's surface by conduction. However, conduction is slow, and the atmosphere is not a good conductor of heat energy.

Convection

The atmosphere distributes energy over Earth's surface mostly by a third form of heat flow. You have learned that all matter above a temperature of absolute zero (0 K) has energy. When matter moves, the energy it contains goes with it. Because the atmosphere is made up of gases, it can flow freely from one place to another, carrying its energy with it. Like liquids, gases are considered **fluids** because they can flow from place to place.

What makes air flow? When most substances are heated they expand and become less dense. Air is no exception. Under the influence of gravity, cooler air, which is more dense, flows under warmer air, replacing it. Therefore **convection** can be defined as a form of heat flow that moves both matter and energy as density currents under the influence of gravity. You will learn more about how convection distributes energy over Earth in Chapter 21.

Without heat flow by convection, differences in temperature over Earth's surface would be much greater than they are now. The oceans and the atmosphere distribute energy by convection currents. Energy is carried from tropical regions toward the poles. Cold fluids (water and air) move from the polar regions to the tropics where they can absorb solar energy. This cycle tends to even out the extremes in temperature on the planet.

Earth Loses Energy

If Earth is always absorbing energy from the sun, why is the planet's temperature relatively constant? Insolation should cause Earth to become warmer, unless somehow the planet loses as much energy as it absorbs. If the inflow of energy is equal to the outflow of energy, the energy of the object is in *equilibrium*.

Earth cannot lose energy by conduction because no other object in space is in contact with Earth. Heat loss by convection would require the flow of matter into space. However, very little matter escapes from Earth. We do not see radiation leaving the planet because Earth does not give off visible light like the sun. So how does Earth give off energy to maintain an energy balance?

You learned that most forms of electromagnetic energy are not visible. Experiments show that all objects above a temperature of absolute zero radiate electromagnetic energy. The form of energy radiated depends on the temperature of the object. The sun is hot enough to give off light energy. Earth is much cooler than the sun, so it must radiate energy in long wavelengths. In fact, the planet loses its energy as heat in the infrared portion of the spectrum. If our eyes were sensitive to infrared (like some military devices used to detect warm objects), we would see the planet glow with a constant flow of escaping heat.

There have been times in Earth's history when it was not in energy equilibrium. The ice ages occurred when the planet radiated more energy than it received from the sun. This imbalance led to colder temperatures and the growth of glaciers. Since the glaciers melted, Earth's average temperature has risen. The end of the ice ages occurred when the planet absorbed more solar energy than it radiated into space. Imbalances in the flow of heat energy are also responsible for daily and seasonal cycles of change in temperature.

 ## The Greenhouse Effect

Just as the atmosphere is not transparent to all forms of solar energy, it is only partly transparent to heat energy given off by the ground. Carbon dioxide and water vapor absorb infrared (heat) radiation. These gases absorb enough energy to elevate the kinetic energy of the atmosphere to a temperature at which the energy escaping from Earth equals

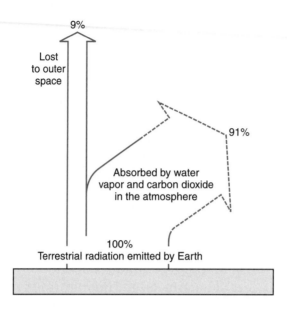

Figure 19-22 Humans are unable to see infrared (heat) radiation given off by Earth. Most of the energy is temporarily trapped in the atmosphere by carbon dioxide and water vapor.

the energy absorbed from insolation. Figure 19-22 illustrates what happens to the heat radiation given off by Earth.

The process by which carbon dioxide and water vapor in the atmosphere absorb infrared radiation is known as the **greenhouse effect**. Perhaps you have seen a greenhouse used to grow flowers or vegetables in a cold location. It may seem strange that a transparent glass building can support plant growth when the greenhouse has little or no heating and the ground may be covered with snow. To understand how a greenhouse works, you need to know that short-wave visible sunlight passes through the glass and heats the inside of the greenhouse. The warm objects in the greenhouse give off long-wave infrared radiation, and glass is not transparent to infrared radiation. Thus these rays are reflected back into the greenhouse. By letting short-wave sunlight enter and not allowing the longer heat rays to escape, the greenhouse traps heat energy.

The atmosphere affects Earth in a similar way. However, Earth's atmosphere absorbs and reradiates energy rather than simply reflecting it. Figure 19-23 illustrates the greenhouse effect. Although carbon dioxide and water vapor are the principal greenhouse gases, they are not the only greenhouse

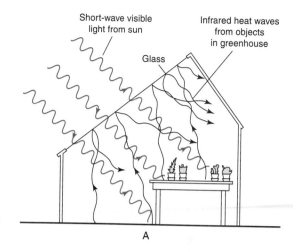

Figure 19-23 Part A. The glass of a greenhouse allows visible light to enter and heat objects inside the greenhouse. The heated objects give off infrared (heat) radiation, which cannot pass through the glass. The trapped energy warms the inside the greenhouse.
Part B. Like a greenhouse, Earth's atmosphere is transparent to visible light. However, the atmosphere absorbs long-wave terrestrial radiation, elevating the temperature of the planet.

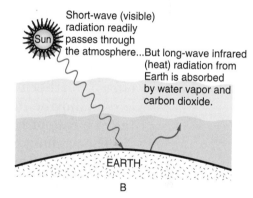

gases. Methane, nitrous oxide, and several other gases contribute to the greenhouse effect. Scientists estimate that the combined effect of these gases is to make Earth's surface about 35°C warmer than it would be without them. If it were not for the greenhouse effect, Earth would be too cold for life as we know it.

 Global Warming

The carbon dioxide content of Earth's atmosphere is only a small fraction of 1 percent. However, of all Earth's greenhouse gases, carbon dioxide worries scientists most. Carbon dioxide

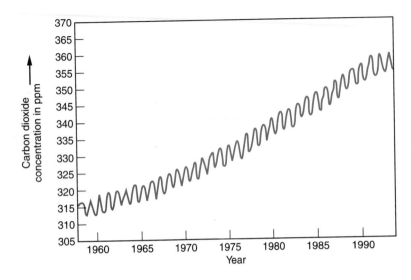

Figure 19-24 An increase in the carbon dioxide content of Earth's atmosphere has been measured over the past half century. The primary cause is the burning of fossil fuels. Yearly cycles resulting from seasonal changes in plant growth are also evident.

alone is responsible for more than half of Earth's greenhouse effect. Furthermore, measurements of the concentration of carbon dioxide in Earth's atmosphere show that it is increasing at a rate of 1 percent every 2 years. (See Figure 19-24.)

The primary cause for the increase in carbon dioxide is the burning of carbon-based fossil fuels, such as gasoline, oil, and coal. At the same time, Earth's forests are being destroyed to supply wood and clear agricultural land. As a result, the ability of forests to absorb carbon dioxide is being reduced. At the current rate of change, the carbon dioxide content of the atmosphere could double by the end of the next century.

The addition of greenhouse gases to the atmosphere could result in an increase of several degrees Celsius in Earth's average temperature. Although this may not sound like a dramatic change, this is comparable to the average global temperature change that has taken place since the end of the last ice age. The predicted rise in Earth's average temperature over the next century or more is called **global warming**.

How would global warming affect humans? Atmospheric circulation is a very complex system. The specific effects for any particular location are difficult to predict. Some locations that are now too cold for agriculture might become important food producers. However, other regions that are now productive could become too hot or too dry for efficient crop growth. Melting of glaciers and polar ice has also been observed. This could raise sea level enough to cause flooding and increased storm damage in coastal locations, including many of the world's largest cities.

The United Nations has convened international conferences to discuss global warming. Most industrialized nations have adopted the goal of stabilizing or even reducing the production of greenhouse gases. However, the United States has not. Our government has pointed to uncertainties about global warming. It is not clear whether global warming is taking place or the temperature change is related to natural cycles of temperature. Our government has been unwilling to accept limitations on economic growth when the benefits are uncertain. The global warming debate between environmental and economic interests seems likely to continue for many decades.

A LESSON FROM VENUS The surface of Venus is hot enough to melt lead. Why is Venus so hot? It is not just because it is closer to the sun than Earth. The surface of Venus is actually hotter than the hottest places on the surface of Mercury, even through Mercury is much closer to the sun than Venus. The atmosphere of Venus is 96 percent carbon dioxide and 100 times as thick as Earth's atmosphere. Some planetary scientists call Venus the planet with "the greenhouse effect gone wild."

TERMS TO KNOW

angle of insolation	global warming	relative humidity
barometer	greenhouse effect	saturated
climate	humidity	scatter
duration of insolation	insolation	temperature
fluid	nuclear fusion	thermometer
fog	precipitation	weather
frequency	radiation	

CHAPTER REVIEW QUESTIONS

1. Which instrument makes use of the fact that most liquids expand when they are heated?

   (1) thermometer
   (2) barometer
   (3) anemometer
   (4) rain gauge

2. What is the temperature at which water freezes on the Kelvin scale?

   (1) 0
   (2) 32
   (3) 100
   (4) 273

3. If atmospheric pressure is 30.40 inches of mercury, what is the air pressure in millibars (mb)?

   (1) 1016 mb
   (2) 1028.3 mb
   (3) 1029 mb
   (4) 1029.5 mb

4. Which list contains electromagnetic energy arranged in order from longest to shortest wavelength?

   (1) gamma rays, ultraviolet rays, visible light, X rays
   (2) radio waves, infrared rays, visible light, ultraviolet rays
   (3) X rays, infrared rays, blue light, gamma rays
   (4) infrared rays, radio waves, blue light, red light

5. Which form of electromagnetic energy has a shorter wavelength than red light?

   (1) radio waves
   (2) microwaves
   (3) infrared radiation
   (4) ultraviolet radiation

6. What prevented life from developing on land before Earth's atmosphere contained a significant portion of oxygen?

   (1) All land locations were too hot for life to exist.
   (2) All terrestrial environments were too cold for life to exist.
   (3) All forms of life require atmospheric oxygen for their vital processes.
   (4) The land received too much harmful solar radiation.

7. In recent years, scientists have been concerned about a decrease in the ozone content of Earth's upper atmosphere. What is the principle threat to humans if this decrease in ozone continues?

   (1) decreasing precipitation leading to crop failure and possible human starvation
   (2) a drop in surface temperatures causing the growth of continental glaciers
   (3) increasing oxygen content of the lower atmosphere causing forest fires
   (4) an increase in dangerous radiation that reaches Earth's surface

8. Which characteristic of a building material would provide the most energy-absorbing exterior covering for a house?

   (1) dark-colored and smooth textured
   (2) dark-colored and rough textured
   (3) light-colored and smooth textured
   (4) light-colored and rough textured

9. Compared with dull and rough rock surfaces, shiny and smooth rock surfaces are more likely to cause sunlight to be

   (1) reflected
   (2) refracted
   (3) scattered
   (4) absorbed

10. Which diagram best represents visible light rays after striking a dark, rough surface?

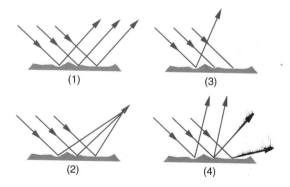

(1)   (3)

(2)   (4)

11. When does the sun heat the ground most strongly in New York State?

    (1) at noon on the first day of spring, about March 21
    (2) at noon on the first day of summer, about June 21
    (3) about 3 P.M. on the first day of spring, about March 21
    (4) about 3 P.M. on the first day of summer, about June 21

12. Base your answers to questions 12 and 13 on the diagram and table below, which represent a laboratory experiment. Hot water at 90°C is poured into cup A. Cool water at 20°C is poured into cup B. Styrofoam covers are placed on top of the cups. An aluminum bar and a thermometer are placed through the holes in each cover as shown. Points X and Y are on the aluminum bar. The data table shows temperature readings taken from the two thermometers each minute for a period of 20 minutes.

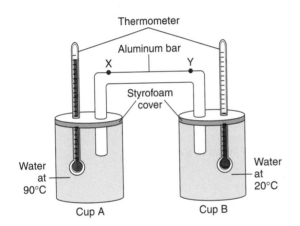

Minute	Temperature of Water (°C)	
	Cup A	**Cup B**
0	90	20
1	88	20
2	86	20
3	85	21
4	83	21
5	82	22
6	81	22
7	80	22
8	79	22
9	78	23
10	77	23
11	76	23
12	75	23
13	74	23
14	73	23
15	72	24
16	71	24
17	70	24
18	69	24
19	68	25
20	67	25

Which change in the experiment would increase the heating rate of the water in cup B?

(1) making the aluminum bar shorter between points X and Y
(2) making the aluminum bar longer between points X and Y
(3) keeping cup A covered, but uncovering cup B
(4) keeping cup B covered, but uncovering cup A

13. The rate of temperature change for the water in cup A for the first 10 minutes was approximately

(1) 0.77°C/minute   (3) 7.7°C/minute
(2) 1.3°C/minute   (4) 13°C/minute

14. How can we tell that an object is absorbing and releasing kinetic heat energy in equal amounts?

(1) There is no change in mass.
(2) There is no change in temperature.
(3) There is no change in volume.
(4) There is no change in physical dimensions.

15. Scientists have become concerned about how humans are causing changes in Earth's atmosphere. An increase in which gas is most likely to result in flooding of coastal locations around the world?

(1) oxygen   (3) nitrogen
(2) carbon dioxide   (4) argon

Open-Ended Questions

16. Explain what a meteorologist means when he reports that the relative humidity is 50 percent.

17. A class of Earth science students recorded the air temperature in a variety of locations around their school property on a sunny afternoon. The coolest location was in a grove of trees. The highest temperatures were over a blacktop parking lot. Explain how a characteristic of the parking lot surface could cause these higher temperatures?

18. A scientist collected hourly temperature readings at an outdoor location near her home. After a week of measurements she noticed that the tem-

perature usually increased between 10 A.M. and 12 noon. Explain why temperature readings normally increase between 10 A.M. and 12 noon.

19. Carbon dioxide is one of the greenhouse gases. These gases cause Earth's climate to be warmer than it would be if they were not present. How do the greenhouse gases elevate the average temperature of planet Earth?

20. Environmentalists say that we should "think globally but act locally." What can you do as an ordinary citizen to reduce the threat of global warming?

Chapter 20

Humidity, Clouds, and Atmospheric Energy

 LET IT SNOW!

Most students like "snow days." The opportunity to sleep late, have some free time, and even play in the snow can be a welcome change from one's usual school routine. The likelihood of winter storms leaving a thick blanket of snow depends on where you live. People who live near the eastern shore of Lake Erie and Lake Ontario receive more snow than any other locations in New York State.

The Tug Hill Plateau, east of Lake Ontario, holds the state record for snowfall. On January 11–12 of 1997, an observer for the National Weather Service measured a snowfall of more than 6 feet in 24 hours. This storm dumped an additional 2 feet of snow over a period of 4 days. Fortunately, the Tug Hill Plateau is one of the least populated areas of New York State and local residents have learned to cope with these monster storms.

The Buffalo area, near the eastern end of Lake Erie, also receives more than its share of the winter snow. A December 25, 2001, snowstorm in Buffalo practically shut down the city when 6 feet of snow fell over a 4-day period. Unlike the Tug

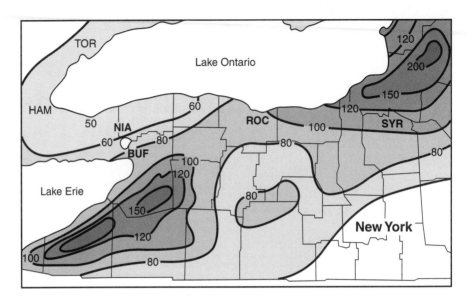

Figure 20-1 Two locations with the greatest average winter snowfall in New York State are located just east of Lakes Erie and Ontario.

Hill Plateau region, the Buffalo metropolitan area has a population of more than 1 million people. There was too much snow to push aside on many streets, so snow had to be transported out of the city in trucks. Fortunately, this storm was localized, and snow removal equipment was rushed in from other towns and cities in Western New York. Figure 20-1 shows the average annual snowfall in inches for parts of western New York State. Notice the two areas of maximum snowfall just east of each of the two Great Lakes.

Lake-Effect Storms

Why do these areas receive so much snow? Both locations are in the path of weather events known as **lake-effect storms**. These are mainly winter-weather events that result from a combination of atmospheric phenomena and geographic factors. The most common wind direction in New York State is from the west. Air traveling more than 320 km (200 miles) over the Great Lakes picks up large amounts of moisture. In the winter, lakes do not cool as quickly as the air, and generally remain free of ice cover. Therefore, in early winter, air blowing over the warm water readily absorbs moisture.

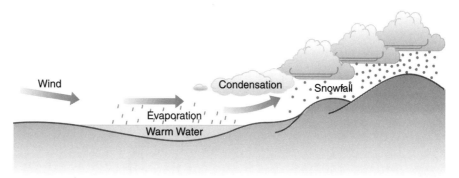

Figure 20-2 Lake-effect storms are most common in the winter when the lakes' water is warmer than the surrounding land areas. Sustained winds blowing over the Great Lakes absorb moisture. When the air is cooled over land, clouds and precipitation form. Additional cooling caused by the moist air rising over the hills increases precipitation.

However, as this moist air moves over land, it cools. When air cools, it loses its ability to hold water, causing clouds and precipitation. Precipitation increases when air rises over the hills at the ends of the lakes. If the westerly winds continue, the process is like a conveyor belt, picking up moisture from the lakes and dumping it on the land over a period of days. Figure 20-2 shows how lake-effect storms form.

| ACTIVITY 20-1 | RATE OF EVAPORATION |

A small sponge cube can be used to investigate the variables that affect evaporation. The cube should be mounted on a string or thread that has been drawn through the cube.

After allowing the cube to absorb water, excess water that might drip should be squeezed out. The mass of the wet cube should be measured with an electronic balance at the beginning and the end of a uniform evaporation time.

Each group will select a location for its cube where the cube will experience evaporation. Make sure cubes are placed throughout the classroom. Consider limitations such as requiring that cubes be placed at least a predetermined distance from electrical devices.

Evaporation is indicated by the percent of original mass lost. Nothing other than air should touch the wet cube, which must be gently handled by its string.

What factors were most important in maximizing evaporation?

HOW DOES THE ATMOSPHERE STORE ENERGY?

In Chapter 19, you learned that most of the atmosphere's energy comes from the sun. You also learned that insolation warms the air and causes increased motion of molecules. We can feel and measure this motion as temperature. However, kinetic heat energy is not the principal reservoir of atmospheric energy. The energy released by a thunderstorm can be greater than the energy of the first nuclear bombs dropped over Japan in 1945. It is clear that large quantities of energy must be stored in the atmosphere and released in weather events. That potential is stored in a special form of heat energy called latent heat.

Has anyone ever suggested that you have hidden talents? Most people have some kind of special ability that enables them to do something better than most other people. For some individuals, school brings out those abilities. For others, unfortunately, they remain hidden. If you have undiscovered abilities, you possess latent talents. Latent means hidden or not apparent. As you will soon learn, heat energy can also be latent, or hidden.

 States of Matter

Water, like other forms of matter, exists in three states: solid, liquid, and gas. Solid water is ice. Liquid water is simply called "water." In the form of a gas it is known as water vapor. The temperatures that occur on our planet allow water to exist in all three states in the natural surface environment.

The expression **phases of matter** is sometimes used as a synonym for states of matter.

When water changes among these states, a great deal of energy is involved. Heating that causes a change in temperature shows up as increased kinetic energy of molecules. However, energy that transforms solids to liquids and liquids to gases does not show up as a temperature change. In this sense, it is hidden, or latent energy. **Latent heat** is therefore energy that is absorbed or released when matter changes state.

A variety of units can be used measure energy. It is convenient to use calories. A **calorie** can be defined as the energy absorbed when the temperature of 1 gram of water increases 1 Celsius degree. That could be from 20 to 21°C or 86 to 87°C. One calorie is also released when 1 gram of water cools 1 Celsius degree.

Different substances absorb different amounts of energy as they change temperature. The reason is that they have different specific heats. **Specific heat** is the energy needed to raise the temperature of 1 gram of a substance 1 Celsius degree, without changing its state. The higher the specific heat of a substance, the more energy needed to raise its temperature. For example, it takes more energy to heat 1 gram of water (specific heat = 1 calorie/gram · °C) 1 degree Celsius than it takes to heat 1 gram of lead (specific heat = 0.03 calorie/gram · °C) the same amount. Figure 21-3 on page 523, from the *Earth Science Reference Tables*, lists the specific heats of some common materials.

Changes in State

For now, you will deal only with heating and cooling of water. Consider what happens to 1 gram of water that starts as ice at a temperature of −20°C and absorbs energy. Figure 20-3 on page 488 graphs the process. Because ice has a lower specific heat than liquid water, ice warms more readily. It takes only 10 calories to warm the ice to the melting temperature of 0°C. The ice stays the same temperature as it continues to absorb energy. In fact, it takes 80 calories to melt a single gram of ice with no change in temperature. That is why

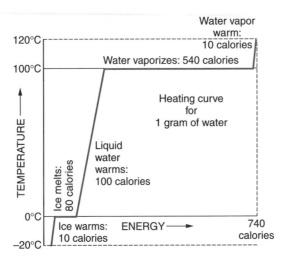

Figure 20-3 This graph is called a heating curve. It shows what happens as water absorbs energy, changing from ice at −20°C to water vapor at 120°C. Notice that most of the energy is used to change liquid water into water vapor with no change in temperature. The level parts of this graph indicate absorption of latent heat.

the name "latent" is applied to energy used for a change in state.

MELTING The change from a solid to a liquid is called **melting**. Once the ice has melted, its temperature can increase. The energy used to melt a gram ice at a constant temperature of 0°C (80 calories) is nearly as much energy as it takes to heat the gram of water from its melting temperature of 0°C to the boiling temperature of 100°C (100 calories).

VAPORIZATION The change in state from liquid to gas (vapor) at any temperature is **vaporization**. **Boiling**, a form of vaporization, is the change in state from a liquid to a gas at the boiling point of the substance. Of all its changes in state, vaporization of water requires the most energy: 540 calories. Again, during boiling there is no change in temperature. After all the water becomes vapor, the addition of a final 10 calories, heats the gram of water vapor to a temperature of 120°C. Of the total energy involved in these five changes, vaporization alone took 73 percent. The evaporation of water is the principle way that the atmosphere absorbs solar energy as shown in Figure 20-4.

CONDENSATION You have read that vaporization is the most important way that insolation adds energy to the atmosphere. How does the atmosphere release that energy to make wind and weather? The energy absorbed by the atmosphere in

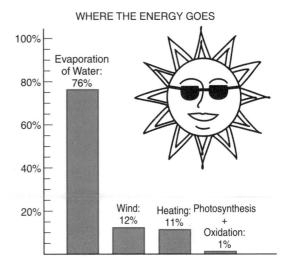

WHERE THE ENERGY GOES

Figure 20-4 More than three-quarters of the solar energy absorbed by Earth evaporates water into the atmosphere. Latent heat is the major reservoir of atmospheric energy.

phase changes is released when water returns to lower energy phases. For example, the 540 calories used to vaporize each gram of water is released when water vapor changes back to a liquid. **Condensation** is the change from a gas, such as water vapor, to a liquid, such as water.

Have you ever noticed that grass is sometimes wet on a cool, moist, but clear summer morning? Water that collects on outdoor surfaces when moist air is cooled is called dew. Where did the moisture come from? **Dew** is moisture that condenses from water vapor in the atmosphere onto cold surfaces. Dew does not fall from the sky, so it is not considered a form of precipitation.

FREEZING Energy absorbed to melt ice is also released when water freezes. **Freezing** is the change in state from a liquid to a solid. Although freezing releases only 80 calories of heat energy per gram (compared with 540 calories per gram for condensation), this is still an important contribution to heating the atmosphere.

Frost is similar in origin to dew. It is composed of ice crystals that form directly from water vapor in contact with surfaces below 0°C. The change of a vapor directly to a solid, skipping the liquid phase, is generally referred to as deposition. This change releases latent heat of vaporization (520 calories per gram) and latent heat of freezing (80 calories per gram). Figure 20-5 on page 490, from the *Earth Science Refer-*

Properties of Water

Energy gained during melting.................80 calories/gram	
Energy released during freezing............80 calories/gram	
Energy gained during vaporization.......540 calories/gram	
Energy released during condensation..540 calories/gram	
Density at 3.98°C..............................1.00 gram/milliliter	

Figure 20-5 Properties of water.

ence Tables, shows these properties of water. (The density of water is listed at 3.98°C because at this temperature water is most dense. Consequently, it is generally cold water that sinks to the bottom of the ocean.)

Ice is the lowest energy phase of water. When ice is made of delicate crystals, it is known as snow. Ice must absorb energy

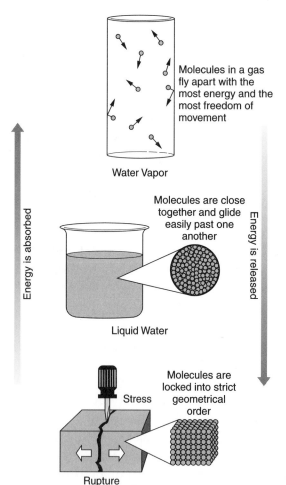

Molecules in a gas fly apart with the most energy and the most freedom of movement

Water Vapor

Energy is absorbed

Molecules are close together and glide easily past one another

Energy is released

Liquid Water

Stress

Molecules are locked into strict geometrical order

Rupture

Ice

Figure 20-6 When water changes state, latent potential energy is absorbed or given off. During melting and vaporization water absorbs energy. Condensation and freezing release energy.

LAB 20-1: Observing Latent Heat

Materials: safety goggles, 250-mL beaker, hot plate or burner, thermometer about 15 cm long (0–120°C), stirring rod, timer or clock, snow or crushed ice, scale, graph paper, heat protection (such as a fire-retardant glove)

SAFETY PRECAUTIONS:

1. Wear safety goggles whenever you heat anything in the lab or whenever you mix two liquids and at least one is not water.

2. Use care around flames or heat sources. The teacher should carefully supervise all lab group setups and procedures.

3. At the end of the procedure, let hot objects cool before they are cleaned and put away.

Procedure:

1. Place fine ice chips or snow in a 250-mL beaker. Add just enough water so that it is visible under the snow near the bottom of the beaker. A total of 200 mL of ice and water works well.

2. Take the initial ice-water temperature and place the beaker on a hot plate or over a medium burner flame. Stir the water-ice mixture and record the temperature every 30 seconds. Continue until the water has been at a full boil for 5 minutes or until half the water has boiled away. (Whichever comes first.) CAUTION: Use heat protection for your hand while stirring. The air above the beaker and the stirring rod may become very hot.

3. It is important to stir the ice-water mix with the stirring rod until all the ice has melted. DO NOT use the thermometer as a stirring rod. The thermometer can rest in the water. Be careful not to hit the thermometer with the stirring rod. Note on your data sheet when the ice has completely melted and when the water comes to a full boil.

4. Graph your data with time on the horizontal axis. What section(s) of the graph indicate latent heat?

to change to liquid water. Liquid water must absorb energy to change to water vapor. Water vapor is the phase of water with the highest energy because of the latent heat it contains. Water vapor can release energy to become liquid water and then ice. Figure 20-6 shows the three phases of water relative to their latent energy. It also shows when energy is absorbed or released during changes in state.

What happens when energy is released by condensation or freezing? That energy warms the air, or at least prevents it from cooling as much as it would without the addition of latent heat. For example, as air rises, expansion causes it to cool. If a cloud is forming as water vapor condenses into cloud droplets, the cooling is slowed. This can keep air in a cloud warmer than its surroundings and keep the air rising. The release of latent heat causes rising air currents that can lead to violent weather, such as thunderstorms and hurricanes.

HOW DOES THE ATMOSPHERE ABSORB WATER VAPOR?

Evaporation is the most important way that moisture enters the atmosphere. If you have ever observed puddles disappear after a rain storm, or dew vanish in the morning sunlight, you have observed evaporation. **Evaporation** can be defined as the vaporization of water when the temperature is below the boiling point.

Evaporation and Kinetic Energy

How can water change to water vapor below the boiling temperature? You may recall that temperature is defined as a measure of the average kinetic energy of molecules. This indicates that within a substance some molecules have more energy than others. Within liquid water, only those molecules that absorb enough energy to transform into water vapor escape as a gas. The cooler molecules are left behind. This is why your body often feels cooler when you get out of the water at a beach or swimming pool. It is like separating the tallest individuals from a crowd of people. With the tallest people gone, the average height of people in the crowd would be less than it was when the tall people were still there. Evaporation is a similar process. The escape of the most energetic molecules reduces the average energy of the molecules remaining.

Sources of Water

Evaporation from the oceans is the primary source of water vapor in the atmosphere. Covering almost three-quarters of our planet, the oceans provide a vast area for evaporation. On land, freshwater and groundwater also contribute to atmospheric humidity through evaporation.

Moisture also enters the atmosphere from plants. Plants draw water from the soil through their roots. Plants use water for a variety of functions, such as transporting nutrients from the soil throughout the plant. Plants also need water to carry on photosynthesis. Carbon dioxide is combined with water in the presence of sunlight to produce glucose for energy.

Water evaporates from plants, especially from their leaves, in a process known as **transpiration**. In some places on land, more water enters the atmosphere by transpiration than by evaporation from surface water and from the ground.

Rate of Evaporation

The rate at which water evaporates depends on four factors:

- *Availability of water*—The greater the exposed surface area, the more water that can evaporate. This is why oceans are the primary source of atmospheric water vapor. It is also the reason the atmosphere is relatively dry in desert regions. Here there is little surface water, and the ground is usually dry.

- *Energy to support evaporation*—You have learned that tropical regions receive stronger sunlight than polar areas do. Therefore, evaporation is active in the tropics for two reasons. Solar energy (insolation) is strongest when the sun is high in the sky, as it usually is in the tropics, and the energy of warm air itself also supports evaporation.

- *Ability of air to hold moisture*—The atmosphere has a limited ability to hold moisture. You have learned that warm air can absorb much more moisture than cool air. The present water vapor content of the air is also important. If the air is holding as much moisture as it can hold at its present temperature, the air is *saturated*. A sponge is a useful model. Sponges are good for picking up liquid spills. If the sponge is full, or saturated with water, it will not pick up any more water. In a similar way, dry air can absorb water, but air that is saturated cannot.

- *Wind*—This factor is related to the ability of water to hold moisture. If the air over a water surface does not move, it becomes saturated, and evaporation stops. The movement of air is needed to bring in a constant supply of dry air. You may have noticed that water evaporates faster on a windy day than it does when the air is calm. As wind speed increases, so does the supply of drier air, and thetefore, evaporation also increases.

HOW DO WE MEASURE WATER IN THE ATMOSPHERE?

The *humidity* of air is an indication of its moisture content. Humid air contains more water vapor than dry air. Humidity can be described in terms of absolute humidity and relative humidity. **Absolute humidity** is the mass of water vapor in each cubic unit of air. It is sometimes measured in grams per cubic meter. However, absolute humidity is seldom reported because the ability of air to hold water vapor changes so much with temperature. Figure 20-7 shows that for each 10°C rise in temperature, air can hold roughly twice as much water vapor. Knowing the absolute humidity does not indicate whether the air is close to saturation, unless the air temperature is also indicated.

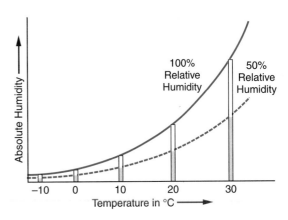

Figure 20-7 For every 10°C increase in temperature, the atmosphere can hold about twice as much moisture. The dotted line shows how much water vapor air is actually holding at any given temperature when the relative humidity is 50 percent.

Dew Point

The dew-point temperature may be used as an indication of absolute humidity. Figure 20-7 shows that as air is cooled, its ability to hold moisture decreases. If moisture is not taken out of the air, it can be cooled until the air is saturated. The temperature at which this occurs is known as the *dew-point temperature*. Sometimes this temperature is just called the dew point. It is called the dew point because when the outside temperature falls below the dew point, dew (or frost) can form on exposed surfaces.

The dew-point temperature can be demonstrated with a glass of ice water. If the glass cools the adjacent air below its dew-point temperature, a film of water will form on the cold outside surface of the glass. In theory, you can measure the dew point with the glass of water. As the surface of the glass is cooled very slowly, the dew point is the temperature of the water when a film of water starts to condense on the outside of the glass. In practice, it is difficult to observe just when condensation begins. Therefore, the results of this method tend to be inconsistent.

ACTIVITY 20-2 EXTRACTING MOISTURE FROM AIR

Place a mixture of ice and water in a glass. (If available, a shiny metal container works even better.) Determine the mass of the glass and mixture with an electronic scale. Then let the glass sit until the ice has melted. You may be able to observe moisture that has condensed on the outside of the glass. Measure the mass a second time. Can you see a change in mass?

Where did the moisture on the outside of the glass come from? Does this procedure work better on some days than others? Why?

A better way to find the dew point is to use an instrument called a **psychrometer** (sigh-CRAH-met-er). A psychrometer is a form of **hygrometer**, an instrument used to measure humidity. Figure 20-8 illustrates a sling psychrometer. It is made from two thermometers mounted side by side on a narrow frame. The sling is the cord or chain that is used to suspend the instrument. One thermometer records air temperature. This is the **dry-bulb temperature**. A wet cloth called a wick covers the bulb of the second thermometer. This

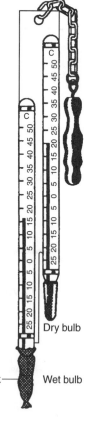

Dry bulb

Wick —— Wet bulb

Figure 20-8 A sling psychrometer consists of two thermometers mounted side-by-side, one with its bulb covered by a wet wick. When the instrument is moved through the air, the dry-bulb thermometer records air temperature. The wet-bulb thermometer is used to measure cooling caused by evaporation. The difference between these temperatures is the key to finding the dew point.

thermometer gives us the **wet-bulb temperature**. The wet bulb is usually mounted lower than the dry bulb to keep the water from splashing on the dry bulb.

When the psychrometer is swung though the air, evaporation from the wet wick causes the wet bulb to register a lower temperature. Depending on the design of the psychrometer, it may be necessary to swing the instrument for a minute or so to get accurate temperature readings. The dryer the air, the more evaporation from the wet wick and the greater the cooling effect. It is the amount of evaporation and cooling that allows us to determine the dew-point temperature. Unless they are equal, the wet-bulb temperature is always lower than the dry-bulb temperature.

Dew point can be confusing. Although it is expressed as a temperature in degrees Celsius (or Fahrenheit), it is not the actual temperature of the air. Instead, it tells you to what temperature the air can be cooled before condensation will begin. Dew point is a temperature, but it is used to indicate how much moisture is in the air. A table can be used to translate dew-point temperatures into absolute humidity, measured as the mass of water vapor per cubic meter. That is why dew point is a good indicator of absolute humidity.

Figure 20-9 on page 498 is taken from the *Earth Science Reference Tables*. This table is used with readings from a sling psychrometer to find the dew-point temperature of the air. A common mistake students make in using this table is to read the top scale of numbers as wet-bulb temperatures. The label clearly indicates that this scale is the *difference* between the thermometer readings. Remember that there is no wet-bulb scale on this table.

SAMPLE PROBLEMS

Problem 1   When you swing a sling psychrometer, the air temperature (dry bulb) is 18°C while the wet-bulb thermometer records 12°C. What is the dew point?

Solution   The difference between 18°C and 12°C is 6°C. Use the dew-point temperature chart. Follow the vertical column down from 6 until it meets the horizontal row from the dry-bulb (air) temperature of 18°C. They meet at 7, therefore the dew-point temperature is 7°C.

Dew-point Temperatures

Dry-Bulb Tempera-ture (°C)	Difference Between Wet-Bulb and Dry-Bulb Temperatures (C°)														
	1	2	3	4	5	6	7	8	9	10	11	12	13	14	15
−20	−33														
−18	−28														
−16	−24														
−14	−21	−36													
−12	−18	−28													
−10	−14	−22													
−8	−12	−18	−29												
−6	−10	−14	−22												
−4	−7	−12	−17	−29											
−2	−5	−8	−13	−20											
0	−3	−6	−9	−15	−24										
2	−1	−3	−6	−11	−17										
4	1	−1	−4	−7	−11	−19									
6	4	1	−1	−4	−7	−13	−21								
8	6	3	1	−2	−5	−9	−14								
10	8	6	4	1	−2	−5	−9	−14	−28						
12	10	8	6	4	1	−2	−5	−9	−16						
14	12	11	9	6	4	1	−2	−5	−10	−17					
16	14	13	11	9	7	4	1	−1	−6	−10	−17				
18	16	15	13	11	9	7	4	2	−2	−5	−10	−19			
20	19	17	15	14	12	10	7	4	2	−2	−5	−10	−19		
22	21	19	17	16	14	12	10	8	5	3	−1	−5	−10	−19	
24	23	21	20	18	16	14	12	10	8	6	2	−1	−5	−10	−18
26	25	23	22	20	18	17	15	13	11	9	6	3	0	−4	−9
28	27	25	24	22	21	19	17	16	14	11	9	7	4	1	−3
30	29	27	26	24	23	21	19	18	16	14	12	10	8	5	1

Figure 20-9 Use this table with wet-bulb and dry-bulb temperature readings to determine the dew point. Follow the vertical column below the difference between wet- and dry-bulb temperatures until it meets the horizontal row to the right of the dry-bulb (air) temperature. Where they meet, read the dew-point temperature.

Problem 2　A student used a sling psychrometer to obtain readings of 8°C and 11°C. What was the dew point?

Solution　This problem introduces two new issues. First, although it is not stated, the dry-bulb temperature must be the higher temperature: 11°C. (Evaporation always makes the wet-bulb temperature the lower of the two temperatures.) Second, 11°C is not listed on the dry-bulb scale. It will be necessary to estimate its position between 10°C and 12°C. The difference between 11°C and 8°C is 3°C. The vertical column below 3°C intercepts the two closest dry-bulb temperatures at 4°C and 6°C. Therefore, you estimate the dew point at an intermediate value of 5°C.

Problem 3　If dry-bulb temperature is 24°C and the wet-bulb temperature is also 24°C, what is the dew point?

Solution　This can be solved logically. If there is no difference between the two temperatures, it is because water was not evaporating from the wet bulb. When no evaporation takes place, the air must be saturated. If the air is saturated (humidity is 100 percent), the temperature is at the dew point. Consequently, the dew point is also 24°C.

 Practice Problem 1

If the dry bulb on a psychrometer reads 24°C while the reading on the wet bulb is 12°C, what is the dew point? What does this indicate about the humidity?

| **ACTIVITY 20-3** | **A STATIONARY HYGROMETER** |

Materials: Two thermometers, ring stand or frame, 30 cm of string, 2- to 4-cm section of a hollow, cotton shoelace (the wick), small dish, fan made of cardboard or any other convenient material.

An instrument to measure humidity can be set up on a counter or table. Suspend the two thermometers next to each other from the ring stand or frame. Place one thermometer bulb in the wet, hollow shoelace. The other end of the shoelace should rest in water in the dish.

Before you use the fan, record the two temperatures . Use your readings with the Dew-point Temperature Table to determine a dew point.

Repeat the procedure, but this time fan the thermometers for one minute before taking wet- and dry-bulb readings. Use this data to determine a second dew point.

Which determination of dew point is more accurate? Why?

 Relative Humidity

Weather reports often include relative humidity rather than absolute humidity or even dew point. This is because relative humidity is easier for most people to understand. Relative humidity is a measure of how full of moisture the air is. *Relative humidity* can be calculated by dividing the absolute humidity by the capacity of the air to hold water vapor. This ratio is expressed as a percent of saturation:

$$\text{Relative humidity} = \frac{\text{amount of water vapor in the air}}{\text{maximum water vapor the air could hold at this temperature}}$$

For example, if the relative humidity is 20 percent, the air is holding 20 percent, or one-fifth, of its moisture capacity. That is quite dry, and evaporation can readily add moisture to the air. However, if the relative humidity is 80 percent, the moisture content of the air is near saturation. Evaporation will be slow and the humidity is likely to be uncomfortable. Sweat may remain on your skin, and clothing does not dry well. These conditions apply at a constant relative humidity regardless of air temperature. That is the advantage of reporting relative humidity.

Relative humidity is not likely to remain the same if the air temperature changes. Consider a parcel of air that is cooling. If there is no evaporation or condensation, the absolute humidity (amount of moisture) in this body of air will remain constant. Remember that relative humidity is a ratio between two values. The ability of air to hold moisture decreases with decreasing temperature. As the air becomes cooler, it is closer to saturation, and the relative humidity increases. This frequently occurs at night when temperatures decrease, leading to the formation of dew or frost.

Figure 20-10 shows typical changes in air temperature, dew point, and relative humidity over a period of one day. The thick black line indicates changes in air temperature. The changes in temperature are the result of overnight cooling and daytime heating by insolation. Over this period there is little change in the dew point, indicating that the absolute humidity is relatively constant, as shown by the dotted line. However, changes in relative humidity depend on air temperature and dew point. At night, the thin line indicates that as the air temperature cools to the dew point, the relative humidity increases to 100 percent. This situation changes when the sun warms the air in the morning. Relative humidity drops as the air is warmed and its ability to hold water vapor increases.

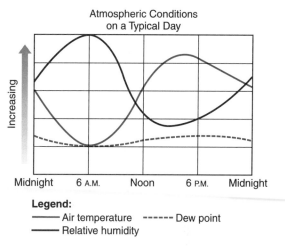

Figure 20-10 The daily cycle of temperature and humidity shows that if the dew point stays nearly constant, changes in air temperature will cause the opposite changes in relative humidity. Although the absolute humidity has changed very little, the relative humidity is affected by air temperature.

Atmospheric Conditions on a Typical Day

Midnight   6 A.M.   Noon   6 P.M.   Midnight

Legend:
—— Air temperature   ------ Dew point
—— Relative humidity

Relative humidity can be determined by using a table found on the same page of the *Earth Science Reference Tables* as the dew point table. This table is reproduced in Figure 20-11. The dry- and wet-bulb temperatures from a sling psychrometer also are used with this table.

Relative Humidity (%)

Dry-Bulb Temperature (°C)	Difference Between Wet-Bulb and Dry-Bulb Temperatures (C°)														
	1	2	3	4	5	6	7	8	9	10	11	12	13	14	15
−20	28														
−18	40														
−16	48	0													
−14	55	11													
−12	61	23													
−10	66	33	0												
−8	71	41	13												
−6	73	48	20	0											
−4	77	54	32	11											
−2	79	58	37	20	1										
0	81	63	45	28	11										
2	83	67	51	36	20	6									
4	85	70	56	42	27	14									
6	86	72	59	46	35	22	10	0							
8	87	74	62	51	39	28	17	6							
10	88	76	65	54	43	33	24	13	4						
12	88	78	67	57	48	38	28	19	10	2					
14	89	79	69	60	50	41	33	25	16	8	1				
16	90	80	71	62	54	45	37	29	21	14	7	1			
18	91	81	72	64	56	48	40	33	26	19	12	6	0		
20	91	82	74	66	58	51	44	36	30	23	17	11	5	0	
22	92	83	75	68	60	53	46	40	33	27	21	15	10	4	0
24	92	84	76	69	62	55	49	42	36	30	25	20	14	9	4
26	92	85	77	70	64	57	51	45	39	34	28	23	18	13	9
28	93	86	78	71	65	59	53	47	42	36	31	26	21	17	12
30	93	86	79	72	66	61	55	49	44	39	34	29	25	20	16

Figure 20-11

SAMPLE PROBLEMS

Problem 4   If wet- and dry-bulb temperatures recorded with a sling psychrometer are 14°C and 24°C, respectively, what is the relative humidity?

Solution   The difference between these temperatures is 10°C. The column leading down from 10°C meets the row extending to the right of the dry-bulb temperature, 24°C, at the number 30. Therefore, the relative humidity is 30 percent.

Problem 5   If both thermometers on a sling psychrometer record −4°C, what is the relative humidity?

Solution   This is similar to Sample Problem 3. The table can be used but it is not needed. If there is no cooling of the wet-bulb thermometer, there was no evaporation, and the air is saturated: 100 percent humidity.

 Practice Problem 2
What is the relative humidity when the wet-bulb temperature is 11°C and the dry-bulb temperature is 15°C?

 Predicting Precipitation

One of the reasons meteorologists measure air temperature and determine the dew point is to predict precipitation. When the air temperature and dew point are close, the chance of precipitation increases. Figure 20-12 illustrates atmospheric conditions over a period of one day at a specific location. The top graph shows the temperature dropping in the late afternoon until it is close to the dew point. This is when precipitation is most likely. The bottom graph makes the prediction more clear by the addition of a line indicating relative humidity. Rising relative humidity indicates that the air is coming closer to saturation in the late afternoon. Again, this is a strong indicator of precipitation.

Predictions are based on weather observations from the past. Precipitation predictions are often expressed as a per-

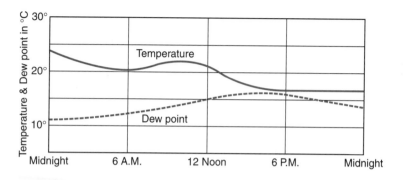

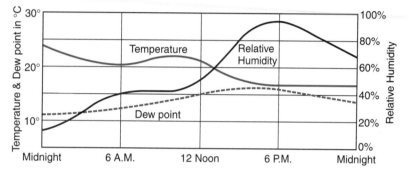

Figure 20-12 The top graph shows air temperature drawing close to the dew-point temperature in the late afternoon. This is a sign that the likelihood of precipitation is greatest in late afternoon. The bottom graph has a third line to show relative humidity during the same period. The chance of precipitation is greatest when relative humidity is high.

cent chance of rain (or snow, etc.). For example, an 80 percent chance of precipitation means that during the past, of 10 days with similar conditions, there was precipitation on eight of the 10 days. A 20 percent chance of precipitation means that precipitation is possible, but not likely. A 50 percent chance of rain indicates that the chance of having rain equals the chance of not having precipitation.

 ## HOW DO CLOUDS FORM?

You have learned that most of the energy for weather events comes from sunlight, and that energy enters the atmosphere primarily through evaporation. You also know that condensation (and deposition) release latent energy as water vapor changes into tiny water droplets and ice crystals to form clouds. Too small to fall as precipitation, the cloud droplets remain suspended in the air. A large body of these tiny droplets

Figure 20-13 Clouds are large regions of ice crystals or water droplets too small to fall through the air. Cloud formation releases energy that warms the air, creating updrafts. This often results in more cloud formation.

or ice crystals is known as a **cloud**, such as the cloud pictured in Figure 20-13. For clouds to form, several conditions must be met.

Cooling of Air

For condensation to occur, air must be cooled below its dew point. At ground level, air is cooled by contact with cold surfaces. How can air be cooled higher in the atmosphere? To understand how this occurs, it is useful to observe what happens when air is compressed. Have you ever felt a pump used to force air into automobile tires? The compressor on the pump gets warm. Some of this energy comes from friction. However, most of it is the result of compressing the air. It takes energy to compress air, and that energy turns to heat, which makes the air and the pump warm.

The opposite change occurs when air expands. Allowing air to expand takes energy out of the air. This causes the temperature of the air to lower. If you depress the pin in a tire valve, air rushes out. A thermometer held in the rushing air shows that the expanding air is cooler than the air in the tire. When air rises into the atmosphere, it expands due to reduction of atmospheric pressure with altitude. Expansion of a gas causes it to become cooler. If the air cools below its dew point, condensation can create a cloud.

ACTIVITY 20-4 **THE EFFECT OF COMPRESSION AND EXPANSION ON AIR TEMPERATURE**

If you place a plastic, liquid-crystal, digital thermometer in a transparent 1-liter soda bottle you can observe small changes in the temperature of air in the bottle. (These thermometers are often sold in aquarium supply stores.) With the thermometer inside, tightly cap the empty soda bottle . (The cap must be airtight.) If you compress the soda bottle by pushing in on its sides, you should be able to see the temperature inside change. How did the temperature change? You can also increase the pressure in the bottle with a special pump cap used to maintain the carbonation in opened soda bottles. These caps are sold in supermarkets.

Whichever way the pressure in the bottle is increased, when the pressure is released, the air temperature inside the bottle will change again. How does the temperature change when the pressure is released? Why do these changes in temperature occur? (Please note that this is not an easy phenomenon to explain. You may need to do some Internet or library research.)

What causes air to rise? Sometimes it is the result of winds blowing over mountains. Sometimes warm, moist air is pushed up by cooler, dryer air that is more dense. However, once the process of cloud formation gets started, it tends to keep going. Remember that condensation and deposition release energy that warms the surrounding air. This causes the air in the cloud to continue rising, and results in more cloud formation. When does it stop? Eventually the cloud runs out of moist air. Without water vapor to feed cloud formation, the atmosphere stabilizes and cloud formation stops.

Condensation Nuclei

When dew or frost forms on grass on a cold night, the air must be below its dew-point temperature. Nevertheless, a cloud (fog) is seldom observed. This is because water must condense on a surface. In the atmosphere, those surfaces are provided by tiny particles of solids suspended in the air. They are called **condensation nuclei**. These particles are added to the atmosphere by dust storms, fires, and the exhaust from automobiles, homes, factories, and power plants. You have probably seen the white trails left by jets flying at high altitude. Actually, the trails are clouds that form on the exhaust particles from the jets' engines. At that altitude, air is usually below the dew point, but the lack of suspended particles prevents cloud formation. After a few minutes, the cloud is scattered, and the trail is no longer visible.

You may have noticed how clear the atmosphere looks after a rainstorm. Precipitation not only takes moisture out of the atmosphere, but it also brings down condensation nuclei. In addition, the falling droplets pick up other suspended particles. In this way, precipitation cleans the atmosphere.

How do clouds disappear? Clouds lose some of their moisture by precipitation. However, precipitation usually stops before the cloud is gone. Furthermore, most clouds do not produce precipitation. Clouds disappear when solar radiation vaporizes the cloud faster than cloud formation takes place. The tiny ice crystals and water droplets change back to water vapor and the cloud disappears.

ACTIVITY 20-5 HOMEMADE CLOUDS

Materials: large, wide-mouthed, container (a transparent glass or plastic jar or a large beaker), 1-gallon plastic food storage bag, match, 300 mL of ice, hot and cold tap water

Run a few centimeters of hot tap water into the bottom of the container. Place the crushed ice cubes and about 100 mL of cool water in the plastic bag. (Crushed ice or snow works better than big ice cubes.) Cover the mouth of the container with the bag of ice and water. Do you see a cloud?

Try this again. This time, hold a lit match in the container and blow it out. Allow some smoke from the extinguished match to circulate in the container. Place the bag of ice and water over the mouth of the container. This time you should clearly see cloud inside the container.

What is the purpose of the hot water, the ice and water, and the smoke?

 Altitude of Cloud Formation

Have you ever noticed that bottoms of clouds are often relatively flat while the tops of the clouds are often puffy and irregular? **Cloud-base altitude** is the height at which rising air begins to form clouds. To determine the altitude at which the air could form a cloud use the air (dry-bulb) temperature, the dew-point temperature, and a cloud-base altitude graph (Figure 20-14 on page 508).

Notice that there are two sets of slanted lines on this graph. The slanted lines labeled "dry-bulb temperature" show the rate at which air cools as it rises and expands. This is sometimes called the temperature lapse rate. The dew-point temperature also decreases with altitude, but not as quickly. The rate of change in dew point is shown by the slanted, dotted lines. Where the air temperature line meets the dew-point line, the height at which cloud formation (condensation) begins can be read from either the left or the right side of the graph. The companion diagram in 20-14 may illustrate this more clearly than the graph alone.

SAMPLE PROBLEMS

Problem 6   If the temperature at ground level is 20°C and the dew point is 12°C, how high must the air rise to begin cloud formation?

Solution   Follow the slanted solid line above 20°C and the dotted line up from 12°C until they meet. (*Suggestion:* Lay the edge of a ruler or a blank sheet of paper to the right and along the 20° temperature lapse rate line to follow it more clearly.) The altitude of the cloud base in kilometers can be read on either of the vertical scales. In this case, it is 1 km.

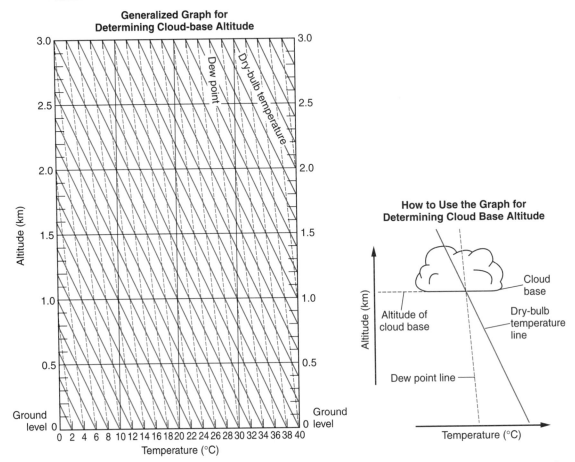

Figure 20-14 If you know the Celsius temperature and the dew point you can determine the elevation at which clouds begin to form in rising air. Follow the slant of the solid line above the dry-bulb temperature. Above the dew point follow the dotted line. They meet at the altitude (in kilometers) of the cloud base.

Problem 7   If you use a psychrometer to find dry-bulb and dew-point temperatures of 23°C and 14°C, respectively, how high must the air rise to form a cloud base?

_____

Solution   This air temperature, 23°C, is not printed on the graph. However, it is half way between 22°C and 24°C. Follow the angle of the slanted solid lines in the middle of the space between the lines for 22°C and 24°C. This imaginary line meets the dotted line from 14°C, the dew point, at a height of 1.1 km.

_____

Problem 8   The air temperature is 90°F. The dew point is 50°F. How high must the air rise to form a cloud?

Solution In this problem, you have been given temperatures in degrees Fahrenheit. The temperatures must be converted to Celsius temperatures to use the graph. That conversion can be done using Figure 19-3 or the "Temperature" table in the *Earth Science Reference Tables*. While 90°F is approximately 32°C, 54°F converts of 12°C. On the cloud-base graph, the lines for 32°C air temperature and 12°C dew point converge close to an altitude of 2.5 km. That is the cloud-base altitude.

 Practice Problem 3

One day, the air temperature was 77°F and the dew-point temperature was 60°F. What was the cloud-base altitude?

| ACTIVITY 20-6 | **THE HEIGHT OF CLOUDS** |

Materials: hygrometer or two 30-cm, alcohol-filled glass thermometers mounted on a ring stand with one thermometer's bulb covered by a wet wick, file folder or similar object to fan the thermometers

Use this apparatus to find the dry-bulb (air) temperature and the wet-bulb temperature. (To get an accurate wet-bulb temperature you will need to fan the wet bulb until the temperature stabilizes. This may take 30 seconds or so.)

Use these temperatures and the "Graph for Determining Cloud-Base Altitude" (Figure 20-14) to find the height at which a cloud could begin to form if the air in the room moved upward into the atmosphere.

Repeat the procedure on days you think the humidity is especially high or low. What would be the most likely sources of error in your determination of cloud base altitude?

TERMS TO KNOW

absolute humidity	dew-point temperature	melting
calorie	evaporation	phase of matter
cloud	freezing	psychrometer
cloud-base altitude	frost	specific heat
condensation	hygrometer	transpiration
condensation nuclei	lake-effect storm	vaporization
dew	latent heat	

CHAPTER REVIEW QUESTIONS

1. Winds from the west often produce winter snowstorms at the eastern end of lakes Erie and Ontario. These storms are known as lake-effect storms. What causes these winter storms that often occur in the Buffalo and Tug Hill Plateau regions?

   (1) increased evaporation over the warm lakes followed by warming of the air as it travels over the land
   (2) increased evaporation over the warm lakes followed by cooling of the air as it travels over the land
   (3) decreased evaporation over the warm lakes followed by warming of the air as it travels over the land
   (4) decreased evaporation over the warm lakes followed by cooling of the air as it travels over the land

2. Which process occurring within Earth's atmosphere causes the most heating of air?

   (1) liquid water changing to ice
   (2) ice changing to liquid water
   (3) liquid water changing to water vapor
   (4) water vapor changing to liquid water

3. Which process cools the atmosphere by absorbing latent vaporization energy?

   (1) evaporation from the oceans   (3) cloud formation
   (2) melting glaciers   (4) precipitation

4. During which phase change of water is the most energy released into the environment?

(1) water freezing   (3) water evaporating

(2) ice melting   (4) water vapor condensing

5. An Earth science class left pan of water outside in the open air to measure the rate of evaporation. For water to be lost by evaporation, the air temperature must be above

(1) the dew point of the atmosphere.

(2) the freezing point of water.

(3) the boiling temperature of water.

(4) the average annual temperature.

6. When the dry-bulb temperature is 22°C and the wet-bulb temperature is 13°C, the dew point is

(1) −10°C.   (3) 14°C.

(2) 5°C.   (4) 33°C.

7. The air outside a classroom has a dry-bulb temperature of 10°C and a wet-bulb temperature of 4°C. What is the relative humidity of this air?

(1) 1%   (3) 33%

(2) 14%   (4) 54%

8. The graph below is a forecast of air temperature and dew point over a period of a little more than two days.

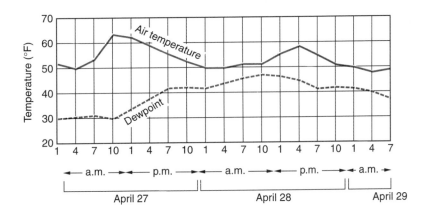

At what time is the rate of evaporation expected to be highest?

(1) April 27 at 10 A.M.
(2) April 28 at 10 A.M.
(3) April 28 at 4 P.M.
(4) April 29 at 4 P.M.

The graph below shows air temperature and relative humidity at a single location during a 24-hour period. Use this graph to answer the next two questions.

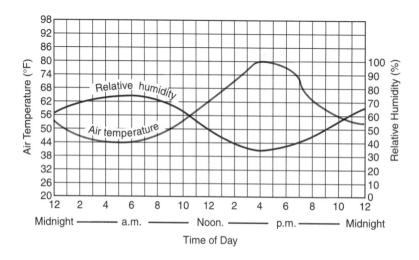

9. What is the approximate change in relative humidity from 12 noon to 4 P.M.?

(1) 10%   (3) 20%
(2) 15%   (4) 30%

10. At which time would the rate of evaporation most likely be greatest?

(1) 6 A.M.   (3) 4 P.M.
(2) 10 A.M.   (4) 11 P.M.

11. What was the average rate of temperature change from 9 A.M. to 3 P.M. on this day?

(1) 2°C/hour   (3) 6°C/hour
(2) 4°C/hour   (4) 8°C/hour

12. Which graph below best shows the relationship between the probability of precipitation and the difference between air temperature and dew point?

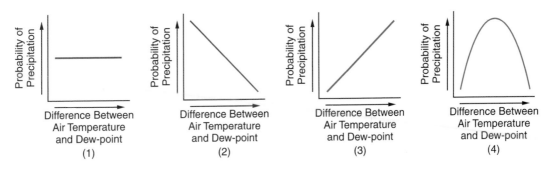

(1)

(2)

(3)

(4)

13. Which weather change usually occurs when the difference between the air temperature and the dew-point temperature is *decreasing*?

(1) The amount of cloud cover decreases.
(2) The probability of precipitation decreases.
(3) The barometric pressure increases.
(4) The relative humidity increases.

14. The cross section below shows how prevailing winds have caused different weather on opposite sides of a mountain range.

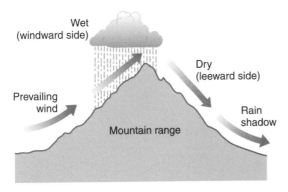

Why does the windward side of this mountain have weather that is more moist?

(1) Rising air compresses and cools, causing the water droplets to evaporate.
(2) Rising air compresses and warms, causing the water vapor to condense.
(3) Rising air expands and cools, causing the water vapor to condense.
(4) Rising air expands and warms, causing the water droplets to evaporate.

15. Clouds can form when

(1) the air temperature cools below the dew point.
(2) sinking air is warmed by compression.
(3) the relative humidity is 0%.
(4) condensation nuclei have been removed from the air.

Open-Ended Questions

Base your answers to questions 16–18 on the diagram below, which shows a hygrometer located on the wall of a classroom. Temperature readings on the hygrometer are used by students to determine the relative humidity of the air in the classroom.

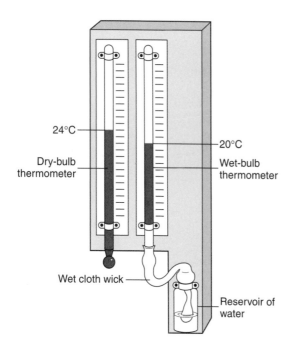

16. Based on the temperature readings shown in this diagram, determine the relative humidity of the air in the classroom.

17. Besides relative humidity, identify another variable of air in the classroom that may be determined only by using both temperature readings on the hygrometer.

18. Describe how water evaporating from the wick attached to the wet-bulb thermometer lowers the temperature reading of that thermometer.

19. A student using a sling psychrometer obtained a dry-bulb reading of 20°C and a wet-bulb reading of 16°C for a parcel of air outside the classroom.

(a) State the dew point.
(b) State the change in relative humidity as the air temperature and the dew point get closer to the same value.

20. Clouds droplets form around small particles suspended in the atmosphere. Describe how clouds form from water vapor. Include the terms "dew point" and either "condensation" or "condense" in your answer.

Chapter 21

Air Pressure and Winds

 FAST AS THE WIND

Establishing a world's record for wind speed is not a simple matter. On April 12, 1934, an anemometer on the summit of Mount Washington in New Hampshire registered a sustained wind speed of 373 kilometers per hour (km/h) (231 miles per hour [mph]). At 1910 meters (6262 feet), Mount Washington is the highest summit in the northeastern United States. This record has stood for 70 years. Prevailing westerly winds are forced up the mountain. The mountaintop and the overlying layers of the atmosphere squeeze the winds, increasing their speed. The average wind speed at this location is 57 km/h (35 mph) making it the windiest surface location in the United States.

A slightly faster wind speed of 380 km/h (236 mph) was reported during a typhoon (hurricane) on the island of Guam in 1997. This report led to an investigation by the National Climate Extremes Committee. The committee found that the anemometer used in Guam was not properly calibrated for winds in excess of 274 km/hr (170 mph). The committee concluded that the combination of high wind and heavy rain caused the Guam instrument to malfunction and that the true wind speed was probably less than 322 km/h (200 mph).

Was the wind speed recorded on Mount Washington the fastest surface wind ever to occur? It is not likely. Winds in the strongest tornadoes are estimated to exceed 483 km/h (300 mph). Scientists have tried to measure tornado winds with ground-based instruments. However, the difficulty of placing instruments in the narrow path of a tornado, flying debris, and damage done by strong tornadoes makes this nearly impossible. Teams of "storm chasers" have tried to put instrument packages where a tornado would envelope them, but none have succeeded in obtaining anemometer measurements of the strongest tornado winds.

A new tool has become available for measuring extreme winds. Meteorologists can now use radar to measure wind speed and direction from a distance. **Radar**, a name taken from the terms "*ra*dio *d*etection *a*nd *r*anging," was developed during the Second World War primarily to observe enemy aircraft. It works by bouncing long-wave radio signals off distant objects. The distance is determined by how long the signals take to return as reflected energy. Advancing technology has enabled engineers to develop radar that can measure the speed with which objects or winds are moving toward or away from the radar station. This is called **Doppler radar**. Doppler radar was used to record a wind gust of 512 km/h (318 mph) in a tornado in Oklahoma in 1999. However, this wind speed does not have the accuracy of the 1934 measurement on Mount Washington.

WHAT CAUSES WINDS?

Surface winds blow in response to differences in air pressure. Winds always move from places of higher pressure to places of lower pressure. When you exhale, you do so by squeezing the air in your lungs, increasing the pressure. Air escapes from your body to equalize the pressure inside and outside your lungs. An air pump works in a similar way. By compressing the air inside the pump, air is forced out of the pump to where the pressure is lower.

You learned earlier that atmospheric pressure is caused by the weight of the atmosphere. Earth's atmosphere is not confined the way air is in your lungs or in an air pump. The atmosphere has a relatively uniform depth near Earth's surface. Differences in the density of air cause changes in the weight of the air. Primarily, temperature and humidity determine the density of air. (As temperature and humidity increase, air becomes less dense.) When air density increases, so does air pressure at Earth's surface, forcing the air to move to places with a lower surface pressure.

DEMONSTRATION #1

THE WEIGHT OF AIR

Use a sensitive balance to measure the weight of a deflated playground ball. Then pump up the ball and determine its weight again. The difference in weight is the weight of the air inside the ball.

DEMONSTRATION #2

THE FORCE OF AIR PRESSURE

This demonstration should be performed over a sink or a large container to catch spilled water. Fill a small glass with water and place an index card over the top. Carefully invert the glass while holding the index card to maintain an airtight seal. Remove your hand from the index card, air pressure will hold the card and the water in place until the wet card loses its stiffness.

DEMONSTRATION #3

AIR PRESSURE AND A SODA CAN

Materials: empty 12-oz soda can, hot plate or lab burner, ring stand, tongs, ice water

Pour about half a centimeter of water in the bottom of the empty soda can. Heat the can of water on a hot plate or over a burner flame until water vapor fills the can and drives out the air. Using the tongs, quickly invert the soda can and place it in the ice water. As the water vapor in the can suddenly condenses, atmospheric pressure will crush the can. Although the soda can has an opening, the change in air pressure inside the can is so rapid and so strong that the can suddenly collapses.

DEMONSTRATION #4

PRESSURE AND DEPTH

Differences in air pressure at different depths within the atmosphere can be modeled with a 2-liter plastic soda bottle or a similar tall plastic container. Three holes are pierced in the bottle at different heights. With the holes covered by plastic tape, fill the bottle with water. Hold the jar over a sink or a container to catch the water. Remove the tape from the holes; notice that the water emerging from each hole travels a different distance. This illustrates that pressure increases with depth in a fluid. Each hole represents a different level in the atmosphere.

Temperature, Air Pressure, and Winds

Heating increases the motion of air molecules and pushes them apart. If you have observed air rising over a campfire, you have observed convection currents in the atmosphere caused by density differences. The fire heats the air, causing it to expand. The low-density air floats higher into the atmosphere, and is replaced by cooler air that flows in from the surrounding area. This cooler air is then heated by the fire and expands to keep the air constantly flowing upward, carrying the heat of the fire into the atmosphere. Expansion by heating and contraction by cooling cause changes in atmospheric pressure.

Humidity, Air Pressure, and Winds

The role of humidity is not as obvious as that of temperature. Under the same conditions of temperature and pressure, the same number of molecules of any gas occupy the same volume. Therefore, if lighter gas molecules are substituted for heavier molecules, there is no change in volume, but the density of the gas decreases. The mass of the individual molecules determines the density of any gas.

You usually think of water as a substance that is more dense than air. Although it is true that liquid water is far more dense than air, this changes when water becomes water vapor. Dry air is 78% nitrogen. If you look at a periodic table of elements you will see that each atom of nitrogen has a mass of 14 atomic mass units (amu). Like many other gases, nitrogen exists in molecules of two atoms (N_2). Therefore, the mass of a molecule of nitrogen is 28 amu. Oxygen (O_2), which makes up most of the rest of dry air, has an atomic mass of 16 amu and a molecular mass of 32 amu. Oxygen is just a little more dense than nitrogen. Therefore air is composed mostly of molecules with a mass of about 28 amu, as shown in Figure 21-1.

Water vapor is a compound made of two atoms of hydrogen and one atom of oxygen (H_2O). The water molecule has three

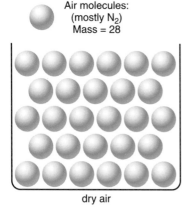

Air molecules:
(mostly N_2)
Mass = 28

dry air

Total mass of air molecules
= 30 × 28 = 840

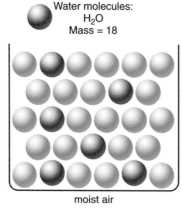

Water molecules:
H_2O
Mass = 18

moist air

Air molecules = 24 × 28 = 672

Water molecules = 6 × 18 = 108

Total mass = 780

Figure 21-1 When water vapor is added to air, the air becomes less dense. Water molecules have less mass than molecules of nitrogen, which make up most of the atmosphere. Therefore, substituting water vapor for dry air makes the air less dense. (The units of mass in these diagrams are atomic mass units.)

atoms: one more than either nitrogen or oxygen. However, the hydrogen atoms are light. They have an atomic mass of just 1 amu. Recall that the oxygen atom has a mass of 16 amu. Therefore, the mass of the water molecule is 18 amu (1 + 1 + 16). This is considerably less than the molecular mass of nitrogen (28 amu) and oxygen (32 amu), which make up 99% of dry air. Therefore, if water vapor molecules replace dry air molecules, the air becomes less dense. Figure 21-1 models dry air as 30 molecules of nitrogen with a total mass of 840 amu. In the second part of this diagram, six water molecules have been substituted for the same number of nitrogen molecules. The total mass decreases to 780 amu. Therefore, adding water vapor to the atmosphere makes air less dense.

The effects of temperature and humidity are confirmed when you use a barometer to measure air pressure in different weather conditions. As temperature and humidity increase, the barometric pressure decreases. Conversely, a change to cooler and dryer weather results in increasing barometric pressure.

WHY DO LOCAL WINDS OCCUR?

There are two categories of wind currents. Regional winds extend over a large area, such as several states of the United States. Local winds are those that extend only for a few miles before they die out.

Convection Cells

Whenever air is heated in one place and cooled in another, circulation tends to occur. Consider a room with a heater on one side of the room and a cold window on the other. Air near the heater absorbs energy. This causes the air to expand and rise. At the far side of the room, air is cooled as it loses its energy. Heat is lost by contact with the cold window and the

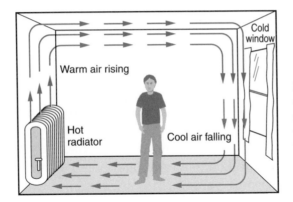

Figure 21-2 Air heated on one side of this room expands and rises. Cooling air on the opposite side contracts and sinks. This energy exchange maintains a flow of air and heat energy called a convection cell.

wall (conduction) as well as by radiating heat toward these surfaces. Air near this end of the room contracts and sinks to the floor where it flows toward the heater. As long as the air is heated in one place and cooled in another, circulation will continue. This pattern, shown in Figure 21-2, is called a **convection cell**. The air currents in the convection cell carry energy from the heater to the cold side of the room and the window.

Winds on Earth are not confined to a closed space the way air is in this diagram. Convection cells do occur within the atmosphere. Rising air in some locations must be balanced by sinking air in other places. Winds that blow in one direction at Earth's surface must be balanced by a return flow somewhere else. The return flow usually happens in the upper atmosphere.

ACTIVITY 21-1   OBSERVING CONVECTION

You can use smoke from an extinguished match or a stick of incense to show convection currents in a classroom. This works best in very cold weather when strong downdrafts overpower the heating effect of the match or incense. The smoke is used to locate places in the room where the air is moving in different directions. If people do not move around, you may be able to map complete convection cells with updrafts, downdrafts, and horizontal air flow.

Can you identify the net flow of energy within the classroom?

Land and Sea Breezes

The wind-producing effects of temperature changes can often be observed at the shore. During stable summer weather in coastal regions such as Long Island or along the Great Lakes, the wind direction can reverse on a daily cycle.

On a sunny day, the land heats up more than the water. To understand why it may help to look at Figure 21-3, from the *Earth Science Reference Tables*. This table shows the specific heat of seven common substances. *Specific heat* is the ability of a substance to absorb or release heat energy. Notice that water in liquid form has a specific heat of 1 calorie/gram · C°. In the form of ice or water vapor, its specific heat is only half as great. This means that a unit of heat energy absorbed by a given mass of ice or water vapor will cause twice the temperature rise it causes in liquid water.

The difference in specific heat is even greater for basalt and granite, which would heat up five times as much as liquid water. Since most beach sand is similar in mineral composition to these two igneous rocks, the sand on the beach heats relatively quickly. Metals, such as iron, copper, and lead, have even lower specific heats. Therefore, they heat up still faster when they absorb energy. The bottom line is, water heats more slowly than most other materials when it absorbs sunlight.

During the day, the land heats up more than the water. Radiation and conduction from the land's surface heat the air over the land. This heated air expands and becomes less

Specific Heats of Common Materials

MATERIAL		SPECIFIC HEAT (calories/gram · C°)
Water	solid	0.5
	liquid	1.0
	gas	0.5
Dry air		0.24
Basalt		0.20
Granite		0.19
Iron		0.11
Copper		0.09
Lead		0.03

Figure 21-3

dense, causing it to rise. The result is a breeze that comes from the water to replace the rising air over the land. **Sea breezes** are light winds that blow from the water to the land. They usually develop in the late morning or afternoon when the land becomes warm. These breezes continue into the evening until the land cools. (See Figure 21-4.)

Sea breezes provide relief in hot summer weather. There are two benefits: The breeze keeps people cool by replacing humid air that builds around the body, allowing sweat to evaporate, and it transports the cooler air over the ocean onto the beach, resulting in relief from summer heat at the hottest time of day.

The wind reverses direction at night and through the early morning, becoming a **land breeze**. Land not only warms faster than the ocean, but it also loses its heat more quickly. The lower specific heat for rock materials means that at night the same amount of energy lost has a greater cooling effect on the land than on water. When the land cools at night, so does the air over it. The air over the water is now warmer than the air over the land. Instead of the air rising over land, air begins to rise over the water during the evening. This causes the

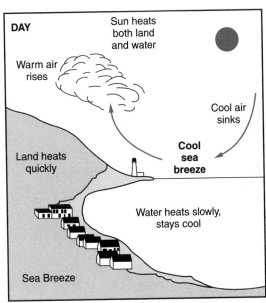
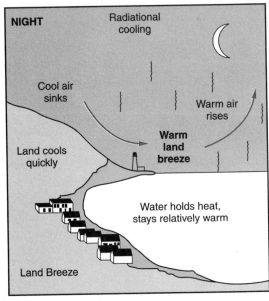

Figure 21-4 Rapid heating and rising air over land areas are responsible for sea breezes that occur during the day, blowing cool air from the ocean. At night, when the water is warmer than the land, the breeze blows from the land.

wind to change direction, blowing from the land to the water. The conditions that lead to a land breeze also are shown in Figure 21-4.

Land and sea breezes do not always occur along the shore. They require large areas of adjacent land and water. Therefore, these breezes do not occur at small lakes or ponds or on small islands. Nor do they develop when daily temperature changes are small, such as during cloudy weather. Strong regional weather events such as the passage of fronts can easily overpower land and sea breezes. However, when these breezes do occur, they can bring welcome relief from summer heat. People who live near the ocean sometimes talk about their "natural air conditioning" from these breezes.

WHAT CAUSES REGIONAL WINDS?

The fastest winds develop in larger and more powerful atmospheric events than land and sea breezes. If you have watched a television weather report you have probably seen maps of the United States with large areas marked "H" and "L." These are regional high- and low-pressure systems.

High- and Low-Pressure Systems

Low-pressure regions are areas where warm, moist air is rising. In the last chapter, you learned that rising air leads to cloud formation when the air is humid. Cloud formation, which occurs by condensation, releases latent energy and warms the air even more. This warming accelerates the updraft. Therefore, once a low-pressure system develops, it tends to strengthen and sustain itself as long as it can draw in moist air. In fact, some low-pressure systems build into major storm events that release great quantities of energy.

High-pressure regions are usually places where cool, dry air is sinking lower into the atmosphere. Although the air gets warmer as it descends, it may still be cooler than the surrounding air. The air spreads out at the surface and makes

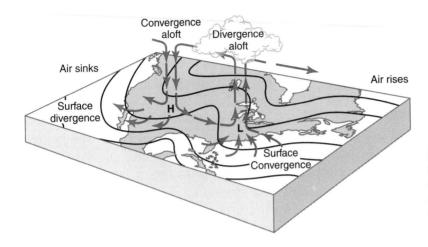

Figure 21-5 Regional high- and low-pressure systems often last for days as they move across the country and determine surface wind patterns.

room for more descending air. High-pressure regions can also remain strong for many days. Rising and falling air are characteristic of convection cells. Therefore, vertical air movements generate surface winds from regions of high pressure to places where the pressure is lower. Figure 21-5 is a diagram of North America that shows regional high- and low-pressure centers.

High-pressure systems are sometimes called zones of divergence. **Divergence** means moving apart. At Earth's surface, descending air spreads as it moves out of a high-pressure system as shown in Figure 21-5. Conversely, winds come together as they blow into regions of low pressure. Rising air at the center of the low sustains these winds. This is a pattern of **convergence**. It is like people converging on an arena where a concert will soon take place. Low-pressure centers are also called zones of convergence.

ACTIVITY 21-2 MOVEMENTS OF PRESSURE SYSTEMS

Use a daily weather map from a newspaper, televised weather report, or the Internet to locate high- and low-pressure centers on a map of the United States. Over the next three days, plot the movements of these pressure systems across the country. Is there a general direction in which they usually move?

The Coriolis Effect

The **Coriolis effect** produces the curved path that objects, including winds and ocean currents, appear to follow as they travel over Earth's surface. It was named after the French scientist who first described it. Consider the three people in Figure 21-6. To conduct an experiment, they are using a rotating platform similar to those often found in playgrounds. In part A, the boy on a rotating platform is about to throw a ball toward the two people opposite him. Part B shows the same people 1 or 2 seconds later. From the point of view of the boy on the ground, the ball travels straight toward him as the two people on the platform move. However, the people on the moving platform see the path of the ball curve to the right. If the platform in Figure 21-6 rotated in the opposite direction, the observers on the platform would see the ball curve left.

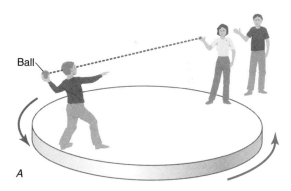

Ball

A

Apparent path as seen by observers on rotating platform

Actual path

Figure 21-6 Whether the ball appears to curve or travel straight depends on whether you are on the rotating platform or standing still. This is why the Coriolis effect produces an apparent curvature.

B

The difference between a turn to the right and a turn to the left can be confusing. For example, if you stand facing another person, what you call the right side of the room will be to the other person's left. If each of you steps to the right, you will be going in opposite directions. Obviously, we need some kind of rule to distinguish which way is "to the right." This is not so different from the way winds are labeled. A wind is named according to the direction from which it comes, not to where it is going.

Right and left curves are determined according to the direction of movement as shown in Figure 21-7. One way to think of this is to imagine that you are walking in the direction of the arrow. A right curve would be to your right only if you are looking forward in that direction. Winds and ocean currents in the Northern Hemisphere appear to curve to the right as they move forward. In the Southern Hemisphere, winds and currents curve to the left.

If Earth did not rotate, patterns of convection on our planet would be relatively simple. Air would descend in high-pressure regions and blow directly toward low-pressure centers. However, the rotation of Earth on its axis causes wind patterns to be more complicated. The winds follow a straight path, but as they blow over long distances, the planet moves under them. The effect is not noticeable over small distances such as those covered by land and sea breezes.

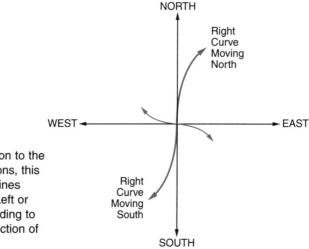

Figure 21-7 In addition to the four compass directions, this diagram shows four lines curving to the right. Left or right is defined according to movement in the direction of the arrow.

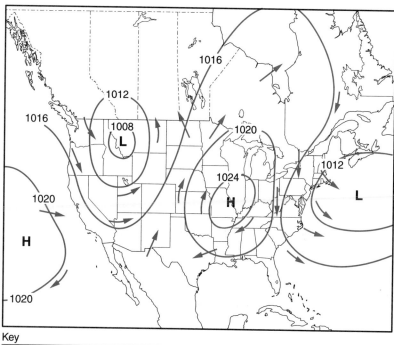

Figure 21-8 Due to the Coriolis effect, regional winds blowing from high-pressure centers to low-pressure regions curve to the right in the Northern Hemisphere. In the Southern Hemisphere, they curve to the left.

Key

H Pressure center   ⌒1008⌒ Isobars   ◄— Wind direction

When you look at larger regional wind patterns, the apparent change in wind direction is very important. The apparent curvature of winds as they move along Earth's surface is the result of the Coriolis effect. Figure 21-8 is a simplified map of high- and low-pressure systems over North America. The isolines, called **isobars**, connect locations with the same atmospheric pressure. These lines highlight the high- and low-pressure centers. Arrows show wind directions. Although the winds do blow out of the high-pressure areas and into the low-pressure systems, the apparent curvature caused by the Coriolis effect swings them to the right of their path in the Northern Hemisphere. In the Southern Hemisphere, the Coriolis effect shifts the winds to the left of their path. In fact, over long distances, the Coriolis effect is so important that winds generally blow almost parallel to the isolines rather than following the pressure gradient from high pressure directly to lower pressure.

If you look at the ocean currents map in the *Earth Science Reference Tables*, you will notice that most or them circle clock-

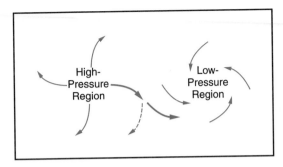

Figure 21-9 In the Northern Hemisphere, winds curve to the right as they exit a high-pressure system. To converge into a low-pressure area, they curve to the left. The dashed arrow shows that without this change the winds would not converge into the low.

wise (to the right) in the Northern Hemisphere and counter-clockwise (to the left) south of the equator. The winds show a similar apparent curvature. In the Northern Hemisphere, winds that flow out of a high-pressure area turn clockwise while winds flowing into a low-pressure area turn counter-clockwise. In the Southern Hemisphere the situation is reversed. Winds that flow out of a high-pressure area turn counterclockwise while the winds flowing into a low-pressure area turn clockwise. Notice that although the winds exiting a high-pressure center in Figure 21-8 curve right, they bend in the opposite direction as they approach low-pressure regions. What could cause them to curve in the "wrong direction" as they blow into a low? The easiest way to explain this change is to point out that if the winds continued to circle to the right, they would move away from the center of the low-pressure region. The dashed line in Figure 21-9 illustrates this. Therefore, to follow the pressure gradient, regional winds change their curvature as they converge into low-pressure systems.

 Prevailing Winds

In New York State, winds blow from the west and southwest more often than they come from any other direction. Remember that winds are labeled according to the direction you face when you look into the wind. **Prevailing winds** refer to the most common wind direction and speed at a particular location and time of year. Figure 21-10 shows two diagrams of Earth. Part A shows how terrestrial winds might blow if

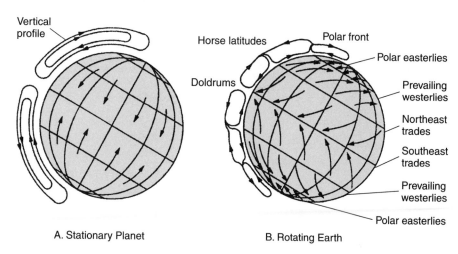

Figure 21-10 If Earth did not spin, wind patterns would be simpler. However, due to Earth's rotation and the Coriolis effect, winds appear to curve, creating prevailing winds from the west and the east and several convection cells in each hemisphere.

A. Stationary Planet

B. Rotating Earth

Earth were not spinning. Cold air would sink at the poles and travel along the surface toward the equator. Strong sunlight heating the air near the equator would cause the air to rise and move back toward the poles. Two large convection cells, as shown in the vertical profile, would dominate planetary winds.

Earth's rotation modifies this motion through the Coriolis effect as shown in part B. Winds curving to the right in the Northern Hemisphere and to the left in the Southern Hemisphere break the two convection cells shown in part A into six convection cells. Within each cell, winds curving to the right in the Northern Hemisphere and to the left in the Southern Hemisphere change the North and South winds into winds east and west winds. Regional weather systems (highs and lows) complicate the pattern even more. Winds can come from any direction depending on changes in the pressure gradient.

Monsoons

Large continents create seasonal changes in the direction of the prevailing winds. These seasonal wind directions are called **monsoons**. They are similar to land and sea breezes; but monsoon winds last for months and move over greater

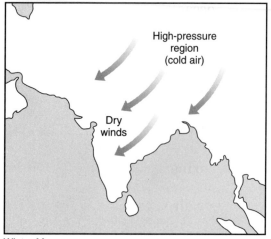

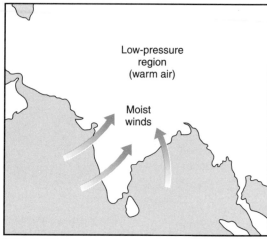

Winter Monsoon

Summer Monsoon

Figure 21-11 Seasonal changes in temperature and atmospheric pressure over the continent of Asia result in seasonal winds called monsoons. Rainfall in India depends on these seasonal changes.

distances. The monsoons of India are a prime example. During the winter, when the sun is always low in the sky, the continent of Asia cools. Cooling of the air creates a long-lasting high-pressure zone over central Asia. Sinking air spreads southward over India, bringing in dry air from central Asia. Rain is very scarce in India during this part of the year. In fact, the dry air becomes warmer as it descends from the high plateaus; so the relative humidity actually decreases.

The monsoon climate is very different in summer. Central Asia becomes warmer as the sun moves higher in the sky. By midsummer, rising air over the continent draws in moist winds from the Indian Ocean. This brings much needed rain to the Indian subcontinent. The rains allow farmers to grow crops. Some years the summer monsoon winds are very weak and the rains come late, if at all. This causes food shortages for the millions of people who depend on the rain brought by the summer monsoon. Figure 21-11 is a simplified map of the seasonal changes in wind direction over India called monsoons.

New York State does not experience dramatic seasonal changes in wind direction and precipitation. However, the southwestern desert of the United States does experience monsoons. The dry conditions of spring and early summer are replaced by moist winds and occasional thunderstorms as summer winds bring moisture off the Pacific Ocean and into the deserts.

WHAT ARE JET STREAMS?

Jet streams were discovered during the Second World War when the pilots of high-altitude aircraft found themselves traveling much slower than their air speed indicated. Today, aircraft will sometimes change their flight paths to take advantage of fast tail winds, or to avoid fighting head winds. Wandering currents of air far above Earth's surface are known as **jet streams**. With wind speeds that can be greater than 160 km/h (100 mph), jet streams circle the globe, usually in the middle latitudes. Jet streams seldom follow surface winds and usually occur where cold polar air meets warmer air in the mid-latitudes. They circle the globe from west to east, usually in the upper part of the troposphere.

Meteorologists need to know the location and speed of these upper atmosphere winds because they influence the development and movements of storm systems. Figure 21-12

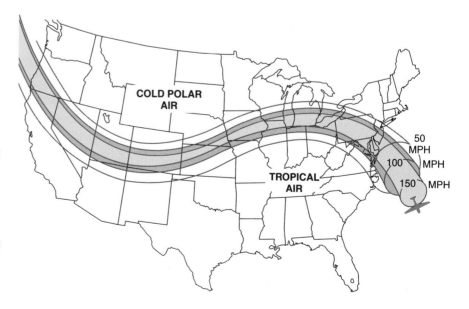

Figure 21-12 The jet stream is a narrow band of high-altitude wind that separates cold polar air from warmer air to the south. The jet stream gives rise to weather systems and steers their movements.

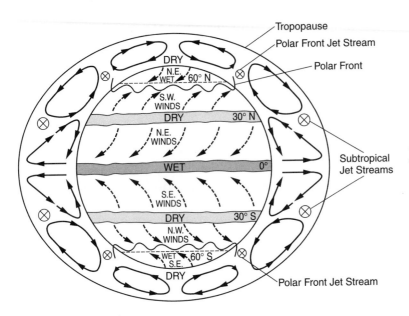

Figure 21-13 The combined effects of uneven heating by the sun and Earth's rotation (the Coriolis effect) set up patterns of atmospheric convection and prevailing surface winds. Zones of moist weather occur where rising air currents cause clouds and precipitation. Deserts are most common in the zones of sinking air.

shows a typical path of the jet stream crossing the United States from west to east.

The path of the jet stream is changeable as it meanders around the globe. In fact, two jet streams sometimes develop in the Northern Hemisphere. They tend to occur at the northern and southern limits of the zone of prevailing westerly winds. Figure 21-13, taken from the *Earth Science Reference Tables*, is a generalization of the pattern of winds on our planet. This diagram shows the large convection cells responsible for prevailing surface winds at various latitudes. Notice how the jet streams generally occur in the regions between the circular convection cells.

Notice in Figure 21-13 how rising and sinking air currents create wet and dry zones at particular latitudes. Where the air is often rising, such as along the equator, the cooling of warm, moist air creates clouds and precipitation. (Remember that air expands as it rises and air pressure is reduced. Expansion causes air to cool below the dew point.) Most of the world's deserts are located approximately 30° north and south of the equator in zones of high pressure. This is where sinking air currents become warmer as they fall through the atmosphere and the relative humidity at the surface tends to be low.

WHAT ARE ISOBARIC MAPS?

Meteorologists draw isoline maps of atmospheric pressure to help them identify weather patterns and predict weather. These maps are based on measurements of barometric pressure taken throughout a large geographic region, such as the 48 contiguous United States. Figure 21-14 is a simplified isobaric map.

Winds can be inferred from an isobaric map based on the following principles.

1. Winds blow out of high-pressure areas and into low-pressure areas.

2. Due to the Coriolis effect, in the Northern Hemisphere, winds circulate clockwise as they diverge from highs. They circulate counterclockwise as they converge into the lows.

3. Winds are the fastest where the pressure gradient is greatest. This is illustrated on the map in Figure 21-14. The fastest winds are in New England, and the far West is relatively calm.

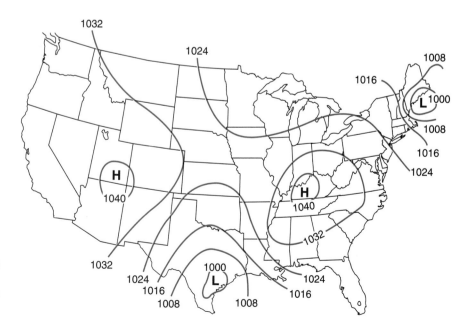

Figure 21-14 Isolines connect locations with the same atmospheric pressure and help to locate areas of high and low pressure. The numbers on the isolines represent barometric pressure in millibars.

| ACTIVITY 21-3 | SURFACE WIND PATTERNS |

Using a copy of Figure 21-14, draw arrows to represent the surface winds at the time this map was drawn. The arrows should show wind directions throughout the map region. Also indicate relative wind speeds by the length of the arrows. (Please do not write in your book.)

TERMS TO KNOW

convection cell	divergence	jet stream	prevailing winds
convergence	Doppler radar	land breeze	radar
Coriolis effect	isobar	monsoons	sea breeze

CHAPTER REVIEW QUESTIONS

1. Which weather variable is a direct result of the force of gravity on Earth's atmosphere?

   (1) barometric pressure   (3) relative humidity
   (2) cloud cover   (4) atmospheric transparency

2. Winds always blow

   (1) from high-temperature locations to low-temperature locations.
   (2) from low-temperature locations to high-temperature locations.
   (3) from high pressure to low pressure.
   (4) from low pressure to high pressure.

3. As air on the surface of Earth warms, the density of the air

   (1) decreases.   (3) remains the same.
   (2) increases.

4. During which process does heat transfer occur because of density differences in a fluid?

   (1) reflection   (3) conduction
   (2) radiation   (4) convection

5. Which atmospheric condition would cause smoke from a campfire on a beach to blow toward the ocean?

(1) warm air over the land and cool air over the ocean
(2) humid air over the land and dry air over the ocean
(3) low-density air over the land and high-density air over the ocean
(4) high pressure over the land and low pressure over the ocean

6. The air near the center of a low-pressure system usually will

(1) evaporate into a liquid.
(2) blow away from the center of the low.
(3) rise to form clouds.
(4) squeeze together to form a high-pressure system.

7. Which of the following has the greatest effect on regional wind patterns at Earth's surface?

(1) charged particles given off by the sun
(2) gravitational force from the moon
(3) Earth's yearly revolution around the sun
(4) rotation of Earth on its axis

8. The diagram below shows some examples of how surface winds are deflected in the Northern and Southern hemispheres because of Earth's rotation.

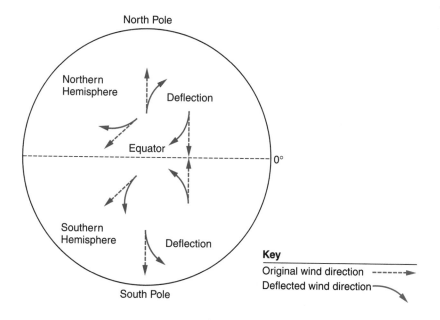

Earth's rotation causes winds to be deflected to the

(1) right in both Northern and Southern hemispheres.
(2) right in the Northern Hemisphere and left in the Southern Hemisphere.
(3) left in the Northern Hemisphere and right in the Southern Hemisphere.
(4) left in both Northern and Southern hemispheres.

9. Which diagram below best shows the circulation of air around a Northern Hemisphere high-pressure center?

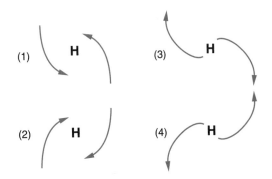

10. What is the most common wind direction 15° south of Earth's equator?

(1) northwest   (3) southwest
(2) northeast   (4) southeast

11. What is the general pattern of air movement on March 21 at Earth's equator?

(1) upward, due to low temperature and high pressure
(2) upward, due to high temperature and low pressure
(3) downward, due to low temperature and high pressure
(4) downward, due to high temperature and low pressure

12. Which kind of wind is best described as a strong west to east current of air high in the troposphere that guides weather systems across North America?

(1) prevailing winds
(2) the jet streams
(3) mid-latitude westerly winds
(4) polar east winds

Use the weather map below to answer the next two questions. Points A, B, C, and D are locations on Earth's surface.

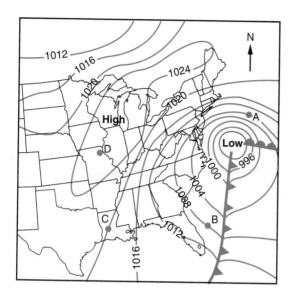

13. The isolines on the map represent values of air

(1) density.   (3) pressure.
(2) humidity.   (4) temperature.

14. The strongest winds are closest to location

(1) A.   (3) C.
(2) B.   (4) D.

15. Which of the following changes is likely to cause an increase in wind velocity?

(1) an increase in cloud cover
(2) an increase in the pressure gradient
(3) a decrease in the rate of precipitation
(4) a decrease in the temperature gradient

Open-Ended Questions

16. Why does the atmospheric pressure usually decrease when the air becomes more humid?

17. Name the instrument used by meteorologists to measure air pressure.

18. The diagram below represents summer afternoon conditions at an ocean shoreline location. Weather conditions are stable with no significant pressure gradient from regional high- or low-pressure systems. On a copy of this diagram draw arrows to show the most likely wind direction at the shoreline caused by the temperature conditions shown in the diagram. Show both horizontal and vertical motion of the air.

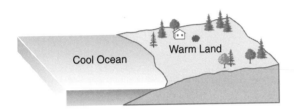

19. The diagram below shows the position of a strong low-pressure system located over central New York State. Make a copy of this diagram and draw three arrows at positions A, B, and C to show the direction of the movement of surface winds outside the center of the low.

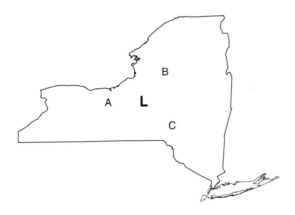

20. What is meant by prevailing winds?

Chapter 22

Weather Maps

 ## WEATHER FORECASTING

Everybody talks about the weather but no one does anything about it. This statement (attributed to Mark Twain) expresses our dependence on weather and our inability to control it. If people cannot change the weather, at least they can improve their ability to predict weather and prepare for weather events. Weather forecasting has improved dramatically over the past century. Meteorologists can now predict the weather several days in advance with remarkable accuracy. However, there are limitations on their ability to predict future weather.

For many centuries, people depended on their limited understanding of weather to forecast weather events. Red sky at night, sailors delight. Red sky in the morning, sailors take warning. These statements are based on the observation that a buildup of clouds in the morning commonly occurs before a storm. However, individual observations at one location may not give enough information to make forecasts accurate for more than a few hours into the future.

That picture began to change about a century ago. Electronic communications, including telegraph and radio, extended awareness of present weather conditions over much larger areas. At the same time, the emerging science of meteorology enabled scientists to see how regional weather systems develop. Weather satellites and regional radar have enabled meteorologists to see the broad scope of weather systems.

Electronic computers came into use in the second half of the twentieth century. These machines can quickly analyze huge amounts of data. Meteorologists thought that the increase in data, along with mathematical models of the atmosphere, would greatly increase their powers of forecasting. Some envisioned accurate predictions of weather weeks or even months into the future. Although forecasting has become more accurate, this predicted range has not been achieved.

During the same century, scientists learned that nature is not as predictable as they had hoped. The atmosphere is a complex, fluid system. Within that system, tiny uncertainties grow to major importance through time. It has been suggested that the way a butterfly moves its wings in Japan can influence the development and path of an Atlantic hurricane several weeks later. This is sometimes called extreme sensitivity to initial events. What scientists mean by initial events is not just that butterfly in Japan. It is the millions of such observations that can be made. The complexity of the atmosphere and its motions are too great to permit accurate long-range forecasting.

WHAT ARE AIR MASSES?

Large bodies of air located over a particular region acquire the characteristics of temperature and humidity characteristic of that region. A body of air that is relatively uniform in temperature and humidity is an **air mass**. While temperature and pressure are used to characterize air masses, these measures also influence other weather variables. For example, temperature and humidity affect air density. So air that is cold and dry tends to occur where barometric pressure is high. Winds, in turn, are affected by the air-pressure gradient. Cloud cover and precipitation are related to humidity. Therefore, when meteorologists characterize an air mass by its temperature and humidity they may imply other weather variables as well.

 ## Source Regions

Based on the temperature and humidity of an air mass, the location where it originated can often be inferred. This is known as the **source region** for the air mass. For example, a relatively cold, dry air mass moving from west to east across New York State most likely came from central Canada. In central Canada, there is relatively little open water and cooler temperatures prevail. An air mass that is warm and moist most likely originated over the South Atlantic Ocean or the Gulf of Mexico. This kind of air mass is likely to bring lower air pressure with clouds and precipitation.

Describing Humidity The terms *continental* and *maritime* are used to describe the humidity of an air mass. **Continental air masses** have relatively low humidity because they originate over land. **Maritime air masses** have relatively high humidity because they originate over the ocean or another large body of water. (Maritime refers to the oceans or human activities associated with oceans.)

Describing Temperature The terms *tropical*, *polar*, and *arctic* define the temperature characteristics of an air mass. **Tropical air masses**, because they originate close to the equator, are relatively warm. **Polar air masses** are cool because they originate near one of Earth's Poles. In the winter when an especially cold air mass that originated in the Arctic enters the United States from Canada, meteorologists call this an **arctic air mass**. Figure 22-1 on page 544 shows the source regions for air masses moving into the United States.

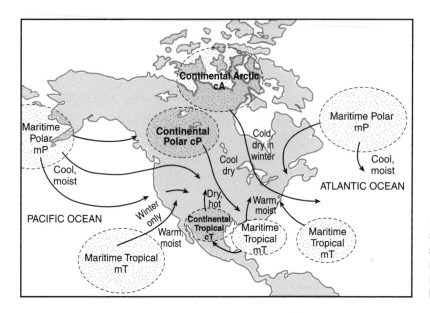

Figure 22-1 This map shows the most common source regions for the air masses of North America. Notice that each air mass has a two-letter code.

 Air Mass Codes

Each type of air mass has a unique two-letter code that includes a small letter followed by a capital letter. These codes are listed in the *Earth Science Reference Tables*. Five air mass types are listed below along with their most common North American source regions.

mT   maritime tropical: A warm, moist air mass that probably came from the South Atlantic Ocean or the Gulf of Mexico

cT   continental tropical: A warm, dry air mass that may have originated in the deserts of the American Southwest or the land regions of Mexico.

mP   maritime polar: A cold, moist air mass that most likely moved in from the northern parts of the Atlantic or Pacific Ocean.

mT   maritime tropical: A warm and moist air mass from the Gulf of Mexico or the Atlantic Ocean south of the United States.

cA   continental arctic: An unusually cold and dry air mass from Arctic Canada.

Note that the letter that designates temperature is capitalized while the letter designating humidity is lowercase. Temperature has a much greater influence on air pressure than does moisture content.

ACTIVITY 22-1 | **IDENTIFYING AIR MASSES**

Use weather maps from a newspaper or from the Internet to identify different air masses invading the United States and infer their source regions. You may decide to use temperature isoline maps to help you locate regions with relatively uniform temperatures.

WHAT ARE MID-LATITUDE CYCLONES AND ANTICYCLONES?

The term **cyclone** has many meanings, even in meteorology. In general, it means a region of relatively low atmospheric pressure. The same term is also applied to hurricanes in the Indian Ocean, and occasionally is used as a synonym for tornado. These are two kinds of intense low-pressure weather features. In this chapter you will learn about **mid-latitude cyclones**, areas of low pressure or storm systems, such as those that usually move west to east across the continental United States.

In Chapter 21, you learned that when air is heated or when moisture is added, the air becomes less dense. This results in lower barometric pressure and causes the air to rise. Rising air is part of heat flow by convection. Rising air draws air from all sides into the low-pressure center.

A low-pressure system is known as a cyclone. A high-pressure system is the opposite of a low-pressure system. Therefore a high-pressure system is called an **anticyclone.**

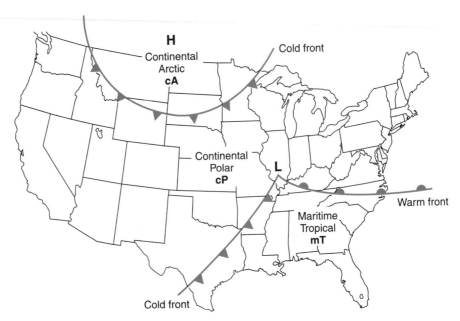

Figure 22-2 The three air masses on this weather map are separated by interfaces called fronts. The symbols on the frontal lines show the direction of wind movement, spreading outward from the high and converging in a counterclockwise circulation around the cyclone (low). The arctic air mass following the polar mass identifies this as a winter weather map.

Unlike cyclones, anticyclones usually bring clear weather. Within an anticyclone, cool, dry air descends and spreads over the surface. Anticyclones do not draw in air masses. A high-pressure system is a single, spreading air mass that generally covers the surface with cool, dry, and stable weather.

It is important to recall that although winds generally blow from high-pressure areas to low-pressure areas, the Coriolis effect makes them appear to curve along Earth's moving surface. Consequently, we see them curve to the right moving out of a high-pressure zone and then turn left into a cyclone. You will see this wind pattern around all Northern Hemisphere, mid-latitude cyclones. (In the Southern Hemisphere, winds curve left when moving out of a high-pressure zone and turn right into a cyclone.)

Figure 22-2 is a weather map of the United States that shows the location of three air masses in the winter. The boundaries, or interfaces, where air masses meet are known as weather **fronts**. Fronts are important because weather changes occur rapidly along fronts as one air mass replaces another. The symbols on the fronts that separate these air masses will be explained later. Remember that winds diverge from a high-pressure center and circulate counterclockwise as they converge into the center of the low. The symbols along the front lines indicate the direction of the winds.

HOW ARE WEATHER FRONTS IDENTIFIED?

Like other interfaces in nature, fronts are places where energy is exchanged. Weather changes generally occur along weather fronts. For example, stormy weather is usually associated with the passage of fronts and strong cyclonic systems. When meteorologists keep track of weather fronts, they observe and forecast weather over broad regions. Meteorologists recognize four kinds of weather fronts: cold fronts, warm fronts, stationary fronts, and occluded fronts.

Figure 22-3 from the *Earth Science Reference Tables* shows the standard symbols that are used to label fronts on weather maps. For example, a cold front moving south is represented by triangular points on the south side of the front line. The "swirl" symbol on the right side of Figure 22-3 is used to mark the center of a hurricane.

 Cold Fronts

Cold fronts are so named because they bring cooler temperatures. Figure 22-4 on page 548 is a three-dimensional view of a cold front. Because cold air cannot hold as much moisture as warm air, it is also likely to be less humid. Cold and dry air is more dense than air that is warmer and more moist. Therefore, the denser cold air will wedge underneath the warmer air it is replacing and force the warm air mass to rise. Rising air expands and becomes cooler. If the warmer air mass being

Air Masses	Front Symbols	Hurricane
cA continental arctic	Cold ▲▲▲▲	
cP continental polar	Warm ⌒⌒⌒⌒	
cT continental tropical		❟
mT maritime tropical	Stationary ⌒▼⌒▼	
mP maritime polar	Occluded ▲⌒▲▲	

Figure 22-3 Weather map symbols

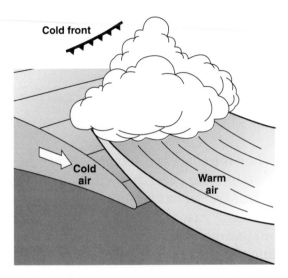

Cold front

Cold air

Warm air

Figure 22-4 Cold, dense air wedges under warmer air and pushes it upward. The rising air often causes cloud formation and precipitation along the front. Following the passage of a cold front, the weather is cooler and usually more dry. On weather maps, cold fronts are labeled with triangles pointed to show the direction of frontal movement.

pushed up is humid, there will be probably be cloud formation and precipitation along the cold front. Dark, puffy cumulus clouds such as those in Figure 22-5 are typical along cold fronts.

Cold fronts usually pass quickly, typically in an hour or two. Precipitation tends to be brief followed by clearing skies and falling temperatures. Cold air is more dense than warm air, so the passage of a cold front usually results in higher barometric pressure. In the summer, cold fronts often bring thunderstorms and cool relief from hot, humid weather. In winter, the passage of a cold front usually results in crisp, clear conditions.

Figure 22-5 A bank of puffy cumulus clouds often announces an approaching cold front.

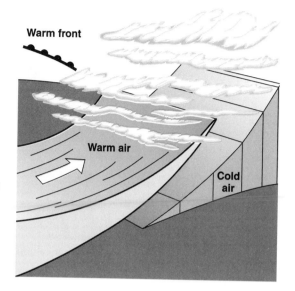

Warm front

Warm air

Cold air

Figure 22-6 A wedge of warm air forced up over cooler air exerts the gentle push of a warm front. The half circles used to label warm fronts are drawn on the side of the front to which it is advancing.

Warm Fronts

Just as cold fronts cause the temperature to fall, warm fronts bring warmer temperatures. Figure 22-6 is a three-dimensional representation of a warm front. Unlike the typical passage of cold front, warm fronts may take a day or two to pass through. Warm air is lighter and less dense than cold air, so warm air does not have the ability push aside cold air. As a result, the warm air drags the cold air mass. Therefore, the warm front moves slower than the cold front. A wedge of warm air rides up and over the air mass it is replacing. The approach of a warm front can be predicted when you see high, thin cirrus clouds that gradually become thicker and lower. As the warm front approaches, stratus clouds cover with sky like a blanket. Figures 22-7 and 22-8 on page 550 show clouds that may appear as a warm front approaches.

The rising air along the warm front often leads to thick layered clouds and gentle but long-lasting precipitation. As the front passes, the temperature gradually increases. More humid conditions commonly follow the frontal interface, as well. Since warm air is lower in density, atmospheric pres-

Figure 22-7 High thin clouds like these may signal an approaching warm front.

sure generally decreases with the passage of a warm front. On weather maps, half circles along the front line indicate advancing warm air.

The weather map in Figure 22-2 on page 546 shows two cold fronts and a warm front. The cold front that extends southwest from the low-pressure center is the advancing edge of a cold, polar air mass. The second cold front moving south from Canada defines an outbreak of especially cold arctic air. Meanwhile, circulation about the low is drawing in a warm air mass and slowly pushing a warm front to the north over the eastern seaboard.

Figure 22-8 As you may have learned in Chapter 5, strata are layers. Stratus clouds like these are typical of warm fronts.

Occluded Fronts

Because cold fronts move faster than warm fronts, a cold front sometimes overtakes a warm front. When this happens, the interface between warm and cold air masses is pushed off the ground and up into the atmosphere. A body of warm air is held aloft by two cooler air masses that merge beneath it. This is called an occluded front as shown in Figure 22-9.

Passage of an occluded front produces a spell of cloudy weather that may or may not bring precipitation. Because the warm air-cold air interface is isolated above the surface, people at ground level might not notice any significant change in temperature as an occluded front passes through. On a weather map, alternating triangles and half circles show the direction in which an occluded front is moving.

Stationary Fronts

The word *stationary* means not moving. Compared with warm and cold fronts, which advance relatively predictably, stationary fronts can remain in one location, keeping skies cloudy for hours or even days. The eventual direction of movement of the frontal interface is difficult to anticipate. There-

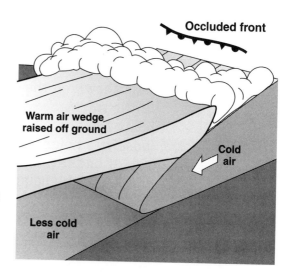

Figure 22-9 When a cold front overtakes a warm front and pushes the warm mass high above Earth's surface, an occluded front is produced.

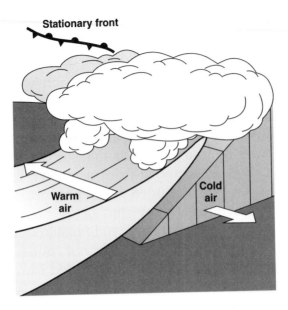

Stationary front

Warm
air

Cold
air

Figure 22-10 Winds blowing in opposite directions on opposite sides of an interface between air masses often characterize a stationary front.

fore, temperature changes along the front are not as predictable as they are when a cold or a warm front passes through. Winds blow in opposite directions on either side of the front as shown in Figure 22-10.

On a weather map, triangles and half circles are drawn on opposite sides of the front. A stationary front can become a warm front if it advances toward the side of the half circles. If the stationary front advances toward the triangles, it becomes a cold front.

The Polar Front

The polar front is the boundary between two great convection cells shown on the Planetary Wind and Moisture Belts chart in the *Earth Science Reference Table* and Figure 21-13. The polar front is also the most common path of the upper atmosphere polar jet stream. The jet stream begins the formation of mid-latitude cyclones and steers them across Earth's surface. If the polar front moves south, it advances as a cold front. Moving to the north it becomes a warm front. In theory, the polar front is a continuous boundary that circles Earth at the northern edge of the zone of prevailing westerly winds. However, the polar front is not very stable and it is often broken apart by cyclonic circulation.

In the summer, the polar front is usually located far north of New York State. In warm weather, New York is more often affected by weather systems that develop along the subtropical jet streams at lower latitudes. However, in the winter, the global wind belts shift to the south. This brings the polar front closer to New York. Outbreaks of arctic air push the polar front through New York and bring winter storms. In spring, the polar front returns to arctic Canada as summer weather patterns settle in.

HOW DO MID-LATITUDE CYCLONES EVOLVE?

Low-pressure systems often develop as swirls along the polar front. Figure 22-11 shows the evolution of a typical mid-latitude cyclone. This sequence is a cycle in which the last

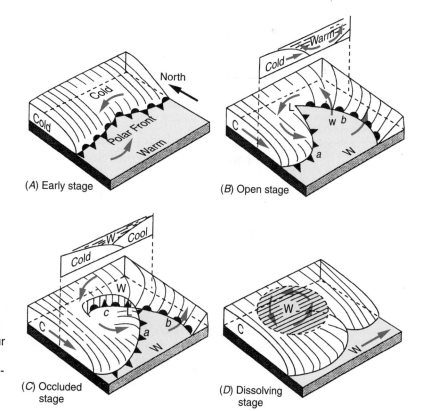

Figure 22-11 Eddies in the polar front can develop into low-pressure systems known as mid-latitude cyclones. Four types of weather fronts take part in this process that eventually returns to a stationary part of polar front.

(A) Early stage

(B) Open stage

(C) Occluded stage

(D) Dissolving stage

stage returns to the initial pattern. Not all weather systems develop in this clear sequence, but the model is useful in understanding the structure and formation of mid-latitude cyclones. As development occurs, low-pressure systems seldom stay in one place. They generally follow prevailing wind directions. In the United States, most low-pressure systems move across the country toward the east or northeast. For this reason, people sometimes say that if you want to know tomorrow's weather in New York, look at today's weather in the Midwest.

The early stage (*A*) shows the polar front as a stationary front. Winds blowing in opposite directions along the front create friction at the interface, and a swirl starts to develop. In the open stage (*B*), winds advancing toward the front change the stationary front to a cold front at *a* and a warm front at *b*. Notice the symbols for these two kinds of fronts, which match the key to front symbols in the *Earth Science Reference Tables*. The profile at the top of the diagram shows cold air starting to push the warm air mass upward.

The third stage (*C*) is called the occluded stage because the cold front is overtaking the slower moving warm front. This creates a new kind of front at *c*: an occluded front. The profile above shows the warm air now pushed completely aloft above the occluded front. In the dissolving stage (*D*), the warm air pushed aloft eventually loses its identity as it mixes with the air around it. Although this is the final stage of the mid-latitude cyclone, a stationary front is forming, which returns to first-stage conditions.

These diagrams do not show cloud formation and precipitation. Wherever warm, moist air is pushed upward, expansion and cooling cause clouds and precipitation. This occurs primarily along the fronts.

ACTIVITY 22-2   STAGES OF CYCLONIC DEVELOPMENT

Find weather maps from newspapers or other media sources to illustrate each of the four stages of cyclonic development shown in Figure 22-10. Label each with the proper name of the stage.

HOW ARE WEATHER DATA RECORDED?

When meteorologists make weather maps, they need to consider many factors over a broad geographic region. It is helpful to show many observations for each weather station in the map area. The time of these simultaneous observations is usually indicated somewhere on the weather map. To place as much information as possible at each observing station, a standard system is used to display abbreviated weather data. It is important to understand how to extract the information from weather stations shown on the map.

The first rule is that each weather variable has a unique position around a circle that indicates the location of the weather station. For example, barometric pressure in millibars is always indicated as a three-digit number above and to the right of the circle. The dew-point temperature in degrees Fahrenheit is always found at the lower left. Figure 22-12 is the weather station model printed in the *Earth Science Reference Tables*. It is common for weather stations to leave out some information, often because it is not important. However, the information shown will always occupy the standard position with respect to the station circle.

Air temperature   Also known as dry-bulb temperature, it is always recorded in degrees Fahrenheit. This temperature can be converted to a Celsius temperature using the temperature conversion scale on the same page of the *Earth Science Reference Tables* as the station model. (For example, the 28°F shown here is approximately –2°C.)

Present weather   This term describes the atmospheric conditions such as precipitation or limited visibility. The present weather symbols shown in the box below the station model are examples of the kinds of conditions that can be indicated at this place on the weather station model. The United States Weather Service uses many more symbols than are found in the *Earth Science Reference Tables*. This weather station in Figure 22-12 on page 556 shows that it was snowing at the time of the observations.

Weather Map Symbols

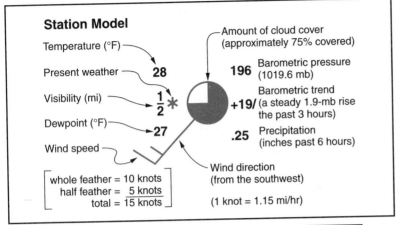

Station Model

Temperature (°F)

Present weather

Visibility (mi)

Dewpoint (°F)

Wind speed

28

$\frac{1}{2}$ *

27

whole feather = 10 knots
half feather = 5 knots
total = 15 knots

Amount of cloud cover
(approximately 75% covered)

196   Barometric pressure
(1019.6 mb)

+19/   Barometric trend
(a steady 1.9-mb rise
the past 3 hours)

.25   Precipitation
(inches past 6 hours)

Wind direction
(from the southwest)

(1 knot = 1.15 mi/hr)

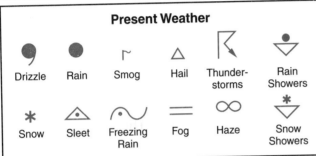

Present Weather

Drizzle	Rain	Smog	Hail	Thunder-storms	Rain Showers

Snow	Sleet	Freezing Rain	Fog	Haze	Snow Showers

Figure 22-12 Meteorologists use a system of placement and coding to show a variety of weather observations at a single location on a weather map.

Visibility This indicates the greatest distance, measured in miles, at which objects can be identified. Visibility is reduced by such conditions as precipitation, fog, or haze. In Figure 22-12 the half-mile limit of visibility is probably caused by snowfall.

Dew point You may recall that the dew point is the temperature at which the air would be saturated if it were cooled. Although dew point is a temperature, it is also an indicator of the absolute humidity. In this station model, the dew point of 27°F is very close to the air temperature of 28°F. This indicates a high relative humidity, which is not surprising because it is snowing.

Wind direction A line connected to the circle indicates wind direction. Just as north is at the top of most maps, these symbols follow the standard compass directions.

Think of the circle as the head of an arrow. In this case, the wind is blowing out of the southwest. (Wind, you will recall, is always named according to the direction from which it comes.) A line extending directly above the circle would indicate a north wind.

Wind speed   Small lines at the end of the wind direction indicator are called feathers. This example shows a 15-knot wind. (As printed on the chart, 1 knot is 1.15 mph.) Wind speeds are rounded to the nearest 5 knots. A 10-knot wind would be shown a single feather at the end of the wind direction line. Two feather lines of equal length would indicate a 20-knot wind. A half feather indented a short distance from the end of the wind direction line would indicate a 5-knot wind. A circle in a circle indicates a calm area on the weather map. In this way, a clam area is not mistaken for a station in which the data is missing.

Cloud cover   Small circles like those in Figure 22-12 are drawn on the map at the position of the weather observing station. The amount of the circle that is dark is used to indicate the portion of the sky covered by clouds. The example shows that the sky is mostly cloudy. A circle that is all white in the center represents a clear sky.

Air pressure   Air pressure is recorded as a three-digit number. It helps to realize that the normal barometric pressure at Earth's surface is generally about 1000 millibars (mb). So the first digit (or the first two digits) are not shown. Furthermore, barometric readings are recorded to the nearest tenth of a millibar. If the three-digit code is less than 500, add a "10" at the beginning and a decimal point before the last number. If the three-digit code is more than 500 add a "9" before the number and a decimal before the last digit. Do whichever procedure brings you closer to 1000 mb. For example, if the barometric pressure were listed as 162, this would translate to 1016.2 mb. However, 994 would become 999.4 mb. If you do not get a value close to 1000 mb you have probably decoded the number incorrectly. The barometric pressure indicated in Figure 22-12 is 1019.6 mb.

Precipitation   The amount of rain or snow in the past 6 hours is sometimes shown. In Figure 22-12 the weather station has received an amount of snow equivalent to one-quarter of an inch of rainfall. (Precipitation is always shown as its liquid water equivalent.)

Barometric trend   Rising air pressure usually brings clearing weather. Stormy weather is more likely if the atmospheric pressure is going down. A "+" indicates rising air pressure and a "−" indicates falling air pressure over the past three hours. The number is in millibars and tenths of millibars. (You need to insert a decimal between the two numbers.) Next to the two-digit number is a line indicating the pattern of change. The data shown in Figure 22-12 indicate a steady rise of 2.3 mb in the last 3 hours.

Note: There is a guide to the placement on this information, as well as how to interpret weather station data, in the *Earth Science Reference Tables*.

SAMPLE PROBLEMS

Problem 1   What weather conditions does the station model shown in Figure 22-13 indicate?

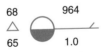

Figure 22-13

Solution   This station model indicates a temperature of 68°F and a dew point of 65°F. Hail is falling. The barometric pressure is 996.4 millibars. Winds are from the east at 5 knots. There has been an inch of rain in the past 6 hours. The sky is half (50%) covered with clouds.

Problem 2   Draw a station model to represent the following conditions: Temperature and dew point are 62°F, and there is fog. Visibility is only one-tenth of a mile. Winds are from the west at 10 knots. The air pressure is 1002.0 millibars and it has gone down steadily 0.9 millibar in the past 3 hours. Cloud cover is 25%.

Solution   Figure 22-14 shows these conditions. The fog symbol and $\frac{1}{10}$ visibility fraction can be shown above or below the wind direction line. The feather indicating wind speed can slant up or down. Otherwise the numbers and symbols must be in the positions shown in the figure.

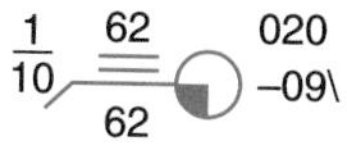

$$\begin{array}{ccc} \frac{1}{10} & 62 & 020 \\ & 62 & -09\backslash \end{array}$$

Figure 22-14

Problem 2   Draw a station model to indicate a temperature of 20°C, a dew point of 11°C, and clear skies.

Solution   Temperatures on the station model are always shown in degrees Fahrenheit. The temperatures in Sample Problem 2 must be converted from Celsius to Fahrenheit before they are placed in their proper positions next to the station circle as shown in Figure 22-15.

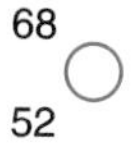

68
52

Figure 22-15

ACTIVITY 22-3   **CURRENT STATION MODELS**

Create station models to describe the local weather conditions as recorded at a particular time over a period of several days. (Your teacher will tell you how long.) Weather data can be obtained from nearby online sources or from your own weather instruments.

 HOW ARE WEATHER MAPS DRAWN AND USED?

You have already seen several weather maps used to illus-trate wind patterns and air masses. In this section you will learn how a weather map is drawn and how weather maps are used to predict the weather for any point on the map. Because weather includes so many variables (temperature, pressure, humidity, wind, sky conditions, and precipitation) it is often useful to show many of these factors over a wide geographic area. A map that combines several weather variables is called a synoptic map because it shows a summary, or synopsis, of weather conditions.

Gathering Data

Remote sensing has become a standard observation technique available to meteorologists. Remote sensing is the use of instruments to gather information at a distance from the instrument. Radar images and photographs from satellites are good examples. You can access current satellite photographs, radar images, and other regional data on the Internet through a local office to the National Weather Service of the National Oceanic and Atmospheric Administration (NOAA) of the United States government. Other organizations such as *The Weather Channel* and news media also provide current online weather maps and images.

ACTIVITY 22-4　DRAWING WEATHER FRONTS

From the Internet, download and print a current satellite weather image of a part of the United States. On that image, label the centers of high- and low-pressure systems. Draw appropriate weather fronts using weather maps or the pattern of development you see. Label air masses by their two-letter codes. Use your drawing to predict tomorrow's weather at your location.

Using Weather Data

Professional meteorologists often have years of experience and access to more information than can be shown on a single weather map. For example, observations of the jet stream help them predict how weather systems will move. They can also use computers to access how weather changed in the past when similar conditions were in effect. Some of the most powerful computers ever built have been used to work with complex equations that represent the atmosphere. Meteorologists sometimes construct mathematical models of the dynamics (changing nature) of the atmosphere to find how far into the future accurate predictions can be made. Currently, forecasts for a few days in advance are quite reliable. Due to the complexity of air movements, it is unlikely that predictions of more than a week or so into the future will ever be truly dependable.

ACTIVITY 22-5   RELIABILITY OF WEATHER FORECASTS

Make a table to record the expected changes in weather conditions over a period of a week. A sample is shown below. You may add more weather factors if you wish. Each day, make your own ("My") predictions of how you expect the weather to change the following day. Also record how these changes are predicted in a particular news media source.

Name _____

Week of _____

CHANGES IN	Monday		Tuesday		Wednesday		Etc. . . .
	My	Media	My	Media	My	Media	
Temperature							
Cloud Cover							
Precipitation							
Winds							

The following day, circle any predictions that are incorrect and record your own predictions as well as the media predictions for the following day. Continue this procedure for at least a week.

How often do you differ from the media predictions?

What is the percentage of correct predictions for you and for the news source?

Figure 22-16 (A) Weather data collected from weather observers over a geographic region can be used to draw a weather map. (B) The isolines of barometric pressure are based on the air-pressure data displayed at each station. (C) The temperature field is apparent from these isotherms. (D) The position of the low and the two weather fronts are based on wind direction and other data from the previous maps. Precipitation usually occurs along weather fronts.

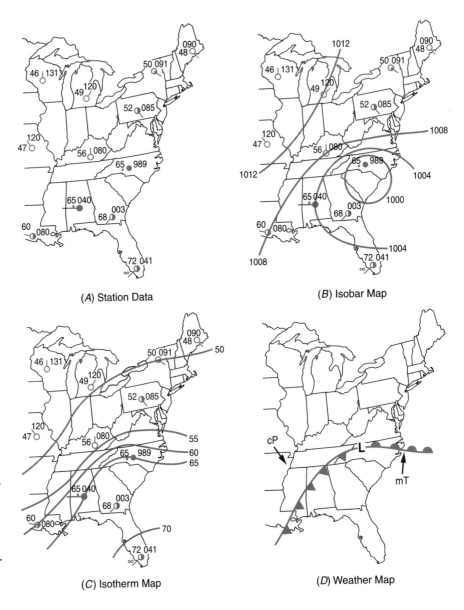

(A) Station Data

(B) Isobar Map

(C) Isotherm Map

(D) Weather Map

Drawing Weather Maps

Figures 22-16A through D illustrate how weather data are transformed into a weather map. Station data is shown in part A, isobars of air pressure in part B, isotherms of temperature in part *C*, and a weather system map in part *D*. Note that isolines are drawn as if the numbers were located at the center of the circles. The pattern of a cyclone (low-pressure system) is apparent from the data and from the circulation of winds around the North Carolina station. The changes in temperature and air pressure along with the winds and precipitation guide the placement of the cold and warm fronts.

Using Weather Maps to Make Predictions

Weather maps are important to meteorologists in forecasting weather. Figure 22-11 shows how low-pressure systems generally develop as they move across the United States from west to east, or toward the northeast. You can use current weather maps to make your own predictions of your local weather for the next day or two. Figure 22-17 is a weather map of the United States. Notice how this map can be used to make the following forecasts for these cities:

New York City   Rain will probably end as the warm front advancing northward passes through the city. Warmer weather and partial clearing will occur as a warm, moist air mass slides over New York. The density of warm,

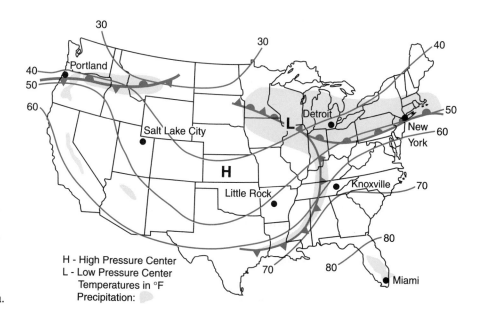

Figure 22-17 Information about weather conditions over a large region can be reported on a synoptic weather map. The map may be used to make local weather forecasts for locations throughout the area.

moist air is low; therefore, the barometric pressure will remain low.

Knoxville   The cold front approaching from the west will bring showers, followed by cooler and clear weather. The temperature and humidity will then drop, and barometric pressure will rise as a cool, dry air mass moves in with the passage of the cold front.

Little Rock   Cool, dry weather is likely to continue. It will probably be a few days before the anticyclone (high-pressure system) that is building into the area no longer dominates local weather.

Detroit   The spell of rainy weather is likely to continue for another day or so until the low-pressure, storm system moves away to the east or northeast.

ACTIVITY 22-6   MAKING DAILY WEATHER REPORTS

Use your own observations as well as information from the media or the weather bureau to make daily weather forecasts. A weather station can be set up at school. Choose a location at which to place weather instruments that will allow accurate readings and be safe from vandals. Your forecast need not agree completely with forecast from the news media if you feel that you can do better.

You may wish to post your predictions, or announce them daily to the class or to the whole school via the public address system.

TERMS TO KNOW

anticyclone	maritime air mass	source region
arctic air mass	mid-latitude cyclone	stationary front
cold front	occluded front	tropical air mass
continental air mass	polar air mass	warm front
cyclone		

CHAPTER REVIEW QUESTIONS

1. Which statement correctly matches each air mass with its usual geographic source region?

   (1) mP is from the North Atlantic Ocean and mT is from the deserts of the southwestern United States.
   (2) mP is from northern Canada and mT is from the deserts of the southwestern United States
   (3) mP is from northern Canada and mT is from the Gulf of Mexico.
   (4) mP is from the North Atlantic Ocean and mT is from the Gulf of Mexico

2. Which type of air mass usually contains the most moisture?

   (1) mT   (3) cT
   (2) mP   (4) cP

Base your answers to questions 3–5 on the weather map of North America below, which shows the location of a front and the air mass influencing its movement.

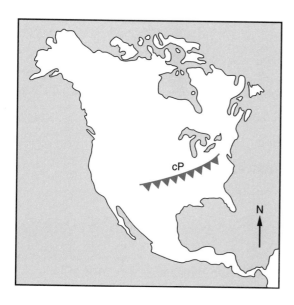

3. Which region is a probable source of the air mass labeled cP on this map?

   (1) central Canada   (3) North Atlantic Ocean
   (2) southwestern United States   (4) Gulf of Mexico

4. Which type of front and frontal movement is shown on the weather map?

    (1) cold front moving northwestward
    (2) cold front moving southeastward
    (3) warm front moving northwestward
    (4) warm front moving southeastward

5. The cP air mass is identified on the basis of its temperature and

    (1) wind direction.   (3) moisture content.
    (2) cloud cover.   (4) wind speed.

6. In which map below does the arrow show the general direction that most low-pressure storm systems move across New York State?

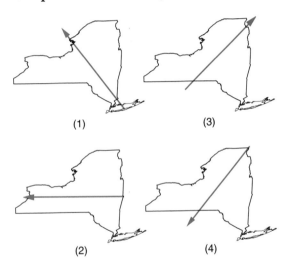

    (1)   (3)

    (2)   (4)

Use the weather station model below to answer questions 7–10.

47   101
∞
36

7. What are the sky conditions shown on this weather station model?

    (1) overcast
    (2) partly cloudy and with clean air conditions
    (3) partly cloudy with haze
    (4) light rain showers

8. What is the atmospheric pressure indicated by the weather station model above?

   (1) 101 mb

   (2) 936 mb

   (3) 1010.1 mb

   (4) 1047.0 mb

9. What is the wind speed at this location?

   (1) calm

   (2) 5 knots

   (3) 15 knots

   (4) 36 knots

10. What is the *Celsius* temperature of the dew point shown on this station model?

    (1) 2°C

    (2) 8°C

    (3) 36°C

    (4) 47°C

11. The station model below shows the weather conditions at Massena, New York, at 9 A.M. on a particular day in June.

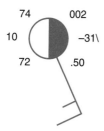

    What was the barometric pressure at Massena 3 hours earlier on that day?

    (1) 997.1 mb

    (2) 999.7 mb

    (3) 1003.3 mb

    (4) 1009.1 mb

12. Which event is most likely to bring increasing temperature and humidity conditions to New York State during the winter?

    (1) a strong anticyclone moving in from the northwest

    (2) the jet stream shifting northward

    (3) the jet stream moving south

    (4) the polar front moving to the south

The weather map below shows a typical mid-latitude, low-pressure system centered over Illinois. Letters A, B, C, and D show four locations on Earth's surface. Use this map to answer questions 13–15.

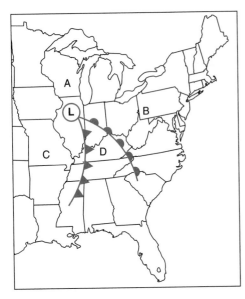

13. Another name that could be applied to regional low-pressure systems like these is

(1) tornado.
(2) thunderstorm.

(3) cyclone.
(4) anticyclone.

14. Which location most like has the highest surface temperature at the time shown on this map?

(1) A
(2) B

(3) C
(4) D

15. What is the most likely direction of movement of this low-pressure system over the next 24 hours?

(1) toward the northeast
(2) toward the southeast

(3) toward the northwest
(4) toward the southwest

Open-Ended Questions

16. The diagram below represents a part of the continental United States. A weather front is approaching a city at location X.

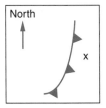

How are *both* the temperature and the barometric pressure likely to change at location X as this front passes?

17. The diagram below is a profile view of a warm front moving across the United States. The arrow shows the front moving toward the north.

Why is the maritime tropical air mass moving above (and not under) the continental polar air mass?

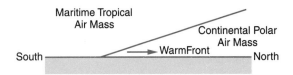

18. The map below shows a low-pressure system over part of North America. Lines AB, BC, and BD represent surface frontal boundaries. Line AB represents an occluded front at the center of a low-pressure system. Symbols cP and mT represent different air masses. Make a copy of this map to complete the following. (Please do not mark in this book.)

a. Place the proper front symbols on lines AB, BC, and BD. Place the symbols on the correct side of each to show the direction of front movement.

b. Name the geographic region over which the mT air mass most likely formed.

19. The diagram below is a station model for a location in New York State.

35   007
*
29

   a. What was the barometric pressure in millibars at this location when the data was recorded?

   b. What was the *Celsius* air temperature at this location when the data was recorded?

20. Using the proper format, draw a weather station model to include the following data collected at Boonville, New York.

Air temperature	65°F
Dew point	64°F
Visibility	2 miles
Present weather	drizzle
Wind direction	from the southeast
Wind speed	5 knots
Amount of cloud cover	100%
Barometric pressure	996.2 mb

Chapter 23

Weather Hazards and the Changing Atmosphere

 THE COST OF NATURAL DISASTERS

There are many ways to measure natural disasters. One measure is the cost to repair damage from the event. Hurricane Andrew crossed southern Florida in 1992, causing an estimated $25 billion in damage. The loss of human lives is an even greater tragedy. A disastrous cyclone (the name then applied to a hurricane in the Indian Ocean) struck Bangladesh in 1970. Bangladesh is a poor nation in the delta region of the Ganges River east of India. Scientists estimate that this storm took the lives of 300,000 people. It is difficult to estimate these grim statistics because such storms wipe out whole families and often destroy public records. Without such documents, the names and numbers of people living in the area at the time of the storm cannot be determined. Among all violent natural disasters, only a few of the largest earthquakes have killed more people than the 1970 Bangladesh hurricane.

In the United States, the greatest loss of life in any natural disaster occurred in 1900. At that time, Galveston, Texas, was the wealthiest city in the state and its major shipping port. A hurricane struck the city in September of that year. Galveston's geographic setting and the lack of advanced warning were the most important factors in this disaster. The city was built on a low-lying barrier island at the entrance to Galveston Bay. The highest point in the city was only 3 meters (10 feet) above sea level.

On the morning of the day the storm struck, no one could predict the events that would unfold in the next 12 hours. The storm came ashore when the storm surge had elevated the ocean nearly 4 meters (13 feet) above the normal high tide level, flooding the city. About 20 percent of the Galveston population of 38,000 people was lost in the 1900 storm. Figure 23-1 is a photograph taken in Galveston just after the 1900 hurricane.

A hurricane of similar strength struck the city 15 years later, but relatively few people were lost in that storm. To protect the city from the storm surge and pounding waves, a 5-meter-high sea wall had been built. Sand was used to raise the level of some parts of the island as high as the new sea wall.

Figure 23-1 Damage caused by the 1900 Galveston Hurricane.

Today, the residents of Galveston would know several days in advance if a major hurricane was approaching. The weather service gathers data primarily from weather radar and satellite images to provide storm warnings. A planned evacuation of the city could also save many lives.

WHAT WEATHER EVENTS POSE HAZARDS?

You have learned that the atmosphere can store great quantities of solar energy, primarily in water vapor. Some weather events concentrate that energy and release it to generate strong winds, excessive precipitation, and other hazards. Understanding storm systems and predicting their movements can save lives and property.

Thunderstorms

You have probably witnessed the build up of clouds leading to a thunderstorm. A **thunderstorm** produces rain, lightning, thunder, and sometimes strong winds and hail. Although thunderstorms can occur at any time of year, they are especially common during humid summer weather. Rapid updrafts caused by an approaching cold front may trigger the formation of massive storm clouds. Thunderstorms also are common in humid, tropical air masses. Some thunderstorm clouds extend to the top of the troposphere. Figure 23-2 on page 574 shows the vertical development and flared top characteristic of clouds that produce thunderstorms.

For reasons that scientists do not yet fully understand, positive and negative electrical charges separate in thunderstorm clouds. Sudden electrical discharges are observed as **lightning**. These electrical currents heat the air to temperatures higher than the surface of the sun. The electromagnetic energy they give off is observed as a flash of light, and the sudden expansion of air causes the sound of thunder.

Figure 23-2 Strong updrafts and flared tops are characteristics of clouds that produce thunderstorms.

Lightning can occur within clouds as well as between clouds, and between clouds and the ground. Lightning strikes cause about 100 deaths each year in the United States. Most lightning deaths are isolated events so they are seldom reported in the news media outside the local area. In many years, lightning is the leading cause of weather-related human fatalities in the United States.

ACTIVITY 23-1	LIGHTNING DISTANCE

(**NOTE:** Thunderstorms can be dangerous. Perform this activity only if you are in a well-protected location and under the supervision of a responsible adult.)

Light travels so fast that you see it essentially the moment it occurs. However, the sound of thunder travels much more slowly. Here is a way to estimate the delay in seconds between the flash of lightning and the sound of thunder arriving at our location. When you see the flash of lightning, begin counting slowly, one thousand one, one thousand two, one thousand three, etc., until you hear the clap of thunder. The distance to the lightning strike is about 1000 feet per second of separation, or about 1 mile for a 5-second delay (1 km for each 3-second delay.)

Flooding, strong winds, and damage from hail are other costly effects of violent thunderstorms. On a summer Saturday in 1976, a strong late afternoon thunderstorm hovered over the Rocky Mountains of Colorado. The storm dropped an estimated 10 to 12 inches of rain. Over a period of three hours the discharge of the Big Thompson River increased by a factor of 200. The water swept into a narrow canyon popular for recreational and camping activities. Along the river, 400 houses were destroyed and 145 lives were lost. Many of the victims had sought refuge from the rain and rising floodwaters in their cars. The flood swept the cars into the river, carrying them and their occupants away. Others, who abandoned their cars and scrambled up the rocks at the side of the canyon, survived and were rescued the next day.

Flash floods are a major hazard throughout the United States. They are a special hazard in the desert southwest where local roads often cross dry streambeds without bridges. Summer monsoon thunderstorms are common in this region. These storms may wash away small bridges, but the flood waters only block these crossings for a few hours at a time. The majority of flood deaths in this country are people who are swept away in their cars when they attempt to drive along roads that cross flooded streams.

The pellets of ice that fall during thunderstorms are called **hail**. Hail begins as small drops of frozen rain high in a thunderstorm cloud. As the ice is blown about within the cloud, it is repeatedly coated by liquid water. Each coating of water freezes and forms a new layer of ice. Eventually, the hailstones become too heavy to be held up by the strong updrafts, and they fall to the ground. Most hail pellets are less than 1 cm (0.4 inch) in diameter but hailstones the size of softballs have been recorded in the midwestern United States. Although rare, such large hailstones can break windows and dent cars. Figure 23-3 on page 576 shows the parts of the United States that have the most frequent thunderstorms.

Hail should not be confused with sleet. **Sleet** forms when drops of rain falling through a layer of cold air near Earth's surface freeze completely. Unlike hail, the ice particles in sleet are always the size of a raindrop or smaller (less than 5 mm), and sleet pellets are transparent. Sleet shows no evidence of

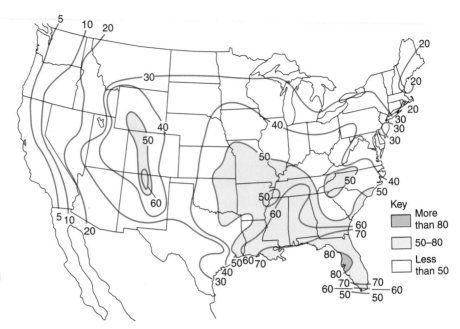

Figure 23-3
Thunderstorms are most common in the southeastern states where warm, humid conditions occur and cool continental air masses clash with tropical air masses.

multiple cycles of coating by water followed by uplift into cold air aloft where each coating freezes. Therefore, the formation of sleet does not require the violent updrafts that produce hail.

 **Blizzards**

A **blizzard** is a winter snowstorm that produces heavy snow and winds of 35 mph (56 km/h) or greater. Blowing snow, limited visibility, and cold temperatures are typical of these storms. People can be stranded by blizzard conditions. Deep snow on county roads or city streets can shut down commerce and services. The wind and snow can break electric and telephone lines. Fire trucks and ambulances may not be able to reach places where they are needed. Some residents, especially the elderly, may be in danger because they run out of food or fuel. People caught outdoors in the storm or those who perform vital emergency services may suffer from frostbite and loss of body heat (hypothermia).

Major lake-effect winter storms were discussed in Chapter 20. Most of them are associated with outbreaks of arctic air masses. These storms usually involve fierce winds and blowing snow that can pile into deep drifts. Away from major lakes, cold continental air masses usually produce limited snowfall amounts. Greater snowfall is often associated with blizzards that draw moist air off the ocean as they move north through the Atlantic coastal states.

The legendary Blizzard of 1888 dumped nearly 2 feet (60 cm) of snow on New York City. The storm brought down recently installed electrical, telephone, and telegraph wires. Public transportation was crippled for several days. Even more snow fell in inland areas, but the "modern technologies" that made the city a marvel of its age also left it more vulnerable to the storm. Figure 23-4 is a weather map of a late winter blizzard that occurred in 1993. This unusual blizzard included snowfalls as deep as 4 feet (125 cm) in some inland areas. Notice the extreme pressure gradient shown by the closely spaced isobars. This pressure gradient caused winds of hurricane force in the 1993 storm.

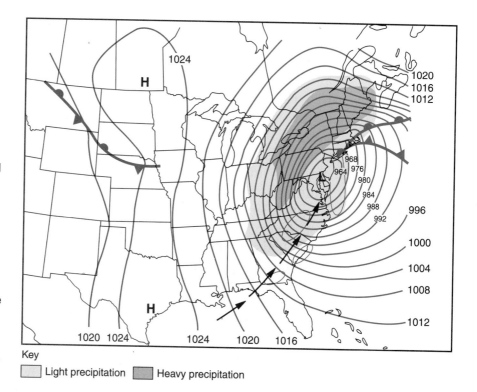

Figure 23-4 A late winter storm moved up the East Coast dumping as much as 4 feet of snow in some areas. The shading shows the most intense areas of snowfall on the morning of March 13, 1993. Notice the tightly spaced isobars, indicating an intense pressure gradient and unusually strong winds.

Key

Light precipitation Heavy precipitation

| ACTIVITY 23-2 | STORM SURVIVAL |

List supplies that can be stored at home to help your family get through a major storm event. When your list is finished, number the items from most important to least important to keep in stock. Compare your list with classmates' lists.

| ACTIVITY 23-3 | A LOCAL WEATHER EVENT |

Use local records such as newspaper articles to investigate and prepare a report on a historical weather event in your community.

 Hurricanes

Hurricanes are the most destructive storms on Earth. Most hurricanes occur in the late summer and early autumn when the tropical ocean's surface water is warmest. Solar energy and the warmth of the ocean support evaporation, which carries water vapor and energy into the atmosphere. Hurricanes begin as tropical depressions (mild low-pressure regions) that become tropical storms when their winds exceed 39 mph (63 km/h). Vertical air currents and energy released by cloud formation (condensation) strengthen the storm. When sustained winds in a tropical storm exceed 74 mph (120 km/h), the storm is classified as a **hurricane**. In the eastern Pacific, these storms are known as typhoons, and in the Indian Ocean they are called cyclones.

Atlantic hurricanes commonly begin as small disturbances off the western coast of Africa. They gather strength as they drift to the west with the northeast trade winds. Some of them turn north before they reach North America and lose strength over the cooler water of the North Atlantic Ocean. Other hurricanes come ashore along the East Coast or move parallel to the coast, causing destruction for hundreds of miles.

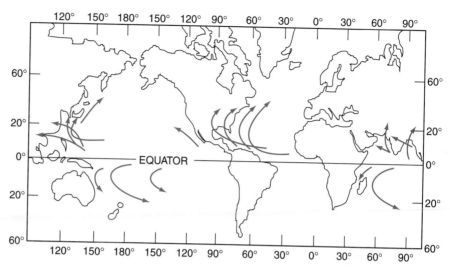

Figure 23-5 Hurricanes in the Atlantic Ocean usually drift westward toward North America. Most turn north until they enter the zone of prevailing west winds and finally drift to the east as they die out over land or cooler ocean water.

Hurricanes also originate over the Caribbean Sea or the Gulf of Mexico. They gather strength over these tropical seas before they hit land. When they move over North America they weaken because they are deprived of their fuel: water vapor from the warm ocean surface. Friction with land features is also an important reason that hurricanes usually weaken after they come ashore. Wind speed quickly decreases as the weakened hurricanes break up into a disorganized group of rainstorms that may cause flooding in inland areas. After turning northward, the remains of the storm are picked up by the eastward flowing jet stream or the zone of prevailing southwest winds. Figure 23-5 shows the most common paths taken by these tropical storms.

Hurricanes are very large. Some are more than 400 miles (600 km) in diameter. As in other Northern Hemisphere cyclones, winds circulate counterclockwise as they converge toward the center. From the outer bands of clouds to the center of the storm, wind speed generally increases, as does the intensity of precipitation. At the center of the strongest hurricanes, there is a small round area of relative calm known as the eye of the storm. In this region, the winds are light and the skies may be mostly clear. However, the calm of the eye is surrounded by the most violent part of the hurricane called the "eye wall" where the fastest winds and most intense rainfall occur. Some people think the storm is suddenly over when

Figure 23-6 This is a satellite image of Hurricane Hugo. The cloud pattern of this major hurricane shows winds circulating counter-clockwise as they converge toward the eye of the storm.

the eye of the storm passes over them. The approaching eye wall quickly brings them back to the reality of the second half of the storm. Figure 23-6 is a satellite image of Hurricane Hugo as it approached the coast of South Carolina in 1989.

The strength of hurricanes is measured on the Saffir-Simpson scale of hurricane intensity as shown in Table 23-1. Category 1 hurricanes are likely to do little damage to homes and other well-constructed buildings as long as they are not built directly along the coast. However, a category 5 hurricane will cause widespread damage and pose danger to nearly anyone caught in the strongest part of the storm.

TABLE 23-1 Saffir-Simpson Scale of Hurricane Intensity Categories

| Category | Sustained Winds | | Storm Surge | | Potential Damage |
	km/hr	mi/hr	m	ft	
1	119–154	74–95	1–2	4–5	Scattered
2	155–178	96–110	2–3	6–8	Moderate
3	179–210	111–130	3–4	9–12	Extensive
4	211–250	131–155	4–6	13–18	Extreme
5	>250	>155	>6	>18	Catastrophic

Hurricanes present the greatest danger to people in low-lying coastal cities, such as New Orleans, Louisiana, and Charleston, South Carolina, and even parts of New York City. Deadly hurricanes have come ashore along the coastline from Texas to New England. While hurricanes bring strong winds and heavy rainfall, the greatest cause of damage and fatalities is the storm surge. A storm surge forms as the result of the combination of very low atmospheric pressure and strong on-shore winds. These conditions can elevate the ocean's level to the point that the beach no longer protects buildings and other structures from the storm waves. The water rises so high that the waves hammer directly on homes and other structures along the oceanfront. People who have not evacuated the area may find themselves with no escape route, stranded for hours by violent seawater in a collapsing building. If the storm comes ashore at high tide, the height of the storm surge presents an even greater danger.

Over the past century, the cost of hurricane damage in the United States has increased, but the death toll has dropped as shown in Figure 23-7. People can move to higher ground, especially when evacuation routes have been established and updated warnings are given on radio and television. The principal cause of the increasing cost of hurricane damage is the development of shoreline areas. Although people can leave the shore for higher ground, the homes and resorts along the ocean remain vulnerable.

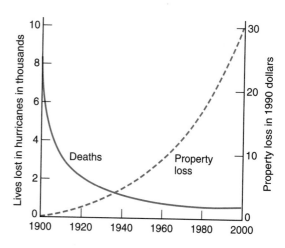

Figure 23-7 Over the past century, the number of deaths and injuries from hurricanes has decreased due to advanced storm warnings and emergency planning. However, property losses have increased due to the development of beachfront areas.

ACTIVITY 23-4 | HURRICANE TRACKING

For this activity you will need an outline map that includes the Atlantic Ocean, the Caribbean Sea, and the Gulf of Mexico. The map must be clearly marked with terrestrial coordinates. Use the Internet or printed references to obtain tracking coordinates of a major Atlantic or Gulf of Mexico hurricane. Plot its daily path over the period of time that it was a strong storm system. Explain why the storm formed where it did and how it died out.

Tornadoes

In terms of area, tornadoes are very small storms, usually less than a third of a mile (0.5 km) in diameter. Most touch down for fewer than 10 minutes although some last for hours and skip from place to place over Earth's surface. Tornadoes are known to have the fastest winds on Earth: sometimes in excess of 300 mph (500 km/h). This is twice as fast as winds in the most violent hurricanes.

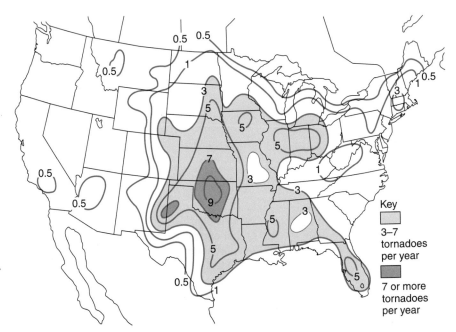

Figure 23-8 This map shows the annual number of tornadoes per 10,000 square miles. Note the similarity to the thunderstorm frequency map in Figure 23-3.

Key

3–7 tornadoes per year

7 or more tornadoes per year

Tornadoes can occur throughout the United States and at any time of year. Nevertheless, they are most common in the Midwest and southern states from March through May as shown in Figure 23-8. At this time, strong cold fronts form in this region at the interface between polar and tropical air masses. Most tornadoes are born as swirling winds within a thunderstorm that intensify and extend upward into the cloud as well as down to the ground. The Fujita scale (F-scale) classifies tornadoes by their wind speed and damage from an F0 which causes light damage to F5 which causes total destruction along its path. Table 23-2 is the Fujita scale of tornado intensity.

Tornadoes can occur in swarms such as the outbreak of April 3–4, 1974. A total of 127 tornadoes, some of them F5, killed 315 people and injured more than 6000 along a path that ranged from Mississippi to western New York State. New York usually gets one or two confirmed tornadoes each year.

TABLE 23-2 Fujita Scale of Tornado Intensity

Scale	Wind Speed		Damage Potential
	km/hr	mi/hr	
F0	68–118	40–73	Light
F1	119–181	74–112	Moderate
F2	182–253	113–157	Considerable
F3	254–332	158–206	Severe
F4	333–419	207–260	Devastating
F5	420–513	261–318	Incredible

Recent research indicates that some of the greatest damage done in strong hurricanes is actually caused by tornadoes that form within the hurricane. Tornadoes are difficult to notice in the fury of the hurricane, but evidence of tightly swirling winds has been found in the debris left by some major hurricanes such as Hurricane Andrew that devastated a region south of Miami in 1992.

ACTIVITY 23-5   A MODEL OF A TORNADO

You will need two 1-liter soda bottles for this activity. Place waterproof, high-tack tape across the top of one bottle and put a small hole in the tape. Fill the second 1-liter soda bottle with water. With the full bottle at the bottom, tape the bottles one on top of the other at their necks. (Some science stores and suppliers sell special caps that connect these bottles, allowing water to flow between them.) When the connected bottles are turned upside down and gently rotated, the water falling into the empty bottle will form a tornado-like vortex.

How is this similar to a real tornado and how is it different?

 **Floods**

In recent years, floods have taken the greatest toll of lives and property in the United States. Some are sudden events such as the flash flood that swept through Big Thompson Canyon in Colorado in 1976. Others are long-term events associated with unusual periods of above average precipitation or spring snowmelt.

The deadliest flood in American history occurred in Johnstown, a small industrial city in western Pennsylvania, in 1889. An earthen dam holding back a private recreational reservoir had been elevated above its original height, and then poorly maintained. The dam failed during an unusually heavy rainstorm. Water surged 24 miles (39 km) along the valley and down 400 feet (120 m) in elevation to the city of Johnstown. Built in a narrow valley, the city was destroyed in minutes and the lives of more than 2000 people were lost.

The Mississippi River and its tributaries have been subject to flooding throughout recorded history. Heavy regional rains and rapid snowmelt have led to spring floods, especially when the ground in the watershed is saturated. In 1993, flood-water covered an area the size of the state of Indiana and set new high-water marks at St. Louis. **Levees** are a high bank along a river of natural or human origin. Artificial levees have been constructed along most of the lower parts of the river to confine the river and prevent flooding of valuable farmland when the river runs high.

However, these structures have prevented floodwater from spreading over the floodplain. Levees also make the river run higher when it floods, thus increasing the danger if the levees break or overflow at flood stage. The levees have also allowed people to live in low areas near the river where breaks in the levees have caused extensive property losses.

Droughts

Many of the same areas subject to sudden flooding also suffer from drought. Rainfall is especially unpredictable in the Great Plains region of the United States. The climate has cycles of wet and dry years. Encouraged by several years of good precipitation, farmers rely on these conditions. However, the droughts return, and without irrigation, farming becomes impossible. Droughts do not lead to sudden loss of life as do storm events, but their long-term economic impact is probably greater.

HOW CAN WE PROTECT OURSELVES FROM WEATHER HAZARDS?

Although major storms cannot be predicted more than a few days in advance, people can make long-range plans to deal with them when they do occur. Advanced planning can prevent property losses. Educational programs about weather hazards help people learn what to do before the storm and how to survive if they are caught in potentially hazardous situations.

Historical Information

It is valuable to know what kinds of destructive weather events have occurred in your area in the past. These are the most likely weather hazards that will occur in the future. For example, hurricanes occasionally pass over New York State, causing damage related to storm surges, wave action, and winds in coastal areas such as Long Island. The greater threat from hurricanes in upstate New York has been flooding of low-lying areas along streams and rivers. Blizzards and thunderstorms are known to cause power and telephone outages and make travel difficult. Damage and danger from these storms can be avoided by careful planning.

Location

People who live in rural regions of New York State know that they can be isolated for days by blizzards and other storms. They need to keep emergency supplies of food and fuel for these events. Floods primarily affect homes and businesses in low-lying areas along streams and rivers. Coastal areas such as Long Island are subject to damage from hurricane storm surges and wave action.

Local Building Codes and Zoning

Perhaps the most important function of government is to protect people from harm. For example, construction codes are designed to require that buildings are made strong enough to withstand most local weather hazards and protect the people inside. Most communities have zoning laws that do not allow the construction of homes in areas at risk from weather hazards such as floods. Steep and unstable land may require special permits and inspections to ensure that the property is not prone to landslides and other forms of mass wasting. Many shoreline communities restrict construction close to the ocean to protect fragile beach environments and prevent storm damage to property.

Emergency Planning

Citizens can learn the best ways to protect themselves in case of weather-related emergencies. When a thunderstorm occurs, people should avoid exposed areas such as hilltops and open fields. It is also important to stay away from tall trees that may be hit or toppled by a lightning strike. Inside buildings, avoid anything that could provide a path for electricity such as electrical wires and water pipes. Automobiles, as long as they are not located in a flood zone, are usually safe locations in a thunderstorm because the metal shell of the vehicle conducts electrical energy around the occupants and into the ground.

Special plans can be made to notify and evacuate people from hazardous areas in case of severe weather. Residents can be informed of impending danger through the news media and loud sirens. Advanced planning helps to coordinate police and emergency services to evacuate people from locations prone to flooding. Roads and bridges in shore areas can be changed to one-way routes leading people to safety and preventing others from entering a danger zone.

ACTIVITY 23-6	COMPREHENSIVE EMERGENCY PLANNING

With the class divided into five groups, each group will select a different weather hazard: thunderstorm, blizzard, flood, tornado, or hurricane. The group will prepare a plan to deal with its weather hazard. The reports should include a section on the greatest dangers, how to prepare for the event, and what people should do (and not do) if caught in a severe weather event.

Early Warning Systems

Hurricanes, blizzards, and floods can usually be predicted several days in advance. The weather service broadcasts its forecast of approaching hazards on special radio frequencies and communicates these reports to the news media. Three levels of alert are used. An advisory indicates conditions that could produce a storm are expected in the near future. The second level is a storm watch, which means that conditions already exist for storm development. A storm warning indicates that a storm has actually been observed in the area, or that a storm is likely to appear at any moment.

Thunderstorms and tornadoes are especially difficult to predict because they are more localized than other weather hazards and they can appear with little or no warning. Doppler radar has become an important tool in identifying these hazards. It enables the weather service to identify swirling winds in storm systems many miles away. In November of 2002, many lives were saved by an early tornado warning in Van Wert, Ohio. In a movie theater, 50 patrons, including a large school group, were moved from the auditorium into bathrooms and interior hallways that were more structurally sound. When a tornado passed through 28 minutes after the warning, these were the only parts of the building left standing. Miraculously, no one in the theater was injured.

| ACTIVITY 23-7 | COMMUNITY PLANNING MAP |

Use a map of your community to identify places where natural hazards are most likely to cause loss of lives and property. Make recommendations about building codes and how residents in those areas should prepare for natural hazards.

HOW IS EARTH'S ATMOSPHERE CHANGING?

Earth's atmosphere is nearly as old at the planet itself. When magma comes to the surface, pressure is reduced and bubbles of hot gas escape into the atmosphere. This process is known as **outgassing**. Gases collected from erupting volcanoes are about 80 percent water vapor, 10 percent carbon dioxide, and 10 percent other gases including nitrogen.

The world's oceans originated from water vapor vented during volcanic eruptions. The water vapor condensed and fell as precipitation. Carbon dioxide and other gases remained as Earth's atmosphere. Objects in space that are much smaller than Earth, such as our moon, do not have enough gravity to prevent gases from escaping into space. However, Earth has been able to hold onto its atmosphere and oceans.

 ## Atmosphere and Life

If carbon dioxide is the most plentiful gaseous component of magma after water vapor, why is the atmosphere composed primarily of nitrogen and oxygen? Only a very small fraction of 1 percent of the present atmosphere is carbon dioxide. Both Mars and Venus have atmospheres that are mostly carbon dioxide.

Unlike our neighboring planets, conditions on Earth are favorable for the development of abundant life on the surface. The history of life and the evolution of the atmosphere are tightly coupled. Earth's earliest life-forms were primitive bacteria that thrived in an atmosphere that had no oxygen. Chemical evidence of these earliest life-forms has been found in rocks 3.5 billion years old. That is only 1 billion years after Earth formed.

The distance between Earth and the sun has played an important part in Earth's development. Consequently, Earth has moderate temperatures. Ocean water protected early life-forms from destructive solar radiation. Early plantlike life-forms in the oceans began to use the energy of sunlight to make food through *photosynthesis*. During photosynthesis, oxygen is produced as a waste product.

In fact, as oxygen built up in the atmosphere, the original life-forms could no longer live at the surface. Their descendants now occupy the deep oceans and subsurface environments where there is little free oxygen. As the atmosphere changed, other organisms, such as animals including humans, evolved that thrived in an environment of abundant

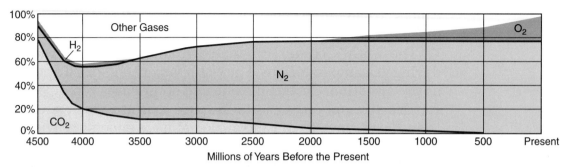

Figure 23-9 Earth's changing atmosphere.

oxygen. Oxygen also supports the ozone layer in the upper atmosphere that protects land dwellers from short-wave ultraviolet radiation. Today's atmosphere is in a dynamic balance sustained by plants that produce oxygen and animals that consume it. Figure 23-9 represents the changing composition of Earth's atmosphere throughout geologic history. The composition of Earth's atmosphere is still changing.

People Affect the Atmosphere

One major factor changing the atmosphere is the production of carbon dioxide through the burning of fossil fuels. As noted earlier, the concern is not simply the increasing carbon dioxide content of the atmosphere. Scientists are more concerned about global warming that could cause the melting of Earth's polar ice caps and other climatic changes.

The fuels that add carbon dioxide to the atmosphere also add oxides of nitrogen and sulfur. These compounds can act as condensation nuclei to form clouds. They also make the cloud droplets acidic. Nitric acid and sulfuric acid are strong acids. Although some of the acidity in the atmosphere comes from volcanoes venting the same substances, the human-generated part of the problem is much more troubling. Figure 23-10 shows smog caused by air pollution over the city of Los Angeles, California. **Smog** is a mixture of fog and smoke.

Acid clouds yield **acid precipitation** when the moisture falls as rain or snow. Some kinds of bedrock, such as lime-

Figure 23-10 The clouds obscuring downtown Los Angeles and the San Gabriel Mountains are smog caused mainly by emissions from burning fossil fuels.

stone and marble that contain calcite, can react chemically with these acids to neutralize them. However, most rocks do not.

The Adirondack Mountains are hit with a three-part problem of acid damage. Air pollution from the industrial sections of the Midwest drifts eastward into the Adirondacks on the prevailing southwesterly winds. Damage to trees from acid clouds is evident in the higher parts of the mountains and other places where trees cling to life in severe environmental conditions. Furthermore, there is little calcite in most of the rocks of the Adirondacks so the acid runs into streams and lakes. Some of the lakes in the Adirondacks are remarkably clear because normal plants and fish cannot live in the acid water. Acid damage is most severe when spring snowmelt brings a sudden rush of acidic water just when fish and plants are most vulnerable.

A historical look at Earth's atmosphere reveals three important facts. First, the composition of Earth's atmosphere has changed through geologic history, primarily in response to living organisms. Second, in the past century, human activities have become the most important cause of change in Earth's atmosphere. Those changes may have even more dramatic results in the next few centuries. Third, people need to understand the changes in the atmosphere and how they are likely to affect them. People may be facing a choice between reducing their effect on the atmosphere and adapting their lives to the changes that occur as a result of human technologies.

TERMS TO KNOW

acid precipitation	levee	rain showers	thunder
blizzard	lightning	sleet	thunderstorm
freezing rain	outgassing	smog	tornado
hazard	rain	snow showers	velocity
hurricane			

CHAPTER REVIEW QUESTIONS

1. The map below shows part of North America.

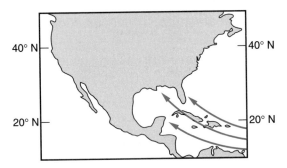

The arrows on this map most likely represent the direction of movement of

(1) Earth's rotation.
(2) prevailing northeast winds.
(3) ocean conduction currents.
(4) Atlantic Ocean hurricanes.

Use the map below to answer questions 2 and 3. The map represents a satellite image of Hurricane Gilbert in the Gulf of Mexico. Each X represents the position of the center of the storm on the date indicated.

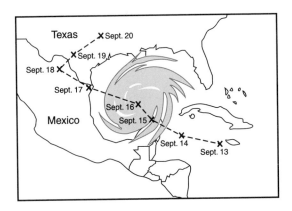

2. Which hazard caused by Hurricane Gilbert was probably the greatest threat to life and property along the coastline at the Texas-Mexico border?

(1) high air temperatures
(2) high barometric pressure
(3) sinking air currents from the upper atmosphere
(4) wave action at a time of high water levels

3. Why did Hurricane Gilbert weaken between September 16 and September 18?

(1) The hurricane changed into an anticyclone.
(2) The hurricane season ended in mid-September.
(3) The hurricane lost its source of warm ocean water.
(4) The pressure gradient within the hurricane increased.

4. Why do hurricane winds usually decrease when a hurricane from the Gulf of Mexico moves over North America?

(1) Land surfaces are always cooler than the ocean surface.
(2) Hurricanes usually turn to the west over North America.
(3) The air pressure gradient is greater over land.
(4) Vegetation and other features on the land surface slow the winds.

5. On a certain day, the isobars on a weather map are very close together over eastern New York State. To make the people of this area aware of possible risk to life and property in this situation, the National Weather Service should issue

(1) a dense-fog warning.
(2) a high-wind advisory.
(3) a heat-index warning.
(4) an air pollution advisory.

6. Venus has a greater concentration of carbon dioxide (CO_2) in its atmosphere than does Earth. As a result, surface temperatures on Venus are

(1) warmer due to absorption of long-wave (infrared) radiation.
(2) warmer due to absorption of short-wave (ultraviolet) radiation.
(3) cooler due to absorption of long-wave (infrared) radiation.
(4) cooler due to absorption of short-wave (ultraviolet) radiation.

7. Scientists believe that Earth's early atmosphere changed in composition as a result of

    (1) the appearance of oxygen-producing organisms.
    (2) drifting of the continents.
    (3) changes in Earth's magnetic field.
    (4) transfer of gases from the sun.

8. An increase in which gas would cause the most greenhouse warming of Earth's atmosphere?

    (1) nitrogen
    (2) oxygen
    (3) carbon dioxide
    (4) hydrogen

9. The graph below shows the change in carbon dioxide concentration in parts per million (ppm) in Earth's atmosphere from 1960 to 1990.

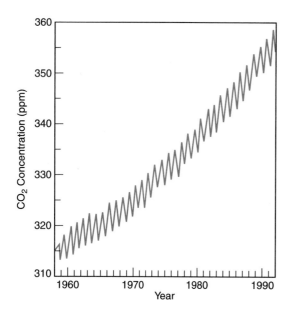

The most likely cause of this overall change in the level of carbon dioxide is an increase in the

    (1) number of violent storms.
    (2) number of volcanic eruptions.

(3) use of nuclear power.
(4) use of fossil fuels.

10. Which area is most likely to have the *least* ecological damage due to acid precipitation?

(1) a region of extrusive volcanic rocks including basalt and rhyolite
(2) a region sedimentary and metamorphic rocks including limestone and marble
(3) a high mountain area where trees begin to give way to other forms of vegetation
(4) a location downwind from a major population and industrial center

Open-Ended Questions

11. Most violent weather gets its energy from water vapor in the atmosphere. What must happen to the water vapor to make this energy available?

Base your answers to questions 12–14 on the two data tables and the hurricane-tracking map below. Location, wind velocity, air pressure, and storm strength are shown for the storm's center at 3 P.M. Greenwich time each day.

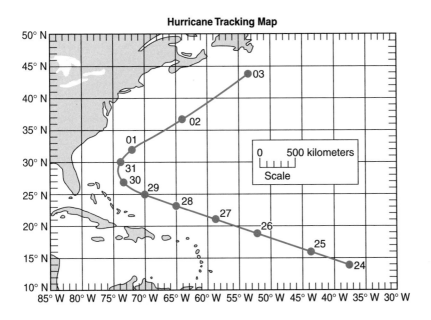

Hurricane Tracking Map

Data Table I

Latitude (°N)	Longitude (°W)	Date	Wind Velocity (knots)	Air Pressure (millibars)	Storm Strength
14	37	Aug. 24	30	1006	Tropical depression
16	44	Aug. 25	70	987	Category 1 hurricane
19	52	Aug. 26	90	970	Category 2 hurricane
21	59	Aug. 27	80	997	Category 1 hurricane
23	65	Aug. 28	80	988	Category 1 hurricane
25	70	Aug. 29	80	988	Category 1 hurricane
27	73	Aug. 30	65	988	Category 1 hurricane
30	74	Aug. 31	85	976	Category 2 hurricane
32	72	Sept. 01	85	968	Category 2 hurricane
37	64	Sept. 02	70	975	Category 1 hurricane
44	53	Sept. 03	65	955	Category 1 hurricane

Data Table II

Storm Strength Scale	Relative Strength
Tropical depression	Weakest
Tropical storm	
Category 1	
Category 2	
Category 3	
Category 4	
Category 5	Strongest

12. Describe two characteristics of the circulation pattern of the surface winds around the center (eye) of a Northern Hemisphere low-pressure hurricane.

13. The hurricane did not continue moving in the same compass direction during the entire period shown. Explain why the hurricane changed direction.

14. Calculate the average daily rate of movement of the hurricane from 3 P.M. August 24 to 3 P.M. on August 28. The hurricane traveled 2600 miles during this four-day period.

    a. Write the equation to determine the rate of change.
    b. Substitute data into the equation.
    c. Calculate the rate and label it with the proper units.

15. a. State two dangerous conditions, other than hurricane winds, that could cause human fatalities as a strong hurricane strikes the coast in North America.

    b. Describe one emergency preparation people could take to avoid a problem caused by either of the two dangerous conditions you stated in response to part a of this question.

16. An Earth science class is preparing a booklet on emergency preparedness. State one safety precaution that should be taken to minimize danger from each of the following hazards. You may not use the same answer twice.

    a. thunderstorm
    b. tornado

Base your answers to questions 17–20 on the two maps below and the reading passage about acid rain.

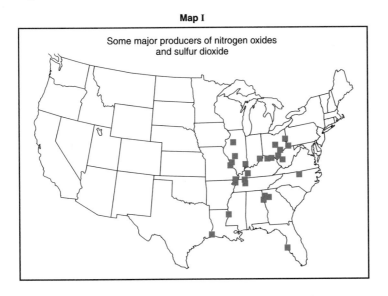

Map I

Some major producers of nitrogen oxides and sulfur dioxide

Map II

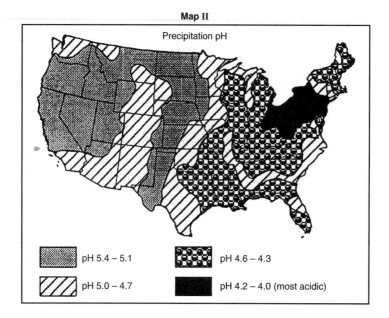

Precipitation pH

pH 5.4 – 5.1   pH 4.6 – 4.3

pH 5.0 – 4.7   pH 4.2 – 4.0 (most acidic)

Acid Rain

Acid deposition consists of acidic substances that fall to Earth. The most common type of acid deposition is rain containing nitric acid and sulfuric acid. Acid rain forms when nitrogen oxides and sulfur dioxide gases combine with water and oxygen in the atmosphere.

Human-generated sulfur dioxide results primarily from coal-burning electric utility plants and industrial plants. Human-generated nitrogen oxides result primarily from burning fossil fuels in motor vehicles and electric utility plants.

Natural events, such as volcanic eruptions, forest fires, hot springs, and geysers, also produce nitrogen oxides and sulfur dioxide.

Acid rain affects trees, human-made structures, and surface water. Acid damages tree leaves and decreases the tree's ability to carry on photosynthesis. Acid also damages tree bark and exposes trees to insects and disease. Many statues and buildings are composed of rocks containing the mineral calcite, which reacts with acid and chemically weathers more rapidly than other common minerals. Acid deposition lowers the pH of surface water. Much of the surface water in the Adirondack region has pH values too acidic for plants and animals to survive.

17. State one reason that the northeastern part of the United States has more acid deposition than other regions of the country.

18. Name one sedimentary or one metamorphic rock that is most chemically weathered by acid rain.

19. Describe one law that could be passed by the government to reduce problems caused by acid deposition.

20. Explain why completely eliminating human-generated nitrogen oxides and sulfur dioxide will not completely eliminate acid deposition.

Chapter 24

Patterns of Climate

ARE CLIMATES CHANGING?

A century ago, 150 glaciers covered the mountains of Glacier National Park in Montana. Today, only 35 glaciers remain. Meanwhile, the remaining glaciers are melting back so rapidly that scientists estimate all of them could be gone in 30 years. From the Himalayas of Asia to the Andes of South America, the shrinkage in glacial ice is a worldwide event. Although a few glaciers are advancing, the great majority of them are clearly melting back.

Measurements of temperatures over the Earth have not been conclusive in showing a warming trend. Natural cycles of temperature change, our limited time frame of accurate temperature readings and the subtle nature of global climate change have made it difficult to measure global warming. Nevertheless, some scientists see the worldwide retreat of glaciers as a clear sign that Earth's climate is getting warmer. Figure 24-1 offers two views of Africa's highest mountain, Kilimanjaro, one taken in 1993 and the other in 2000. Notice the dramatic decrease in the glaciers and snow that occurred in this seven-year period.

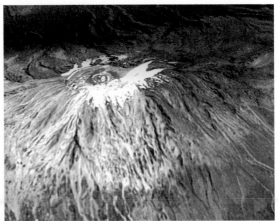

Figure 24-1 Photographs of Mt. Kilimanjaro taken in 1993 (left) and in 2000 (right). The snow and ice cover of Africa's tallest mountain decreased dramatically in the seven years between these images. Many scientists see this, and the worldwide recession of glaciers, as an indication of global warming.

In addition to indicating climatic change, glaciers also hold a record of past climates. The ice in glaciers can be thousands of years old. From the thickness of annual layers and the crystal structure of the ice, scientists can infer conditions of precipitation and temperature that occurred in the distant past. Glacial ice also preserves samples of the atmosphere trapped in the snow before it was buried within glaciers. Like layers of rock, glaciers preserve a record of prehistoric Earth that scientists can use to unravel the past.

 ## WHAT IS CLIMATE?

We usually think of the weather at a particular place and time, or perhaps over a period of hours or days. **Climate** is the average conditions of weather based upon measurements made over many years. Temperature and precipitation are the primary elements of climate, although humidity, winds, and the frequency of storms are also important aspects of climate. The normal seasonal changes in these factors are a part

Figure 24-2 Vegetation is an indicator of climate. These desert plants indicate that the local climate is usually dry with hot summers.

of climate as well. Scientists' understanding of the climate of an area is based primarily on historical records. The more observations that are available and the longer the period during which they have been kept, the more accurately scientists can describe the climate.

Climates can be classified according to temperature, such as tropical climates, which are usually warm; **temperate climates**, which include large seasonal changes; and polar climates, which are usually cold. Humidity and precipitation are often grouped when describing a humid climate or **arid** (dry) **climate**.

Vegetation is sensitive to climatic conditions. The plants found in an area are an indication of the climate, as you can see in Figure 24-2. If you travel through places where the natural vegetation changes you are probably observing the effects of changes in climate. Rain forests, deserts, grasslands, and tundra are terms that describe both vegetation and climate. Figure 24-3 includes two maps of North America. The first map shows zones of similar climate. The second map shows zones of similar vegetation. Notice how closely the zones of climate and vegetation match.

(a) Climate Zones

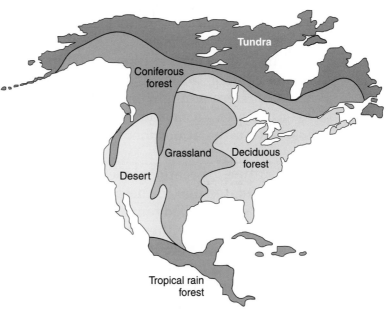

(b) Vegetation Zones

Figure 24-3 The similarity between the boundaries of the North American zones of climate and vegetation illustrates how the climate of a region determines its natural plant community.

HOW DOES LATITUDE AFFECT CLIMATE?

Variation in the intensity of insolation (sunlight) is the major cause of temperature differences over Earth's surface. In the tropics, where the noon sun is always high in the sky, solar energy is strongest. At the poles, where the sun is never high in the sky, solar energy is weakest.

 ## Temperature

TROPICS The tropics are sometimes called the latitudes of seasonless climate. Although the noon sun is a little higher in the sky in some parts of the year than in others, the change is small. The seasonal change in the length of daylight is also very small—in fact, hardly noticeable at all. Therefore the strength of solar energy changes very little throughout the year. Except for mountain locations, the weather is always warm.

The tropics extend from the Tropic of Cancer 23.5° north of the equator to the Tropic of Capricorn 23.5° south of the equator. Here sunlight passes through the minimum thickness of Earth's atmosphere; so relatively little heat energy is lost within the atmosphere.

MID-LATITUDE Locations such as New York State have seasonal climates due to the annual cycle of changes in insolation. These are called temperate climates because the average temperature is neither hot nor cold. The largest seasonal changes actually occur in the mid-latitudes. The seasons in the Northern Hemisphere are the reverse of those in the Southern Hemisphere. When it is summer in the Northern Hemisphere, it is winter in the Southern Hemisphere.

POLAR REGIONS The polar regions are generally cool throughout the year, but they do experience seasonal changes. In the winter, the days are very short and the sun, if it is visible, is always low in the sky. Insolation is extremely weak and tem-

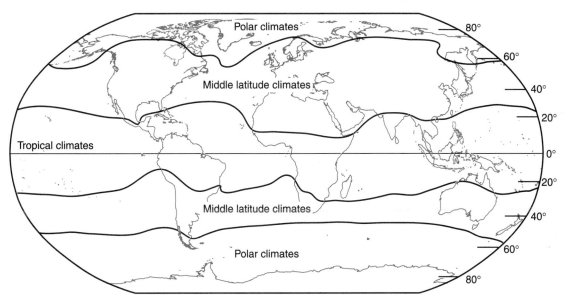

Figure 24-4 The annual cycle of insolation is the primary cause of seasonal temperature changes in climate and the global climate zones.

peratures may stay below freezing for months at a time. Even the summer sun is not very high in the sky, but daylight in the summer is very long. Because there is a large difference in the strength of insolation between winter and summer, polar locations are significantly warmer in the summer than they are in the winter. Figure 24-4 shows these world climate zones.

 **Precipitation**

Latitude also affects patterns of precipitation. These patterns are a result of Earth's rotation acting on terrestrial winds. You learned earlier that instead of one big convection cell in each hemisphere, the Coriolis effect forms three convection cells in each hemisphere.

The three convection cells are shown on the Planetary Wind and Moisture Belts diagram in the *Earth Science Reference Tables*. Figure 24-5 on page 606 is a representation of part of that diagram. The left side of the diagram is a profile of Earth showing convection cells in the Northern Hemisphere. The diagram on the right shows Earth's surface as flat, the

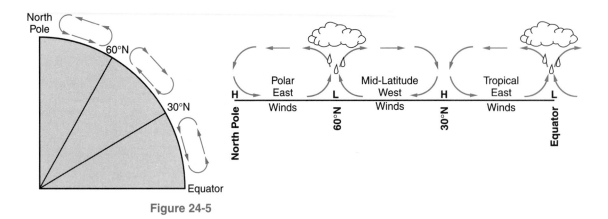

Figure 24-5

way it looks as you stand on it. Along the equator and at about 60°N latitude, air rises, forming low-pressure regions that circle Earth. The rising air causes cloud formation and generous precipitation at these latitudes. However, at a latitude of 30°N and at the North Pole (90°N) are regions of high pressure where sinking air warms and gets drier as it is compressed by atmospheric pressure. These latitudes have relatively little precipitation. The 90° segment shown in Figure 24-5 is one of four similar profiles that circle Earth.

The rotation of Earth and the position of the continents break convection in the Northern Hemisphere into three cells. Rising air near the equator and at about 60°N results in two low-pressure regions of plentiful precipitation that circle Earth. Sinking air at about 30°N and near the North Pole results in high-pressure zones dominated by low relative humidity and little precipitation. Six similar profile sections circle Earth, three in the Northern Hemisphere and three in the Southern Hemisphere.

These high- and low-pressure belts are not stationary. Seasonal changes cause them to shift toward the equator in the winter and toward the poles in the summer. Furthermore, the wandering jet streams move these regions of high- and low-pressure and interrupt them with the passage of storm systems. Other geographic features you will soon read about also influence patterns of precipitation. If you look at the location of the world's rain forests and deserts you will see that they generally occur in the latitude zones indicated in Figure 24-4.

ACTIVITY 24-1   LOCATING DESERTS AND RAIN FORESTS

On an outline map of the world's continents draw east-west lines at the equator, 30° north and south of the equator, and 60° north and south of the equator. On the same map label the major desert regions and rain forests of the world.

Do these features occur where you would expect them according to Figure 24-4 and the *Earth Science Reference Tables?*

WHAT OTHER GEOGRAPHIC FACTORS AFFECT CLIMATE?

Latitude plays a major role in determining the climate of an area. However, other factors, such as elevation, nearness of large bodies of water, winds, and ocean currents also affect climate.

Elevation

Elevation is indirectly related to the average temperature of a location. Recall that air expands as it rises within the atmosphere. As rising air expands, it becomes cooler. Perhaps you have noticed that high mountains are often snow covered, even in the summer, as shown in Figure 24-6 on page 608. Mount Kilimanjaro in Africa and Cotopaxi in South America are near the equator. However, both mountains have permanent snow cover near their summit. Nearby locations at lower elevation have a tropical climate where it never snows. Rising air cools at a rate of about 1°C per 100 meters (4°F per 1000 feet).

People who live in the low desert of Arizona often seek relief from the summer heat by traveling to nearby mountain locations. Within the state, travel north or south has little effect on temperatures. However, the large changes in elevation in that part of the country have major climatic consequences.

Figure 24-6 Due to differences in elevation, high mountains are often covered with snow while nearby lowlands are warmer and free of snow.

 ## Mountain Barriers

Mountain ranges affect patterns of precipitation and temperature. For example, moist winds off the Pacific Ocean cross California and rise into the Sierra Nevada Mountains. Rising into the mountains, the air expands and cools below its dew point. Cloud formation and precipitation create a cool climate with abundant precipitation on the western side of the Sierra Nevada Mountains. When the winds descend on the opposite side of the mountains, the air is compressed by increasing barometric pressure and becomes warmer. As the descending air becomes warmer without picking up moisture, the relative humidity decreases.

The climate on the eastern side of the Sierra Nevada Mountains is very different from that on the western side. Rain forests that exist on the Pacific, or windward, side of the western mountains contrast sharply with inland deserts. Seattle and the coast of northern California have a temperate, moist climate. Inland cities such as Spokane, Las Vegas, and Phoenix are located in the desert climate zone. Climate on the downwind, or leeward, side of mountains is sometimes called a "rain shadow" climate because the mountain barriers rob

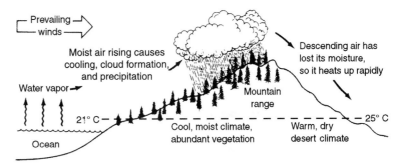

Figure 24-7 Cloud formation and precipitation usually occur on the windward side of a mountain range where moist air rises and cools. On the downwind side, descending air is warmed by compression, so the relative humidity quickly drops, generating an arid climate.

the air of its moisture before the air descends into the valleys. Figure 24-7 illustrates the difference between climates on the opposite sides of a mountain range.

Figure 24-7 shows another difference between climates on the opposite sides of a mountain range. Notice that the temperature near sea level on the left (ocean) side of the diagram is cooler than the temperature at the same elevation on the right (inland) side of the mountains. Why is the temperature warmer on the downwind side of the mountains?

As air rises into the mountains, condensation (cloud formation) releases energy, slowing the cooling. Instead of cooling at the dry air rate of 1 C° per 100 meters, the air cools at a lower rate of about 0.6 C° per 100 meters when clouds form. The descending air is not able to pick up moisture. Therefore, the air heats at the greater, dry air rate of 1°C per 100 meters. Figure 24-8 on page 610 illustrates this difference in the rate of temperature change between moist air moving up the mountain with cloud formation and the descending air, which is dry and heats more rapidly. Changes in climate caused by mountain barriers are called orographic effects.

 ## Large Bodies of Water

The Atlantic Ocean and Long Island Sound moderate the climate of New York's Long Island. Therefore, winters are usually warmer on Long Island and summers cooler than in other

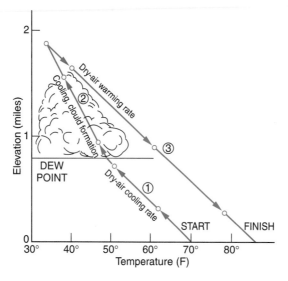

Figure 24-8 (1) As air rises, it expands and cools quickly. (2) When clouds form, condensation releases the energy stored in water vapor, slowing the rate of cooling. (3) When the air sinks to a lower elevation, there is no change in state to slow the rapid warming rate.

parts of the state. This is especially true for places along the coast when the winds are off the ocean. The inland regions of New York State experience the highest and lowest temperatures. The lowest temperature ever recorded in New York State was −52° F (−47°C) in the Adirondack Mountains. The record high temperature in New York State was measured in the capitol district: 108°F (42°C) at Troy. Both places are far from the moderating influence of the Atlantic Ocean and the Great Lakes.

Why do the oceans have such a great effect on climate? Table 24-1 is in the *Earth Science Reference Tables*. This table lists the specific heats of seven common substances. As you learned in Chapter 21, *specific heat* is a measure of the ability of a substance to warm as it absorbs energy or cool as it gives off energy. In general, metals have low specific heats. They heat up rapidly when they absorb energy. Rocks also have relatively low specific heat values. However, water has a very high specific heat.

This means that when the same amount of energy is absorbed by equal masses of water and these other substances, water has the least temperature change. It also means that when equal amounts of these substances cool, water releases the most energy. When water and land receive equal solar energy, the land heats more than the ocean. Large bodies of water change relatively little in temperature. Winds off the oceans or the Great Lakes, such as Lakes Erie and Ontario, moderate the temperature of nearby land areas.

TABLE 24-1 Specific Heats of Common Materials

Material		Specific Heat (calories/gram · C°)
Water	solid	0.5
	liquid	1.0
	gas	0.5
Dry air		0.24
Basalt		0.20
Granite		0.19
Iron		0.11
Copper		0.09
Lead		0.03

There are three other reasons that land areas experience greater changes in temperature than the oceans. First, because water is relatively transparent, sunlight penetrates deeper into water than it does on land. Rock and soil are opaque, so insolation energy is concentrated at the surface. Second, water is a fluid, so convection currents can distribute energy to the interior. Solids have no convective mixing. Finally, evaporation from the oceans uses some of the solar energy that would otherwise heat the oceans. Although there is some evaporation of water from soil, it is far less than evaporation from the oceans. Figure 24-9 on page 612 summarizes these factors.

Most terrestrial climates can be classified as maritime or continental. **Continental climate**, typical of inland areas, is characterized by large seasonal changes in temperature. In-

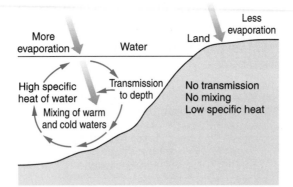

Figure 24-9 Four reasons that the oceans heat more slowly than land: (1) Sunlight can penetrate water but it is concentrated on the surface of the land. (2) Convection currents carry energy into the water. (3) Water has a higher specific heat than rock or soil. (4) Some solar energy is used in the evaporation of water.

land areas often do not experience the moderating influence of large bodies of water. Areas with a continental climate can be arid or moist, depending on the source region of the air masses that move into the area. The **maritime climate**, sometimes known as the marine climate, occurs over the oceans and in coastal locations where water moderates the extremes in temperature. Areas that have a maritime climate experience a consistently moderate to high humidity.

 Prevailing Winds

New York State has greater extremes of climate than many other coastal states. California, for example, is known for its mild climate. Although inland areas of California experience greater ranges of temperature than the coastal locations, even these extremes are not as great as those are in New York State. The reason for this difference is the wind direction. Both states are in the global belt of prevailing west and southwest winds. However, those winds come off the Pacific Ocean in California. In most of New York State, the winds come from inland areas where temperatures are highly changeable. As a result, the nearby Atlantic Ocean has relatively little affect on the climate of most of New York State.

Monsoon climates include an annual cycle of change in weather patterns caused by shifting wind directions. In winter and spring, the wind comes from high-pressure centers over the continents. Spring weather is warm and dry with large changes in temperature on a daily cycle. When summer low pressure builds over the continents, the wind shifts direction. It brings moist air from the ocean. Summer weather generally is more humid with cooler days and warmer nights. The summer monsoons also bring clouds and precipitation, which reduce the temperature as well as the daily range of temperature.

 Ocean Currents

Many tourists are surprised to see palm trees growing in the some parts of England and Ireland. Palm trees are not native to these countries, but these imported plants can survive the mild weather conditions found in some parts of the British Isles. The Gulf Stream and the North Atlantic Current transport warm ocean water from the South Atlantic Ocean to the area surrounding Great Britain. These islands experience more moderate temperatures than does New York State. The British Isles have damp and mild winters in which hard frosts are uncommon. This is true in spite of the fact that Great Britain is roughly 10° of latitude north of New York State. Along the East Coast of North America at the same latitude as Great Britain is the Labrador province of Canada, where the winters are even colder than in New York State.

Other locations are relatively cool because of nearby cold ocean currents. The California Current keeps the coastal city of San Francisco in "sweater weather" throughout the summer. Even in the summer, local residents who visit the ocean may wade in the surf, but the water is too cold for people to swim without a wet suit. People who live just a few miles inland often experience desert heat in the summer while the city of San Francisco is cool and temperate. The Surface Ocean Currents map in the *Earth Science Reference Tables* provides a useful way to tell where warm and cold ocean currents affect the climate of coastal locations.

ACTIVITY 24-2   **CLIMATES AND OCEAN CURRENTS**

Using a political map of the world and the Surface Ocean Currents map from the *Earth Science Reference Tables,* make a list of countries or regions that are affected by warm ocean currents. Make another list of places affected by cold currents. Alphabetizing your list will help you compare your locations with the lists of other students.

Ocean currents also affect patterns of precipitation. Cold air can hold far less water vapor than warm air. In addition, cool air blowing over warmer land surfaces causes the relative humidity to decrease. Decreasing relative humidity makes precipitation unlikely. Therefore, coastal regions affected by cold ocean currents are usually places where rainfall is scarce. A weather station in the Atacama Desert, along the west coast of South America, has been in place for decades without experiencing any measurable precipitation. On the other hand, the relatively warm Alaska current makes coastal Alaska one of the rainiest places in the United States.

 Vegetation

The local climate and soil determine natural vegetation. Therefore, vegetation is a good indicator of the climate. For example, the temperate rain forests along the Pacific coast of the United States and Canada can thrive only in a cool, moist climate. However, vegetative cover also contributes to the climate. Thick vegetation, such as the trees and plants found in a forest, moderate temperature by holding in cool air during the day and preventing the rapid escape of warm air at night. Vegetation slows surface winds. In addition, plants contribute moisture to the air. During precipitation, the plant cover slows runoff and gives water at the surface time to soak into the ground. Groundwater is then absorbed by the roots of plants and rises into the leaves where, over an extended period, it is slowly lost by *transpiration*. Transpiration and photosynthesis also absorb solar energy, which would otherwise heat the land and air during daylight hours. So forest conditions are generally more moderate and consistently more humid than open land in the same area.

Human activities such as cutting wood, plowing fields, mining, or construction remove native plants and replace them with open ground, paved surfaces, or buildings. The human population of the planet has increased and our technology has become more advanced. **Deforestation**, cutting forests to clear land, and **urbanization**, the development of heavily populated areas, have replaced natural vegetation with farmlands and cities at an ever-increasing rate. Bare ground and paved surfaces do not allow evaporation of ground water, and they heat up quickly during the day and cool quickly at night. As a result of urbanization, the local climate becomes more arid and warmer with an increased daily range of temperatures.

 ## Urban Heat Islands

Human activities release heat energy to the environment. Heating and air conditioning release heat to the outdoor environment. Cars, trucks, buses, and other forms or transportation consume fuel and release heat. Businesses and industries, which are concentrated around cities, produce heat in varying amounts. Many of the same activities that take place in rural regions also occur in urban areas, or cities. However, in a city, the high concentration of human activities produces an urban heat island. In general, urban areas warm up more quickly and stay warmer than rural locations.

The effects of urban heat islands are relatively easy to observe. Have you ever noticed how much longer winter snow lasts is the country than it does in nearby urban areas? Even undisturbed parkland in cities will be clear of snow before similar rural land is snow-free. On summer nights, city dwellers often need air conditioning all night while neighboring rural inhabitants can find relief by opening their windows to the cool evening air.

 ## WHAT GEOGRAPHIC FEATURES OF NEW YORK STATE AFFECT THE LOCAL CLIMATE?

Differences in climate throughout New York State are not very large. Many climatologists would classify the whole state as a humid, continental, temperate climate with large seasonal variations in temperature. However, local geographic features do cause some significant differences in climate at various locations in the state.

As noted previously, the Atlantic Ocean and Long Island Sound make the climate on Long Island more moderate in temperature than inland areas of New York State. Winter temperatures do not get quite as cold and summer temperatures do not get as hot as those of inland areas. Winter precipitation that falls as snow upstate is more likely to be rain on Long Island. Breezes off the ocean keep the humidity higher than other parts of the state. Long Island is also more vulnerable to hurricanes and coastal storms.

Winter snow lasts longer in higher parts of the Adirondack Mountains and the Catskills for two reasons. First, the mountains, due to their elevation, are a little cooler than other areas of New York State. Second, mountains also influence patterns of precipitation. Air rising into these two mountain areas expands and cools, causing increased precipitation throughout the year. Mountains also influence the climate on their downwind side. The land around Lake Champlain and the central Hudson Valley are in the rain shadow of mountains and may have as little as half the annual precipitation of the nearby mountains.

You have previously read that the parts of New York State at the eastern end of Lakes Erie and Ontario are subject to increased precipitation from "lake-effect" storms. This is especially true in late autumn and early winter when the lake water is warmer than surrounding land areas. The lakes also moderate temperatures in nearby land areas. The first hard frost of autumn occurs later in these areas. The extended growing season makes land near the lakes valuable for agriculture.

HOW IS CLIMATE SHOWN ON GRAPHS?

Climate graphs are a visual way to show different kinds of climates. On the following graphs, a dark line shows the average monthly temperature and the average monthly precipitation is indicated by monthly bar graphs.

Figure 24-10 is a climate graph for Syracuse, New York. Notice the large seasonal changes in temperature and plen-

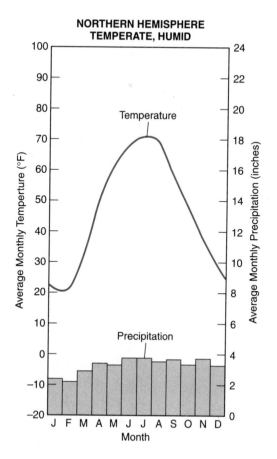

Figure 24-10 Climate graph for Syracuse, New York.

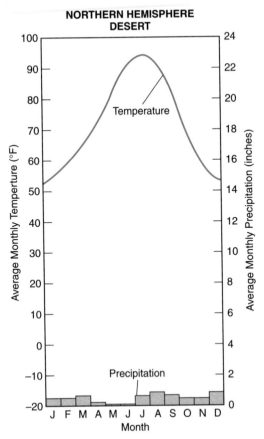

Figure 24-11 Climate graph for Phoenix, Arizona.

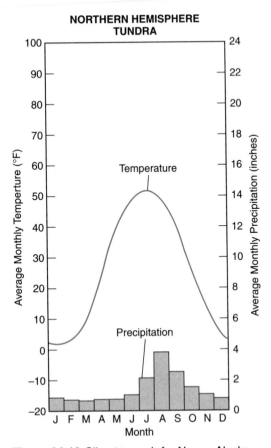

Figure 24-12 Climate graph for Nome, Alaska.

tiful precipitation throughout the year. Remember that these are average conditions over many years. Therefore, unusual events such as droughts do not show on these graphs.

Figure 24-11 is a climate graph for a desert location in the southwestern United States. Although it shows major seasonal changes in temperature like the Syracuse graph (Figure 24-10), this location is warmer in both the winter and the summer. Also notice the limited precipitation throughout the year, especially in the spring before the summer monsoon season.

Figure 24-12 is a tundra climate in arctic Alaska. Temperatures are significantly lower than Syracuse, New York, throughout the year. Although precipitation is low, so is evaporation in this cold climate.

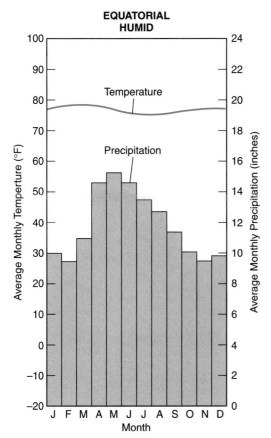

Figure 24-13 Climate graph for Fonte Boa, Brazil.

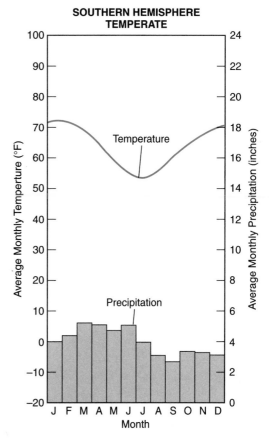

Figure 24-14 Climate graph for Sydney, Australia.

The rain forest of tropical Brazil provides the data for Figure 24-13. The average temperature changes very little throughout the year and precipitation is usually plentiful.

Notice that the highest and lowest temperatures in the Southern Hemisphere location illustrated in Figure 24-14 are off by 6 months from those of the Northern Hemisphere locations. This is also a coastal city so the annual temperature range is not as great as it is at the previous temperate locations.

The last climate graph is a monsoon location in India. (See Figure 24-15 on page 620.) Precipitation is very seasonal. Also notice how the temperatures fall off when the summer monsoons arrive in July.

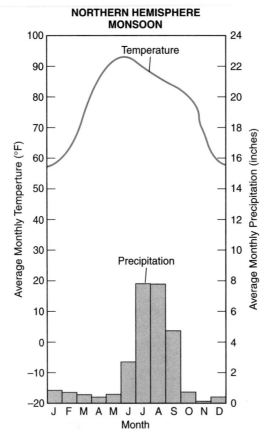

Figure 24-15 Climate graph for New Delhi, India.

TERMS TO KNOW

arid climate	**deforestation**	**temperate climate**
continental climate	**maritime climate**	**urbanization**

CHAPTER REVIEW QUESTIONS

1. A high air pressure, dry climate belt is located at which Earth latitude?

   (1) 0°
   (2) 15°N
   (3) 30°N
   (4) 60°N

2. The average temperature at Earth's North Pole is colder than the average temperature at the Equator because the Equator

   (1) receives less ultraviolet radiation.
   (2) receives more intense insolation.
   (3) has more cloud cover.
   (4) has a thicker atmosphere.

3. Which graph best shows the average annual amount of precipitation received at different latitudes on Earth?

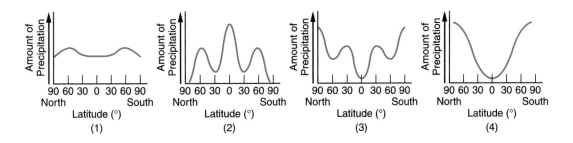

4. Which location probably receives the *least* average annual precipitation?

   (1) Lake Placid in the central Adirondack Mountains of New York
   (2) Buffalo, New York, at the eastern end of Lake Erie
   (3) the South Pole research station in central Antarctica
   (4) Belém, Brazil, along the equator in the Amazon tropical forest

5. Which graph below best shows the general effect that differences in elevation above sea level have on the average annual temperature?

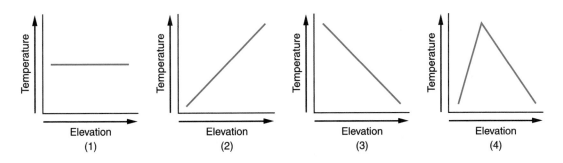

6. The diagram below shows prevailing winds that blow across a mountain range.

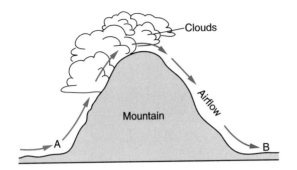

Compared to the temperature and humidity conditions at location A, the conditions at location B are

(1) warmer and less humid.

(3) cooler and less humid.

(2) warmer and more humid.

(4) cooler and more humid.

7. The map below shows the locations of four cities, A, B, C, and D, in the western United States where the prevailing winds are from the southwest.

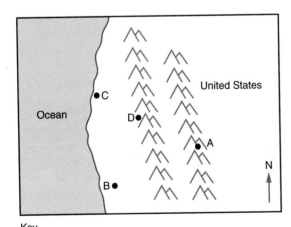

Which city most likely receives the *least* amount of average yearly precipitation?

(1) A

(3) C

(2) B

(4) D

8. Liquid water can store more heat energy than an equal amount of almost any other naturally occurring substance because liquid water

   (1) covers 71 percent of Earth's surface.
   (2) has its greatest density at 4°C.
   (3) has a high specific heat.
   (4) can be changed into a solid or a gas.

9. On a clear summer day, the surface of the land is usually warmer than the surface of a nearby body of water because the water

   (1) receives less insolation.
   (2) reflects less insolation.
   (3) has a higher density.
   (4) has a higher specific heat.

10. The graph below shows the average monthly temperature for two cities, A and B, that are both located at 41° north latitude.

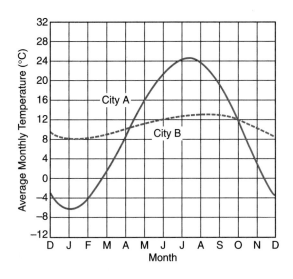

Which statement best explains the difference in the average yearly temperature range for the two cities?

   (1) City B is located in a different planetary wind belt.
   (2) City B receives less yearly precipitation.
   (3) City B has a greater yearly duration of insolation.
   (4) City B is located near a large body of water.

The map below shows the average annual precipitation in New York State. Isoline values represent inches per year. Use this map to answer questions 11 and 12.

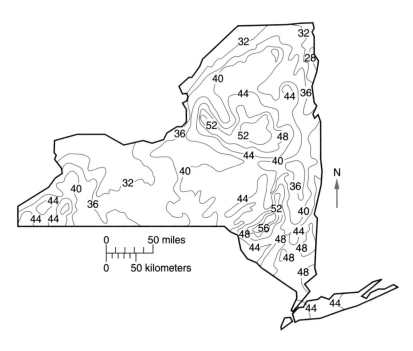

11. Jamestown, New York, receives more rainfall per year than Elmira. The primary reason for this difference is that Jamestown is located

    (1) closer to a large body of water.
    (2) at a higher latitude.
    (3) at a lower elevation.
    (4) in the prevailing southerly wind belt.

12. Which of these locations has the lowest average annual precipitation?

    (1) Kingston   (3) Old Forge
    (2) New York City   (4) Plattsburgh

13. Some housing developments in the barren southwestern deserts of the United States have included large irrigated areas of lawns and trees. As a result, the summer weather conditions within these developments have become more

    (1) warm and moist.   (3) cool and moist.
    (2) warm and dry.   (4) cool and dry.

Use these four graphs to answer the next two questions. Each graph below represents climate conditions for a different city in North America. The line graphs show the average monthly temperatures. The bar graphs show average monthly precipitation.

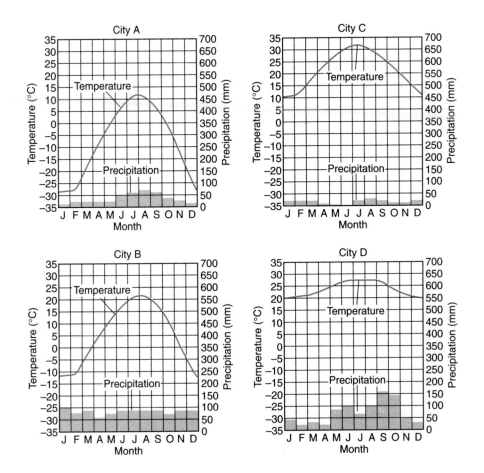

14. At which cities is the winter precipitation most likely to be snow?

    (1) A and B   (3) B and C
    (2) A and C   (4) B and D

15. In which sequence are the cities listed in order of decreasing average yearly precipitation?

    (1) A, B, C, D   (3) C, A, D, B
    (2) B, D, A, C   (4) D, C, B, A

Open-Ended Questions

Use the data below to answer questions 16–18. The table shows the elevation and average annual precipitation at 10 weather stations, A through J, located along a highway that passes over a mountain.

Data Table

Weather Station	Elevation (m)	Average Annual Precipitation (cm)
A	1350	20
B	1400	24
C	1500	50
D	1740	90
E	2200	170
F	1500	140
G	800	122
H	420	60
I	300	40
J	0	65

16. On a copy of the grid below, graph the data from the data table according to the following directions. (Please, do not write in this book.)

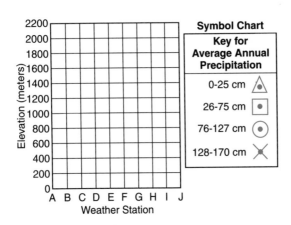

 a. Mark the grid with a point showing the elevation of each weather station above its corresponding letter, *A* through *J*.
 b. Surround each point with the proper symbol from the chart to show the average annual precipitation for the weather station.

17. State the relationship between the elevations of weather stations *A* through *E* and the average annual precipitation at these weather stations.

18. Although stations *C* and *F* are at the same elevation, they have different amounts of average annual precipitation. Explain how the prevailing wind direction might cause this difference.

19. State why locations east of Lakes Erie and Ontario are more likely to receive lake-effect snowstorms than locations west of the these lakes?

20. Westerly winds blowing off Lake Ontario usually drop more precipitation on the Tug Hill Plateau landscape region than they do on the lowlands along the eastern shore of the lake. Why does the Tug Hill Plateau receive more precipitation than the nearby lowlands?

Chapter 25

Earth, Sun, and Seasons

 ## OUR INTERNAL CLOCK

Like most other animals on Earth, people have an internal clock that works on a 24-hour cycle (the circadian rhythm). Scientists have conducted experiments in which people were voluntarily isolated. They received no natural light and no clues to tell day from night. When the subjects of the experiment established their own periods of sleep and activity, they usually settled into a cycle that was about 24 hours in length.

Scientists have learned that this cycle is influenced by brain activity that controls the release of hormones, chemicals that influence bodily functions and behavior. Furthermore, information passed to the brain by the senses, especially the sense of sight, keeps the body's cycles in line with the cycle of day and night. In this way, the cycle of Earth's rotation has become a part of our biological identity.

Our internal clock is noticeable to travelers who journey considerable distances east or west into different time zones. They experience jet lag. This includes difficulty sleeping when the people around them are going to bed, and lapses in concentration at the time that they would have been sleeping at home. After a few days, most people adjust to the new time routine, and the effects of jet lag go away.

Day and Night

Our daily cycle of night and day is so much a part of our lives that we have devised our system of clock time based on the length of the day. Most of the world's nations use the International System of Units (metric system) in which most measures are related to larger and smaller units by a factor of 10. However, the worldwide system of time is still a 24-hour day with 60 minutes to the hour and 60 seconds to the minute. Calendar time, with a year of approximately $365\frac{1}{4}$ days, is based on the cycle of seasons and Earth's revolution around the sun. These cycles are more important to us than a system of time with a base of 10. Even in nations using metric measures, clock time is still set to Earth's cycles of motion.

Two locations on Earth do not have a 24-hour cycle of night and day. The North and South Poles are on a yearly cycle of daylight and darkness. At both poles sunrise occurs on the first day of spring in that hemisphere. The sun remains continuously visible in the sky for 6 months, until the first day of autumn in that hemisphere. Night begins at that time, and continues for the next 6 months. Humans are unable to adapt their activities to a 12-month cycle of day and night.

In addition, there is no natural way to assign clock time at these two locations. Although there is no settlement at the North Pole, the United States does maintain a permanent research station at the South Pole. Clocks at the South Pole are set to the time in New Zealand, the closest inhabited area, and the place from which most people fly to the South Pole research station.

WHAT IS ASTRONOMY?

The study of objects beyond Earth and its atmosphere is known as *astronomy*. Most of our knowledge of the heavens comes through the light (electromagnetic radiation) these objects give off or reflect. In some ways astronomy is one of the oldest sciences. Historians have found written records of celestial observations that go back thousands of years. This branch of science also involves some of the most advanced technology. It has consistently yielded new insights into where we came from as well as the fundamental nature of matter and energy.

In this chapter, it will be important to distinguish between what is apparent and what is real. For example, we observe the daily path of the sun through the sky. Although we know that this cycle is caused by Earth's rotation on its axis, we conduct our lives as if it the sun actually moves around Earth. We talk about the sun rising and setting as if these were real motions of the sun. We call the motion of the sun through the sky an *apparent motion* because this movement of the sun looks real, even though we know it is Earth that is moving. Later in this chapter, you will learn how scientists proved that Earth moves. However, for the sake of describing celestial events, it is convenient to stay with the Earth-centered point of view.

HOW CAN WE DESCRIBE THE POSITION OF CELESTIAL OBJECTS?

To any observer on Earth, the sky looks like a giant dome that stretches across the sky from horizon to horizon. **Celestial objects** are the things seen in the sky that are outside of Earth's atmosphere. These objects include the sun, moon, planets, and stars. You may know that some of these objects are farther away than others. Nevertheless, many appear as if they are points of light on the surface of the sky.

In locating celestial objects, it is convenient to treat the sky as a two-dimensional surface. Two coordinates are all that are needed specify a location on a two-dimensional surface. For example, positions based on the *x*- and *y*-axes of a graph. Angles of latitude and longitude are the coordinates used to locate places on Earth's two-dimensional surface. Like Earth, the sky appears as a curved surface. Therefore, scientists also use angles to locate positions in the sky.

In the sky, you can specify a location using *azimuth* (compass direction) and angular altitude. You learned about compass directions such as North or Southwest when you read about maps in Chapter 3. For locations in the sky, the compass direction refers to the direction along the horizon directly below the position in the sky. Azimuth usually is expressed as an angle measured from North (0°), clockwise to East (90°), South (180°), West (270°) and around the horizon back to 360°. (360° is also 0°, due North.)

Using this coordinate system, altitude is the second coordiante. **Altitude** is the angle above the horizon. Altitude angles start at a level horizon (0°) and increase to 90° at the point directly overhead, sometimes called the **zenith**. The coordinates of compass direction and angular altitude allow us to locate any point in the sky. Figure 25-1 shows the angle of direction and altitude for a star that is high in the southeastern sky.

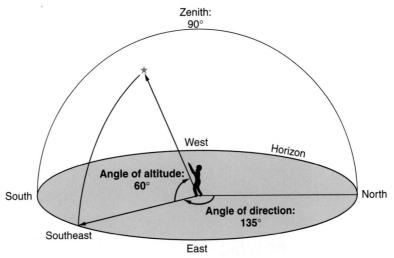

Figure 25-1 Any point in the sky can be located by two angles. The angle of direction is the clockwise angle from due north to a place on the horizon directly below the point in the sky. The angle of altitude is the angle from the horizon up to that point.

WHAT IS THE SUN'S APPARENT PATH ACROSS THE SKY?

For most observers on Earth, the sun rises in the eastern part of the sky. The sun reaches its greatest angular altitude at solar noon. It then moves down to dip below the western horizon. This apparent motion of the sun is the basis for the earliest clocks: sun dials. Figure 25-2 shows a very large sundial.

Local **solar noon** is the time at which the sun reaches its highest point in the sky. For mid-latitude locations in the Northern Hemisphere, such as New York State, the sun at solar noon reaches its highest point when it is due South. The noon sun is never directly overhead in New York State. In fact, the sun is never at the zenith except for observers located between the Tropic of Capricorn (23.5°S) and the Tropic of Cancer (23.5°N).

Figure 25-2 Solar time can be measured with a sundial. The long arm of the sundial (the gnomon) casts a shadow on the dial's horizontal face. The movement of the shadow is used to measure the passage of time. Solar noon occurs when the shadow points exactly north. This is the largest sundial in the United States, located near Phoenix, Arizona.

ACTIVITY 25-1   THE LENGTH OF A SHADOW

Select a place where buildings or trees do not block the sun for an extended period. During the course of one day, take measurements of the changes in length of the shadow of a vertical object 1-meter tall (1 meter = 39.37 inches). Use these measurements to draw a graph of the change in length of the shadow from sunrise to sunset. Measurements must be made on a level surface. Think about the best times to take measurements and which values of shadow length you can determine by logical thinking. Your teacher will collect your data table and your graph.

According to your graph, what was the clock time of solar noon? As an extended activity, make the changing direction of the shadow part of your data and your graph.

ACTIVITY 25-2   CONSTRUCTING A SUNDIAL

A number of sites on the Internet as well as books contain plans for making a simple paper or cardboard sundial. All individuals or groups of students can make the same style, or each can select a different design to construct.

Sundials must be positioned carefully in an open area on a level surface with a north–south alignment. You can check your sundial by comparing its reading with a clock or watch. Solar time and clock time will differ depending on your longitude even if you account for daylight savings time.

HOW DOES THE SUN'S PATH CHANGE WITH THE SEASONS?

The previous section did not specify the exact direction of sunrise and sunset, nor did you learn how high the noon sun reaches in New York State. These observations depend upon the time of year. In fact, the seasons can be defined based on observations of the path of the sun through the sky.

For observers with a level horizon, such as anyone at sea, the sun rises exactly (due) east and sets due west on just two days of the year. People who live on land would also observe this if hills, trees, and buildings did not block a level line of sight to the horizon. These two days of the year are the **equinoxes**. Equinox can be translated as equal night. On these days, daylight and night are approximately equal in length: about 12 hours of daylight and 12 hours of night.

It is useful to remember that the complete apparent path of the sun is a circle. Part of this circle is above the horizon and part is below the horizon. The total angular distance around a circle is 360°. Furthermore, the apparent motion of the sun along that circle is constant. The rate at which the sun appears to move is determined by dividing 360° by 24 hours, which equals 15° per hour. No matter when or where you observe the sun, it always appears to move at a rate of 15° per hour. However, the length of daylight changes. This is because the portion of the circle that is above or below the horizon changes on a yearly cycle.

 ## Spring

Consider the changing path of the sun for an observer facing south at 42° north latitude in New York State. March 21 is the *spring equinox* as shown in Figure 25-3. (This is also called the vernal equinox.) The sun rises due east at 6 A.M. It moves to the observer's right as it climbs to a noon altitude angle of 48°. In the afternoon, the sun continues to the right, setting due west at 6 P.M. Of the sun's circular path of 360°, half is in the sky and half below the horizon, so there are 12 hours of daylight.

 ## Summer

The longest period of daylight in New York State occurs 3 months later, usually on June 21, which is called the **summer solstice**. On that date, the sun rises in the northeast at

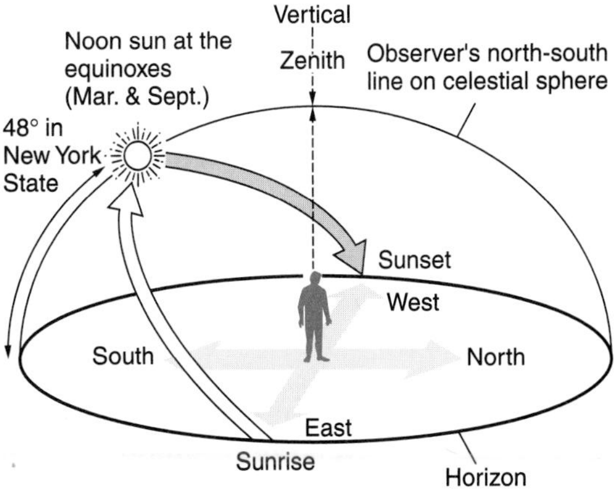

Figure 25-3 In New York State on the spring equinox, the sun rises due east at about 6 A.M. and sets due west at about 6 P.M., giving 12 hours of daylight. The sun reaches its highest point at solar noon when it is due south, 48° above the southern horizon.

about 4:30 A.M. standard time (5:30 A.M. daylight savings time) and sets at about 7:30 P.M. standard time (8:30 P.M. daylight savings time). (See Figure 25-4.)

New York has about 15 hours of daylight at the summer solstice because the sun rises well into the northeast and sets in the northwest. Therefore, more than 60 percent of the sun's circular path is above the horizon. This is also when the noon sun is highest in the sky in New York State—71.5° above the horizon (but still 18.5° below the zenith). Insolation is strongest at this time of the year. The season of summer begins on the summer solstice.

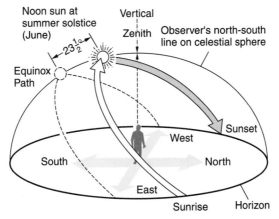

Figure 25-4 In New York State on the summer solstice, the sun rises in the southeast at about 4:30 A.M. (standard time) and sets in the southwest about 7:30 P.M., providing 15 hours of daylight. The sun reaches its highest point at noon when it is nearly 72° above the southern horizon. The sun's path in March is shown by a dotted line.

Autumn

The **autumnal equinox** occurs about September 22. On that day sunrise occurs at 6 A.M. due east, and sunset at 6 P.M. due west. The sun follows the same path it took 6 months earlier at the spring equinox. Figure 25-3 on page 635 is appropriate for both equinoxes. Because there are not exactly 365 days in a year, periodic adjustments must be made in our calendars. Therefore, the dates of the beginning of the seasons vary from year to year.

Winter

The first day of winter occurs each year between December 21 and December 23, which is called the **winter solstice**. On that day, the sun rises in the southeast a shown in Figure 25-5. Its path through the sky is relatively short and the sun sets in the southwest. Less than 40 percent of the sun's circular path is above the horizon. Sunrise does not occur until about 7:30 A.M., and the sun sets at about 4:30 P.M. The length of daylight at 42° north on the winter solstice is about 9 hours.

The maximum altitude of the sun, at solar noon, is only about 24.5° above the southern horizon. The sun is never very high in the sky, even at noon. This is why solar energy is weakest at this time of year. This pattern of change is the same for

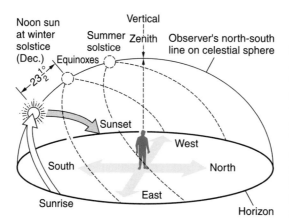

Figure 25-5 In New York State on the winter solstice, the sun rises in the southeast at about 7:30 A.M., and sets in the southwest about 4:30 P.M., giving just 9 hours of daylight. The sun reaches its highest point at noon when it is 24.5° above the southern horizon. Dotted lines show the sun path at the equinoxes and the summer solstice.

all mid-latitude locations on Earth, although the angular altitude of the noon sun does change with latitude.

For mid-latitude locations such as New York State, the seasonal changes can be summarized as follows:

1. The sun generally rises in the east and sets in the west. However, the sun rises north of east when the days are longest in spring and summer and south of east during the short days of fall and winter. The sun always sets in the west, but the pattern of change is similar. The sun sets in the northwest during the long days of spring and summer. When the days are shorter in fall and winter, the sun sets in the southwest.

2. The direction of both sunrise and sunset moves northward along the horizon as the days get longer in winter and spring. During the summer and autumn, when the days get shorter, the sunrise and sunset position moves southward from day to day.

3. The noon sun is highest at the time of the summer solstice and lowest at the time of the winter solstice. At these times, the noon sun is $23\frac{1}{2}°$ higher or lower than its equinox position.

4. When the noon sun is highest in the sky, the days are longest.

ACTIVITY 25-3   **OBSERVING THE SUN**

CAUTION: It is dangerous to look directly at the sun, especially for an extended time because of the potential for damage to your eyes. This danger is much greater if a person is looking through a telescope, which can concentrate the sun's energy. It is best to never look at directly at the sun through any optical device. However, there are two relatively safe ways to observe the sun.

One relatively safe method is to use a small reflecting telescope to project an image of the sun. Direct the sunlight onto a sheet of paper mounted inside a cardboard box. Focus the image

Figure 25-6 A reflecting telescope can be used to project the sun's image onto a sheet of paper mounted in a cardboard box. Dark areas known as sunspots can sometimes be seen on the sun's surface. ***Never look directly at the sun through any optical device.***

on the paper with the eyepiece. The box shields the paper screen from direct sunlight. This procedure is shown in Figure 25-6.

A second method makes use of a special kind of mirror called a front-sided mirror. Most mirrors have the silver reflective coating behind the glass to protect the coating from scratches. Front-sided mirrors can be purchased from scientific supply companies. This activity works best with a tiny piece of mirror about a quarter the size of a small coin.

Mount the small piece of front-sided mirror on a tripod so it reflects a beam of sunlight through a window. Project the reflection onto a light-colored interior wall of a darkened room. Although it can be difficult to aim this reflection, the image projected inside can be used for two purposes. This procedure will create a dim image of the sun revealing details on the solar surface. It can also be used to observe and even measure the sun's apparent motion through the sky.

DOES THE SUN'S PATH DEPEND ON THE OBSERVER'S LOCATION?

At any particular time, half of Earth is lighted and half is in shadow. Whether it is day or night depends upon where you are located. This is a consequence of living on a spherical planet.

 ## Changes with Longitude

If you telephone someone in California, you need to take into account that the sun rises in New York 3 hours before it rises in California. This is due to Earth's rotation at 15° per hour. Local time generally changes by 1 hour for each 15° change of longitude east or west. This is why Earth is divided into time zones that generally are 1 hour apart for each 15° change in longitude.

 ## Changes with Latitude

Although the local solar time does not change with latitude, the path of the sun through the sky does change. You have already read that observers at the North or South Pole experience 6 months of daylight followed by 6 months of night. In fact, every location on Earth has a total of 50 percent of the year when the sun is in the sky and 50 percent when it is below the horizon. However, as a result of Earth's curvature, the cycle of day and night is most changeable at the poles and most constant near the equator.

People at the equator experience approximately 12 hours of daylight every day. Although the position of sunrise along the horizon changes, as it does in the mid-latitudes such as New York, the noon sun is always high in the sky. When it is spring and summer in the Northern Hemisphere, the sun rises to the north of east at the equator. The noon sun is a little north of the zenith (the point straight overhead). From late

September through most of March the sun rises in the southeast and reaches its highest point a little south of the zenith. At the equinoxes the sun rises due east, sets due west, and the noon sun is directly overhead at the equator. Changes from season to season are hardly noticeable. This is why the tropics are sometimes called the zone of seasonless climates.

Moving out of the tropics, you would notice that the noon sun is never directly overhead. You would also notice distinct seasons and changes in the length of daylight. South of the equator, the seasons are the opposite of those in New York. When its summer here and the days are longer, it is winter south of the equator and daylight is shortest. When it is spring in New York, days are getting longer and the noon sun higher in the sky each day. At the same time, is autumn south of the equator where the opposite changes are happening.

WHAT IS REALLY MOVING, SUN OR EARTH?

This was an easy question for our ancestors to answer. They could not feel the ground moving. Their observations fit the idea that the sun circles Earth in a daily path that changes through the year.

The Geocentric Model

For most people that answer was logical and complete. This is called the **geocentric model**. (The prefix *geo-* refers to Earth, as in geology, the study of the solid Earth.) Geocentric means Earth-centered. The geocentric model assumed Earth was stationary and the real motion of celestial objects caused the motions seen in the sky.

As early as 2000 years ago, some scholars made observations that caused them to question this traditional view of Earth. These observations led some scholars to consider the idea that Earth is not a flat surface, but a huge sphere. The geocentric model was revised to include Earth as a sphere, but still at the center of the orbits of the sun, planets, and stars.

 **Heliocentric Model**

Copernicus published his ideas about the sun-centered model in Poland in 1543. It was controversial. The well-known Italian mathematician and inventor Galileo published his own observations in the early 1600s supporting the sun-centered model. Important clues came from observations of the motions of planets in the night sky. German astronomer Johannes Kepler developed the mathematical tools needed to predict the positions of the planets among the stars. However, his theories made sense only if he assumed that the planets, including Earth, orbit the sun. This is the central idea of the **heliocentric model** of the universe. (The prefix *helio-* refers to the sun.) The heliocentric model places the sun at the center of planetary motion.

According to this model, it is mainly the motions of Earth that cause the apparent motions of celestial objects. The motion of the planets in their orbits around the sun is known as **revolution**. The heliocentric model also includes the **rotation**, or spin, of Earth on its axis.

A useful way to remember the difference between rotation and revolution is to recall that a top is a spinning toy. Rotation has two *T*s, the first letter of *top*. The American Revolution started in 1776, which is a year. A revolution is the yearly cycle of Earth's motion in its orbit around the sun. Figure 25-7 on page 642 compares the geocentric and heliocentric models.

Experimental proof of Earth's motion was not discovered until 1851. By then, most astronomers already supported the heliocentric model. Motions of the planets in the geocentric model had become too complicated to explain by the laws of physics.

 Proof of Rotation

Jean Foucault invented a special pendulum that is free to rotate as it swings back and forth. The Foucault pendulum is shown in Figure 25-8 on page 642. If a Foucault pendulum were used at the North or South Pole, it would seem to rotate

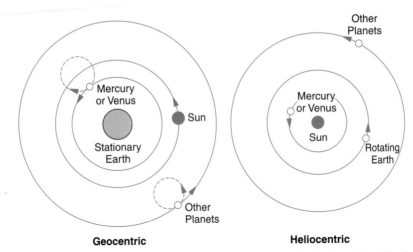

Figure 25-7 Early astronomers thought that the sun and planets orbited Earth. This is called the geocentric (Earth-centered) model. Arrows show the direction of motion in orbits. The complicated paths of the planets among the stars required astronomers to add circles to the orbits as shown by the dashed figures. A major reason that astronomers grew to prefer the heliocentric (sun-centered) model is that it is simpler.

in a complete circle every 24 hours (15° per hour) as Earth spins underneath it. At the equator, a Foucault pendulum does not seem to rotate at all because of its orientation with Earth's axis. It is important to remember that the Foucault

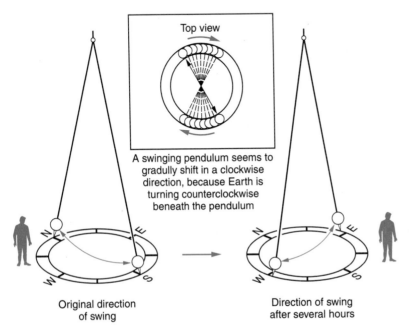

Figure 25-8 The apparent rotation of a Foucault pendulum is experimental proof that Earth rotates. One complete rotation takes 24 hours at the poles. However, the time per rotation increases if the pendulum is moved toward the equator.

pendulum swings in the same plane as the Earth rotates beneath it. A Foucault pendulum in New York State takes about 36 hours to complete one apparent rotation.

ACTIVITY 25-4 LOCATE A FOUCAULT PENDULUM

Some museums and other public places, such as the main lobby of the United Nations in New York City, have a Foucault pendulum. Locate the one nearest to where you live. If you are able to visit it, try to stay long enough to see a noticeable change in the direction of swing.

The Coriolis effect, a second proof of Earth's rotation, was discussed in Chapter 21. Due to the Coriolis effect, winds in the Northern Hemisphere curve to their right as they move out of a high-pressure system. Ocean currents in this hemisphere also curve to their right. The apparent curve to the right is actually a result of the winds and ocean currents trying to follow a straight path on a moving planet. Remember that this apparent curve is to the left in the Southern Hemisphere.

 Earth's Motions as Viewed from Space

If you could view Earth from a stationary position in space beyond Earth's orbit, you might see the planet at one of the positions shown in Figure 25-9 on page 644. This diagram shows Earth's two most important motions: rotation and revolution. Note that this diagram is not drawn to a uniform scale. From outside Earth's orbit, the sun would appear as a tiny circle of light. Except in its closest position, Earth would appear as a tiny dot. In addition, at any given time Earth is at only one position in its orbit. Earth revolves around the sun in an annual cycle. With 365 days in a year and 360° in a circle, the planet revolves a little less than 1° in its orbit each day.

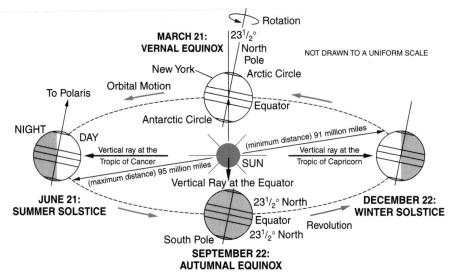

Figure 25-9 The seasons are the result of Earth's daily rotation on an axis that is tilted 23 1/2° and its annual revolution around the sun. Notice that Earth is slightly closer to the sun when it is winter in the Northern Hemisphere. The seasons labeled on this diagram apply only to the Northern Hemisphere.

 ## Earth's Motions and the Seasons

The seasons are caused by a combination of two Earth motions. In addition to its revolution, Earth rotates on an axis that is tilted approximately 23.5° from a direction perpendicular to the plane of its orbit. The direction of Earth's axis is constant throughout the year. Polaris appears very close to the location in space directly above the North Pole. However, for half of the year, our spring and summer, the Northern Hemisphere is tilted toward the sun. For the other half of the year, our autumn and winter, the Northern Hemisphere is tilted away from the sun.

The latitude at which the vertical ray of sunlight strikes Earth determines the beginning of each season. The **vertical ray** is sunlight that strikes Earth's surface at an angle of 90°. It is sometimes called the direct ray of sunlight. This ray strikes Earth at a position where the sun is directly overhead, that is at the zenith. This ray of sunlight always hits Earth within the tropics. However, its position changes in an annual cycle

SPRING   The spring, or vernal, equinox, shown by the top Earth position in Figure 25-9, occurs near the end of March. At that time, the vertical ray is at the equator, and spring begins in the Northern Hemisphere (autumn in the Southern Hemisphere). Earth's axis is pointed neither toward nor away from the sun, and most of the planet receives 12 hours of daylight. The exceptions are the North and South Poles, where the sun circles very near the horizon, resulting in twilight for 24 hours.

SUMMER   Over the next 3 months, Earth moves toward the June solstice, the beginning of summer in the Northern Hemisphere. At this time, the North Pole is tilted toward the sun and the vertical ray reaches its greatest latitude (23.5°N) north of the Equator, the **Tropic of Cancer**. In the Northern Hemisphere, this is when the noon sun is highest in the sky, the sun's path through the sky is longest, and daylight is longest.

It is just after this time that Earth reaches its greatest distance from the sun. However, the change in the Earth-sun distance is very small. In fact, it is too small to affect the seasons. If Earth's distance from the sun caused the seasons, you would observe two major differences from the present seasons. First, summer and winter would be reversed. Our warmest weather would be in January. Second, summer in the Northern Hemisphere would be summer in the Southern Hemisphere instead of the present 6-month difference.

AUTUMN AND WINTER   Over the next 3 months as Earth moves to its September position, the vertical ray moves back to the Equator. This is the autumnal equinox. Our winter begins after another 3 months, near the end of December, when the North Pole is tilted away from the sun. At this time, the vertical ray reaches its most southerly latitude (23.5°S), the **Tropic of Capricorn**.

Two other latitude positions are important. The **Arctic Circle** (66.5°N) is the latitude north of which the sun does not rise on our December solstice. North of this latitude, the sun is in the sky for 24 hours at the June solstice. The **Antarctic Circle** (66.5°S) is the corresponding latitude in the Southern Hemisphere. On the December solstice (summer solstice in

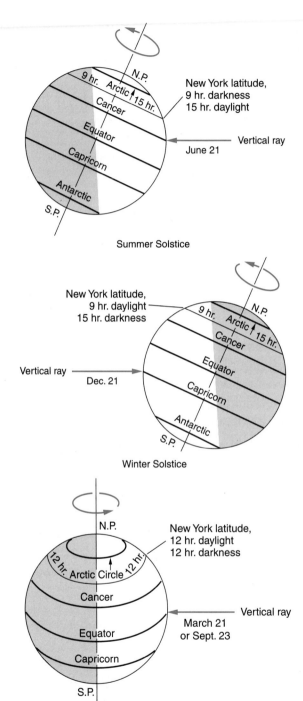

New York latitude,
9 hr. darkness
15 hr. daylight

Vertical ray
June 21

Summer Solstice

New York latitude,
9 hr. daylight
15 hr. darkness

Vertical ray
Dec. 21

Winter Solstice

New York latitude,
12 hr. daylight
12 hr. darkness

Vertical ray
March 21
or Sept. 23

Spring and Fall Equinoxes

Figure 25-10 Notice how the position of the vertical ray and the daily hours of sunlight change with the seasons.

the Southern Hemisphere), the sun is in the sky for 24 hours. On the June solstice (winter solstice in the Southern Hemisphere), the sun does not rise. Figure 25-9 on page 644 should help you understand the seasonal changes in the path of the sun through the sky and why the seasons are reversed north and south of the equator.

Figure 25-10 shows Earth at the solstice and equinox positions. Notice how the latitude at which the vertical ray of sunlight strikes Earth changes as does the length of daylight in New York State.

ACTIVITY 25-5 MODELING EARTH MOTIONS

Position a single bright light, representing the sun, at the center of a darkened classroom. With a globe, illustrate Earth's motions of rotation on its axis tilted 23.5° and revolution around the sun. Notice how the height of the sun in New York State and the length of daylight change as Earth orbits the sun. In what ways is this model unlike the real Earth and sun?

If Earth's axis were perpendicular to the plane of its orbit, there would be no seasons. In places like New York State, there would be no annual cycle of temperature and changes in the sun's path. On the other hand, if Earth's axis were tilted more than 23.5°, New York State would experience greater changes in temperature from winter to summer. More extremes in the path of the sun in its annual cycle would also be observed. The sun's path would be longer and higher in summer, but lower and shorter in winter.

HOW DO EARTH'S MOTIONS AFFECT THE APPEARANCE OF OTHER CELESTIAL OBJECTS?

The night sky was familiar to our ancestors who had no electric lights, television, or computers. Although the stars are randomly distributed throughout the sky, ancient people imagined patterns in the stars.

Constellations

Certain patterns of stars are designated as constellations. Orion the hunter is a constellation prominent in New York's winter sky. Near Orion is a group of stars that represent his dog Sirius and Lepus the rabbit. Others patterns represent objects such as Lyra the harp.

Constellations were often associated with traditions and legends that became a part of the cultural heritage. The constellation called Cassiopeia was named for an Ethiopian queen who proclaimed that she was more beautiful than other women were. Two bright stars Castor and Pollux dominate the constellation Gemini. In mythology, Castor and Pollux were twins.

Different cultures imagined different things about the same group of stars. For example, the group of stars we call The Big Dipper is known as The Plow in Britain. To the ancient Greeks, this was a part of the Great Bear, which we call by its Latin name, Ursa Major. Figure 25-11 shows this constellation with a drawing of a bear superimposed on the star pattern. For many constellations, it takes a good deal of imagination to see in the pattern of stars the objects they represent.

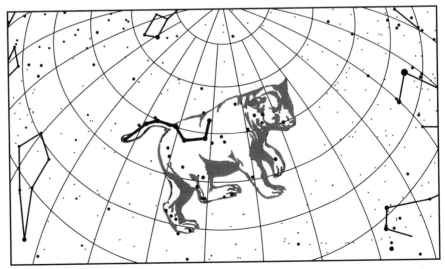

Figure 25-11 The group of stars called The Big Dipper is a part of the constellation Ursa Major (The Great Bear). For observers in New York State, this group of stars is always visible in the northern sky.

ACTIVITY 25-6   THE BIG DIPPER AND POLARIS

On a clear night, find the group of stars called the Big Dipper. Follow the pointer stars to Polaris. Sketch the Big Dipper and Polaris showing their orientation with respect to the Northern Horizon. You may wish to repeat this activity several hours later or in a few months at the same hour to see if their orientation changes.

ACTIVITY 25-7   ADOPT A CONSTELLATION

Find a list of the major constellations in an astronomy book or another source. Select one constellation for which you will do the following:

1. Draw your constellation in white or other light-colored ink on a sheet of black paper.

2. Write a brief summary of the mythology traditionally associated with your constellation and draw the appropriate person or object on the star pattern of your constellation.

3. Tell when and where your constellation can be observed in the sky.

Star Maps

Modern astronomers use the constellations to designate 88 regions of the night sky. This is a convenient way to establish a map of the stars. The stars always occupy the same position with respect to other stars. When a star is part of particular constellation, that association helps observers locate the star in the night sky. Figure 25-12 on page 650 is a map of the evening sky in the month of April.

To use the map, hold a copy of it upside down so that the compass directions printed on the map line up with the true compass directions. If you find the Big Dipper high in the sky, you can use it to locate other stars and constellations as shown by the arrows on the chart. The two stars at the end

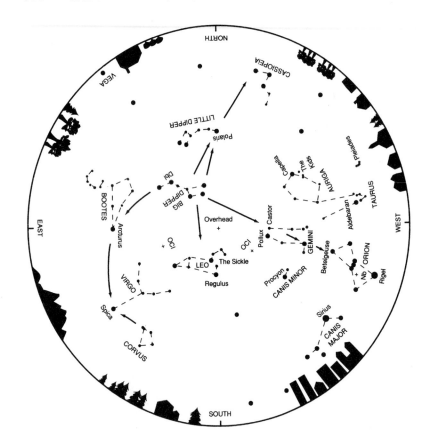

Figure 25-12 This star map shows constellations visible at about 40° North latitude at about 10 P.M. in March and about 8 P.M. to 9 P.M. in April.

of the bowl of the Big Dipper point to Polaris. Binoculars or a small telescope will help you observe objects labeled OCl (open clusters), Dbl (double stars), and Nb (nebulae). The planets wander among the stars, so they are not shown on this chart. Current information from a newspaper or the Internet can help you locate the visible planets.

ACTIVITY 25-8   LOCATING MAJOR CONSTELLATIONS

Use a copy of Figure 25-12 or a similar star map to locate constellations in the night sky. Try to observe on a clear night from a location that is far from artificial lights. The sky should not be obstructed by nearby buildings or trees. Make a list of the constellations you are able to find. Number them according to how easy they were for you to locate in the night sky.

WHY DO THE STARS SEEM TO CHANGE THEIR POSITIONS?

We do not usually see stars shift their positions relative to each other. For that, they are sometimes called the fixed field of stars. However, the cycles of daily rotation and yearly revolution of Earth influence our observations of stars .

Daily Apparent Motions

Earth's rotation makes stars appear to move through the sky just as the sun appears to move. In New York State, most stars rise in the east, travel to the right across the southern sky, and set in the west. Like the sun, their path depends on where they rise. Stars that rise in the northeast travel high in the sky and set in the northwest. Those that rise in the southeast move lower in the sky to set in the southwest.

A different kind of motion can be observed in the northern part of the sky, where the sun is never found. In earlier chapters, you read about Polaris, the North Star. Polaris is not one of the brightest stars, but it appears to be located almost directly above Earth's North Pole near a point in the sky called the celestial north pole. Consequently, Polaris is the only star that does not seem to move. Actually, Polaris is not exactly above the North Pole; therefore, it does make a very small counterclockwise circle in 24 hours.

Other stars that are not as close to the celestial pole make larger circles, all at a rate of 15° per hour, Earth's rate of rotation. Figure 25-13 on page 652 shows the apparent motion of stars in the night sky. The planets and comets also show this apparent motion, although they slowly shift their positions among the stars from night to night.

Figure 25-14 on page 652 illustrates the trails of stars that would be observed in each of the four compass directions by an observer in New York State. These diagrams show the stars rising in the east, moving through the southern sky and setting in the west. Only stars in the northern sky do not seem to rise and set; they appear to circle counterclockwise around Polaris.

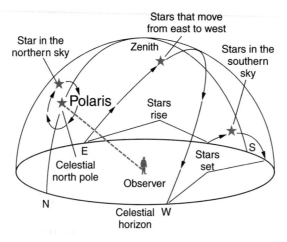

Figure 25-13 Due to Earth's rotation, the stars appear to rotate counterclockwise around Polaris, including those that rise in the east and set in the west. In fact, all star paths can be thought of as circles around the north and south celestial poles.

Looking East...
Stars rise and move to the right.

Looking South...
Stars circle to the right.

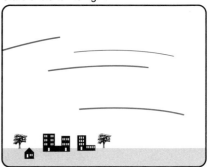

Looking West..
Stars set as they move to the right.

Looking North...
Stars circle counterclockwise around Polaris.

Figure 25-14 These drawings represent time exposure photographs of the night sky. True photographs would show light star trails moving through a dark sky.

ACTIVITY 25-9   PHOTOGRAPHING STAR TRAILS

You can use an adjustable camera that allows you to take very long exposures to take photographs of star trails. The camera must be mounted on a tripod or other object(s) to hold it steady while the shutter is open. A clear, dark sky away from artificial lights is essential. Exposures 15–30 seconds in length show star patterns in constellations. Exposures of 5 minutes to an hour will yield patterns like those in Figure 25-14. For best results, try a few star photographs and have them developed. After viewing your photographs, you can plan changes to improve your pictures in later attempts. If color photography is used, these trails may be of different colors, indicating the surface temperature of the stars.

 Yearly Apparent Motions

In addition to the apparent daily cycle of motion of stars through the sky, the stars in the evening sky also change in a yearly cycle. In the summer, Scorpio, the scorpion, with its red star, Antares, is prominent in the southern sky. At this time, the stars of Scorpio are located on the opposite side of Earth from the sun. This is why they are visible in the night sky at that time of year.

By winter, 6 months later, Earth has revolved half way around its orbit. Orion, the hunter, is now visible in the southern sky in the evening. Scorpio is still in the sky, but it is now located in the direction of the sun. During the day, the sky is too bright for most stars to be visible. Figure 25-15 on page 654 shows Earth's orbit and some of the brighter constellations visible in the evening at different times of the year. Because Earth moves about 1° in its orbit each day, changes in the evening sky are not noticeable from night to night. However, over a period of months this change is obvious.

Not all the constellations are seasonal. Stars and constellations in the northern sky, such as Polaris in Ursa Minor (the Little Bear), Ursa Major (including the Big Dipper), and Cassiopeia, are visible throughout the year. However, the

Figure 25-15 As Earth revolves around the sun, the night side of Earth faces different directions in space. Only constellations opposite the sun are visible because the light of the sun hides stars during the day.

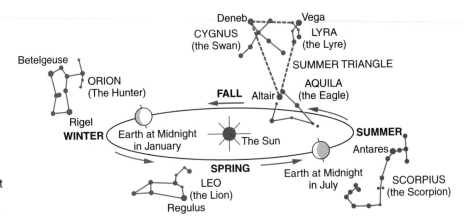

polar constellations that are below Polaris at one time of year are high in the sky, above Polaris, 6 months later.

ACTIVITY 25-10 CELESTIAL OBSERVATIONS

This chapter has introduced a number of changes that you can observe and/or measure quite easily. For example, you can measure the rate at which the sun moves through the sky or the apparent motions of stars.

The topics in this chapter relate to a number of long-range projects that are not difficult to perform. You might use a digital camera to a photograph the horizon, recording changes in the position of sunset or sunrise over a period of weeks. The changing angular altitude of the noon sun can also be documented. Changes in the moon and the stars can also be observed and documented.

TERMS TO KNOW

altitude	equinox	solar noon	vertical ray
Antarctic Circle	revolution	solar time	winter solstice
Arctic Circle	rotation	summer solstice	zenith
celestial object			

CHAPTER REVIEW QUESTIONS

1. In which direction on the horizon does the sun appear to rise on July 4 in New York State?

(1) due north　　　　　　　　　(3) north of due east
(2) due south　　　　　　　　　(4) south of due east

2. The diagram below represents a simple geocentric model. Which object does letter X represent?

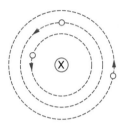

(Not drawn to scale)

(1) Earth　　　　　　　　　　(3) the moon
(2) the sun　　　　　　　　　　(4) Polaris

3. Which observation provides the best evidence that Earth rotates?

(1) The position of the sun changes during the year.
(2) The location of the constellations in relationship to Polaris changes from month to month.
(3) The length of the shadow cast by a flagpole at noon changes from season to season.
(4) The direction of swing of a freely swinging pendulum changes during the day.

4. The apparent rising and setting of the sun, as viewed from Earth, is caused by

(1) Earth's rotation.　　　　　(3) the sun's rotation.
(2) Earth's revolution.　　　　(4) the sun's revolution.

5. The length of an Earth day is determined by the time required for approximately one

(1) Earth rotation.　　　　　　(3) sun rotation.
(2) Earth revolution.　　　　　(4) sun revolution.

6. The length of Earth's year is based on Earth's

   (1) rotation of 15°/h.
   (2) revolution of 15°/h.
   (3) rotation of approximately 1°/day.
   (4) revolution of approximately 1°/day.

7. The diagram below shows the latitude-longitude grid on an Earth model. Points A and B are locations on the surface.

   On Earth, the solar time difference between point A and point B would be

   (1) 1 hour.   (3) 12 hours.
   (2) 5 hours.   (4) 24 hours.

8. Summer days in New York State are likely to be hotter than winter days because in summer

   (1) Earth is closer to the sun.
   (2) the number of sunspots increases.
   (3) Earth's northern axis is tilted toward the sun.
   (4) the sun gives off more energy.

Base your answers to the questions 9–11 on the diagram below, which shows the tilt of Earth's axis and the pattern of day and night on a particular day of the year. Positions A through E are along Earth's surface. Point D is located in New York State.

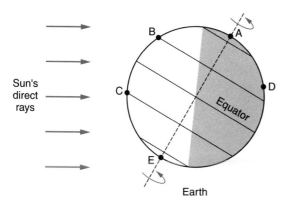

Earth

9. Which diagram best represents the angle of the sun's rays at location C at noon on this day?

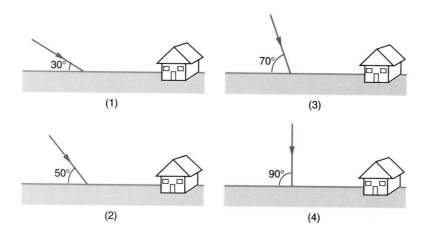

10. On this day, which location had the greatest number or hours of daylight?

(1) B   (3) D
(2) C   (4) E

11. What date is illustrated on the diagram of Earth above?

(1) March 21   (3) September 22
(2) June 21   (4) December 22

Base your answers to questions 12 and 13 on the diagram below, which represents the position of the sun with respect to Earth's surface on certain dates. The latitude of six locations on the same line of longitude is shown.

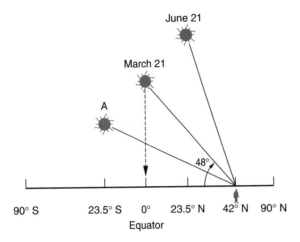

12. When the sun is at position A, which latitude receives the most direct rays of sunlight?

(1) Tropic of Cancer (23.5°N)
(2) Equator (0°)
(3) Tropic of Capricorn (23.5°S)
(4) Antarctic Circle (66.5°S)

13. When the sun is at the March 21 position, New York State will usually have

(1) longer days than nights.
(2) 12 hours of daylight and 12 hours when the sun is below the horizon.
(3) the lowest altitude of the sun at solar noon for the whole year.
(4) the highest altitude of the sun at solar noon for the whole year.

14. As observed in New York State, in which part of the sky do the stars seem to move in small circles?

(1) north
(2) east
(3) south
(4) west

15. The diagram below represents the major stars of the constellation Orion, as viewed by an observer in New York State.

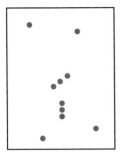

Which statement best explains why Orion can be observed from New York State on December 21, but not on June 21?

(1) Orion has an eccentric orbit around Earth.
(2) Orion has an eccentric orbit around the sun.
(3) Earth revolves around the sun
(4) Earth rotates on its axis.

Open-Ended Questions

16. State two factors that combine to cause Earth's seasons.

Base your answers to questions 17–19 on the diagram below that represents Earth at a specific position in its orbit as viewed from space. The shaded area represents nighttime.

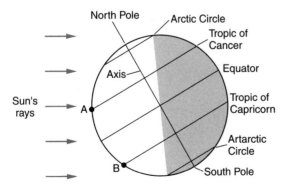

17. *a.* State the month represented by the diagram.

   b. Name the area that receives the most intense radiation from the sun when Earth is at this position in its orbit.

18. Describe the length of daylight at point A compared with the length of daylight at point B on the day represented by the diagram.

19. The diagram below represents the position of Earth in its orbit *6 months later.* Make a copy of this diagram. (Please do not write in this book.)

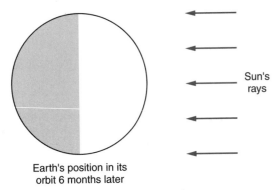

Sun's rays

Earth's position in its orbit 6 months later

   a. Draw the position of Earth's axis and label the axis.
   b. Label the North Pole.
   c. Draw the position of Earth's equator and label the equator.

20. What would happen to the average summer and winter temperature in New York State if the tilt of Earth's axis were to decrease from 23.5° to 20°?

Chapter 26

Earth and Its Moon

THE RACE FOR THE MOON

Ever since Galileo made his first telescopic observations of the moon's surface in 1609, humans have dreamed of visiting Earth's closest neighbor. A series of events in the mid-twentieth century changed that dream into reality. During the Second World War, the German military developed rockets that were capable of long-range guided flight.

Based on this technology, the Russians launched the first artificial satellite into orbit around Earth in 1957. This began the "space race" between the Soviet Union and the United States. In addition to the first artificial satellite, the achievements of the Soviet Union included the first human to travel outside Earth's atmosphere, the first unmanned crash landing on the moon, and the first photographs of the far side of the moon.

 Apollo Program

American politicians and scientists were embarrassed by the early triumphs of their rival. Thus began the American program to explore space, which led to the Apollo program of

Figure 26-1 The last Apollo mission in 1972 carried geologist Harrison Schmitt to the surface of the moon. The six manned moon landings greatly expanded scientists' understanding of Earth's nearest neighbor in space.

lunar exploration. In July of 1969, American astronaut Neil Armstrong became the first human to set foot on the surface of the moon. That mission also returned the first samples of moon rock to Earth. In the next six Apollo missions, a total of 382 kg (842 pounds) of moon rocks were brought back to Earth for scientific study. Figure 26-1 is a photograph of geologist and Apollo 17 crew member Harrison Schmitt investigating the lunar surface. Since that last Apollo mission in 1972, no other humans have visited the moon.

Among the many discoveries of the Apollo program, scientists learned that the mineral composition of the surface of moon is very similar to the composition of Earth's mantle. Plagioclase feldspar, pyroxene, and olivine are the most common minerals in the lunar samples.

Lunar Surface

The surface of the moon has two landscape types, the lunar highlands and the maria (from the Latin word for "seas"). The rocks of the lunar highlands are mostly anorthosite, a rock type widespread in New York's Adirondack Mountains, but not common in most places on Earth. The lunar maria are relatively flat and darker in color than the highlands. They are composed of basalt, a relatively common dark-colored, fine-grained, igneous rock on Earth.

The moon has none of Earth's most common sedimentary rocks because it has no atmosphere or surface water. Chemical weathering does not occur on the moon, although the surface is covered by material from meteorites and moon rocks broken by meteorite impacts. However, breccia, a rock formed from the breakage and welding of rock fragments, is found in the lunar highlands. Radiometric dating has shown the oldest moon rocks are about the same age as Earth, 4.6 billion years.

ACTIVITY 26-1 LUNAR SURVIVAL KIT

Imagine that you are part of a team chosen to explore the moon. You were scheduled to land near the mother ship, which will take you back to Earth. Unfortunately, you were forced to land 100 km from the rendezvous point on the lighted side of the moon. None of your crew is injured, but you need to select the most important items for your survival. Arrange the 15 items below from the most important item (#1) to the least useful (#15). Then, for each item, briefly explain why you gave it that priority.

List of items:
box of matches, dry food concentrate, 15-m nylon rope, parachute silk 6 m by 6 m, portable heater, two .45 caliber pistols, 4 liters of dehydrated milk, three 50-kg tanks of oxygen, chart of the constellations, first aid kit, solar powered receiver–transmitter, 20 liters of water, ocean life raft, five signal flares, flat metal mirror 10 cm by 10 cm

WHAT IS THE HISTORY OF EARTH'S MOON?

A moon is a natural satellite of a planet. A **satellite** is an object in space that revolves around another object under the influence of gravity. The moon is Earth's only natural satellite. (Earth, like the other planets, is a satellite of the sun.) Although Mercury and Venus have no moons, all the other planets do have natural satellites. Earth is the only planet in our solar system with a single moon.

Our moon is the largest compared with the planet it orbits. The moon's diameter is a little more than a quarter of Earth's diameter. In fact, our moon is larger than the outermost planet of our solar system, Pluto.

Another curious feature of our moon is the fact that its period of rotation and revolution are the same, about $27\frac{1}{3}$ days. Consequently, the same side of the moon always faces Earth. This is why features of the far side of the moon were unknown until the Soviets sent a satellite around the moon to take photographs early in the "space race."

The Origin of Earth's Moon

There have been several theories about the origin of the moon. Some astronomers have suggested that the moon and Earth formed as a double planet orbiting the sun. Others have speculated that the moon was an object in space that came close enough to Earth to be captured by Earth's gravity.

Most astronomers now agree that a collision between Earth and a smaller planet probably created the moon shortly after Earth's formation. That impact destroyed the smaller planet and created a debris field in space. The debris came together under the influence of gravity, forming our moon.

Surface Features of the Moon

The cratered surface of the moon contrasts sharply with Earth's surface. Many large objects undoubtedly hit the moon and Earth. The highlands of the moon have been solid for more than 4 billion years. They are so covered with impact craters that the surface seems to be made of crater upon crater.

The moon is small enough and its gravity weak enough that the moon was unable to hold an atmosphere. Consequently, there is no shell of gases surrounding the moon and no water on the moon's surface to cause weathering, erosion, and deposition. Furthermore, unlike Earth, the surface of the moon is not composed of active tectonic plates that create and recycle surface material. Therefore, features on the moon last much longer than they do on Earth. The only active process on the moon is impact by meteoroids and other objects from space. Early in the history of the solar system, there were many more of these objects and impacts were far more common.

It is clear from the dark lava flows that became the maria that the moon once had a molten interior. Most of the moon's volcanic eruptions occurred between 3.8 and 3.1 billion years ago. The maria show less cratering than the highlands because most of the impacts that created the largest craters occurred before the surface of the maria had formed. By 3 billion years ago, the surface of the moon probably looked pretty much as it does today. Whether the moon still has molten rock in its interior is a question that scientists have not yet settled.

 ## HOW CAN WE DESCRIBE ORBITS?

The earliest astronomers believed that the sky was a giant dome that covers Earth from horizon to horizon. The gradual realization that Earth is a sphere led to the idea that the sky is a larger sphere that surrounds Earth. The wandering motion of the planets led to the idea that planets and the moon move in orbits independent of the fixed stars. These orbits were originally thought to be circles. However, careful observations of the planets showed that their motions could be explained best if their orbits were not circles, but ellipses.

An **ellipse** is a closed curve formed around two fixed points such that the total distance between any point on the curve and the fixed points is constant. The fixed points are known as the foci of the ellipse. (The singular of foci is focus.)

If you use pins and a loop of string to draw an ellipse, the position of each pin is a **focus** of the ellipse. The string keeps the total distance to the two foci constant.

In all orbits, the object the satellite revolves around, known as the primary, is located at one focus. For example, the Earth is at one focus of the moon's orbit. The sun is also located at one focus in Earth's orbit. The other focus is a point in empty space. You should also note that Earth is not at the center of the moon's orbit and the sun is not at the center of Earth's orbit.

ACTIVITY 26-2 | **ORBIT OF THE MOON**

Obtain the following materials: one soft board approximately 12 inches (30 cm) square (fiberboard, ceiling tile, or soft pine work well), two straight pins, adhesive tape, one piece of light string about 30–40 cm long, one sharp pencil, one clean sheet of standard size paper ($8\frac{1}{2}$ inches by 11 inches), one metric ruler.

Find the center of the sheet of paper by drawing two lines connecting opposite corners of the paper.

1. Tie the string into a loop that is 10 cm long when fully stretched ($\pm$ 0.5 cm).

2. Tape the paper to the soft board.

3. Stick one pin through the center of the sheet of paper and into the soft board. Place the loop of string around it. Then stretch the loop to its greatest distance with the tip of the pencil. Keeping the loop taut and the pencil perpendicular to the paper, make a circle around the pin. Label it "circle" along its circumference.

4. Place the two pins 4.5 cm from the center of the paper along the paper's long axis. Stretch the string around both pins and then draw out the string to make an ellipse.

5. Make another ellipse by placing the two pins 1 cm apart (0.5 cm each side of the center). Draw this ellipse and label it "Orbit of the Moon."

Is the "Orbit of the Moon" noticeably elongated?

Calculating Eccentricity

The shape of an orbit is described by its **eccentricity**, or how elongated the ellipse is. The following equation, used to calculate the eccentricity of an ellipse, is found in the *Earth Science Reference Tables*:

$$Eccentricity = \frac{distance\ between\ foci}{length\ of\ major\ axis}$$

Whenever you are asked to calculate eccentricity, either the values will be given to you or a diagram will be provided that clearly shows the ellipse and the position of the two foci. If you are given a diagram, you will need to use a centimeter scale to measure these two distances. Remember that a centimeter scale is printed on the front cover of the *Earth Science Reference Tables*. The **major axis** is the distance across the ellipse measured at its widest point. (The smaller axis, the width of the ellipse, is known as the minor axis.) These two features of an ellipse are shown in Figure 26-2.

SAMPLE PROBLEMS

Problem 1   Calculate the eccentricity of the ellipse in Figure 26-2.

Solution   The distance between the foci of this ellipse is 3 cm. The distance across the ellipse (the major axis) is 4 cm. The calculation is shown below.

$$Eccentricity = \frac{distance\ between\ foci}{length\ of\ major\ axis}$$

$$= \frac{3\ cm}{4\ cm}$$

$$= 0.75$$

Figure 26-2 Eccentricity is calculated by dividing the distance between the foci of an ellipse by the length of the major axis. Eccentricity is a ratio without units.

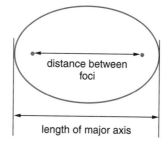

distance between foci

length of major axis

Notice that there are no units of eccentricity. Eccentricity is a ratio between two measured values.

Problem 2   The greatest distance across Earth's orbit is 299,200,000 km. The distance from the sun to the location in space that is the other focus of Earth's orbit is 5,086,400 km. Calculate the eccentricity of Earth's orbit.

Solution

$$Eccentricity = \frac{distance\ between\ foci}{length\ of\ major\ axis}$$

$$= \frac{5{,}086{,}400\ km}{299{,}200{,}000\ km}$$

$$= 0.017$$

Notice that the result is a very small number. Earth's orbit is nearly a perfect circle. That is why the changing distance between Earth and the sun is not a significant factor in seasonal changes in temperature on Earth.

Problem 3   The greatest distance across Mars' orbit is 455,800,000 km. The eccentricity of its orbit is 0.093. What is the approximate distance between the sun and the second focus of Mars' orbit?

Solution   The first step is to rearrange the formula to isolate the distance between the foci. Multiplying both sides of the equation by *length of major axis* does this.

$$Eccentricity = \frac{distance\ between\ foci}{length\ of\ major\ axis}$$

The next step is to substitute the values and solve for the distance between the foci:

Length of major axis × *Eccentricity* = *distance between foci*
455,800,000 km × 0.093 = *distance between foci*
42,389,400 km = *distance between foci*

This value is expressed to a much greater accuracy than the length of the major axis of the ellipse. So it should be rounded off to 42 million km. (Also note that the problem asked for the "approximate distance.")

 Practice Problem 1
Calculate the eccentricity of an ellipse in which the distance between the foci is 15 cm and the length of the major axis is 60 cm.

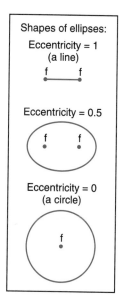

Figure 26-3 The orbits of Earth and the moon look like circles. Only by carefully measuring the length and width of the orbits, or by locating the two foci, can scientists observe that the orbits of the moon and Earth are not quite circles.

 Practice Problem 2

The greatest distance across the moon's orbit is 772,000 km; the eccentricity of its orbit is 0.055. How far apart are the foci of the moon's orbit?

Figure 26-3 shows the range of shape of ellipses. If the foci are at the ends of the ellipse, the ellipse will be a line segment with an eccentricity of 1. If the two foci are located near the ends of the ellipse, the ellipse looks flattened. A circle is the special case of ellipse in which the foci are at a single point. The eccentricity of a circle is 0. When considering the eccentricity of an ellipse, it may help to remember that just as the number "1" can be written as a line, an ellipse with an eccentricity of one is a straight line. Furthermore, just as the number "0" can be written as a circle, an ellipse with an eccentricity of zero is a circle.

WHAT DETERMINES A SATELLITE'S S ORBIT?

The path that Earth takes around the sun is determined by two factors: inertia and gravity. **Inertia** is the tendency of an object at rest to remain at rest or an object in motion to move at a constant speed in a straight line unless acted on by an unbalanced force. If you roll a heavy ball across a flat, hard floor, the ball continues in a straight line until some force causes it to change its speed or its direction. That force could be friction with the floor causing the ball to slow down. It could be a force applied by an object or a wall as the ball collides with it. On the other hand, it could be someone pushing the ball to one side, causing the ball to change direction. Similarly, an object moving through space will move in a straight line unless some force causes it change its speed or its direction.

 ## Gravity

The moon, planets, and other satellites follow curved paths. This tells you that a force must be acting on them. That force is gravity. **Gravity** is the force of attraction between all objects. If the objects are relatively small, such as your body and familiar objects around you, that force is so small you cannot feel it. Although your body's mass is small, the mass of Earth is very large. Therefore, the force of gravity is quite noticeable. **Weight** is the force of attraction between your body and Earth.

When the mass of either or both objects increases, the gravitational force between them also increases. Therefore, a person would weigh more on a more massive planet. Correspondingly, you would weigh less on Mars or the moon than you do on Earth because these smaller celestial bodies have less mass than Earth.

Gravity also depends on the distance between the centers of the two objects. People who climb to the top of high mountains experience a very small but measurable decrease in their weight. This is because they are moving away from Earth's center. In fact, the higher a person goes above Earth's surface, the less the person weighs.

The motion of Earth or any celestial object in its elliptical orbit can be thought of as a combination of two kinds of motion. The first is straight-line motion under the influence of inertia. The second is falling motion toward the primary of

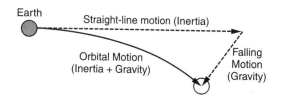

Earth

Straight-line motion (Inertia)

Orbital Motion
(Inertia + Gravity)

Falling
Motion
(Gravity)

Sun

Figure 26-4 The curved motion of the moon in its orbit is the result of straight-line motion caused by inertia, and its falling path toward Earth caused by gravity.

the satellite under the influence of gravity. The result is a curved path as shown in Figure 26-4.

Orbital Energy

The orbital energy that Earth has is a combination of kinetic and potential energy. The potential energy is a result of the distance between Earth and the sun. Kinetic energy is the speed of Earth in its orbit. The combined energy remains constant unless it is reduced by friction. However, there is no friction in space. The balance between the two components of orbital energy, speed and orbital distance, does change. As a satellite moves farther from its primary, its orbital speed decreases. As it moves closer, its speed increases. Figure 26-5 is

Figure 26-5 In this photograph, the lunar module craft is a satellite orbiting the moon. The moon is a satellite of Earth; Earth is a satellite of the sun.

a photograph of Earth taken from a spacecraft in orbit around the moon.

Moving in an ellipse, Earth changes its distance from the sun by a small amount. Therefore, Earth's speed in its orbit changes. In fact, for all planets orbiting the sun, orbital speed is a function of the distance between the satellite and the primary. Earth moves a little faster when it is closer to the sun and slower when it is farther from the sun. Earth is closest to the sun in early January, which is when its orbital speed is greatest (although the change is relatively small). The moon also changes speed in its orbit as it moves slightly closer and farther from Earth.

WHY DOES THE MOON SHOW PHASES?

You have probably noticed that the moon seems to change its shape in a monthly cycle. The apparent shape of the moon, which is determined by the pattern of light and shadow, is called the **phase** of the moon. For example, when the whole side of the moon facing Earth is lighted, we observe the full moon. The gibbous phase occurs when most of the moon appears lighted. When half of the moon appears lighted, it is the quarter moon phase. The crescent phase is a narrow, curved sliver of light. When the whole part of the moon's surface seen from Earth is in shadow, it is known as the new moon. The phase that you see depends on the relative position of the moon with respect to Earth and the sun.

Why is there a monthly cycle of moon phases? This cycle occurs because it takes roughly a month for the moon to orbit Earth. However, the daily rotation of Earth makes the monthly orbiting of the moon more difficult to follow. What you actually observe is the moon rising and setting nearly an hour later each day. Each day for half its cycle the moon moves about 13° away from the Sun's position in the sky. For the other half of the cycle, the moon moves toward the Sun's position in the sky. For this reason, each day or night you see progressively more or less of the moon's lighted surface (the side of the moon facing the sun). The cycle of the moon's phases takes about $29\frac{1}{2}$ days. To fit exactly 12 months into a year, the average calendar month was made a little longer than the full cycle of moon phases.

In a large, darkened room, a friend can help you demonstrate these motions. Due to the dangers posed by a bare light bulb, this should be done under adult supervision. Use a naked light bulb at a distance from you of at least 2–4 meters (7–13 feet) to represent the sun. Your head represents Earth; your eyes represent the position of the observer. A ball can represent the moon. The ball should be much closer to you than the light bulb (the sun). In the most simple representation, stand in one place (although you can turn your head) and observe the lighted part of the ball as your friend moves the ball around (orbits) you. You may want to name each phase as it appears to you. (See Figure 26-6.)

A more realistic, although complicated, demonstration would include you, as Earth, moving in an orbit around the sun while you turn (rotate) 365 times in each orbit. Meanwhile, the moon (the ball held by a friend) needs to move along with you and while orbiting you about 12 times for each time you orbit the more distant light bulb. As you will see, al-

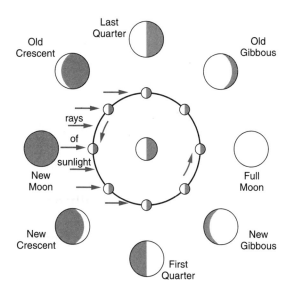

Figure 26-6 The central part of this diagram shows the moon's orbit around Earth as viewed from a position high above the North Pole. The eight outer circles show how a person standing on Earth's surface would see each phase.

though all these motions affect our observations of the Moon, it can be difficult to think about all of them as we observe the changing phases of the Moon.

Earth, Moon, and Sun

Figure 26-6 on page 673 includes two ways of looking at the moon in a single diagram. The central part of this diagram shows the moon orbiting Earth as viewed from a point in space above the North Pole. Notice the arrows showing the orbital motion of the moon as it revolves around Earth. The straight arrows on the left represent rays of light from the sun, which is well outside the diagram to the left.

The eight outer circles show how the moon appears at each position as viewed from Earth. Notice that the first quarter phase at the bottom of the diagram is lit on the left as viewed from above the orbit (inner circle), but lit on the right as viewed from Earth. To understand why the bottom diagram is reversed, turn your book around and look across Earth to the first quarter phase. Then the figure of the moon in its orbit will be lit on the right. Remember that the portion of the moon that is lighted, as well as the side that is lit, depend on the position of the observer.

Earth and Moon

The moon takes about $27\frac{1}{3}$ days to complete one orbit around Earth. However, the cycle of moon phases takes about $29\frac{1}{2}$ days. The reason it takes longer than one revolution of the moon to observe a full cycle of phases is shown in Figure 26-7. Consider Earth and the moon to start at position A. They will be at B $27\frac{1}{3}$ days later. The moon has moved through one complete revolution of 360°. However, to get back to the new moon phase, it must move a little further in its orbit. The reason the cycle of phases takes longer than one revolution of the moon is that Earth is moving in its orbit around the sun.

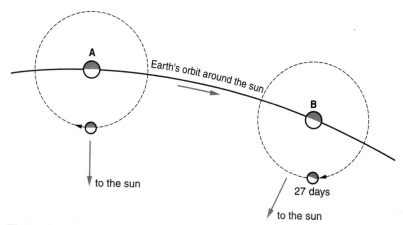

Figure 26-7 Consider Earth and the moon at position A, and 27⅓ days later at position B. The moon has made one complete revolution of 360° from A to B, but the apparent position of the sun has changed. Therefore, the moon must travel another two days to get back to the new moon phase. (This is a view from above the South Pole.)

WHAT IS AN ECLIPSE?

The moon and Earth are not transparent. They cast shadows. Sometimes, the moon's shadow falls on Earth. This may affect the way we see the sun. At other times, Earth's shadow falls on the moon. This may affect the way we see the moon.

 Solar Eclipses

As the moon orbits Earth, it sometimes comes between the sun and Earth, casting a shadow on Earth. When one celestial object blocks the light of another, an **eclipse** occurs. If the shadow is cast on Earth's surface, sunlight is blocked. To a person on Earth, the moon is observed to move in front of the sun and then move away. This is called an eclipse of the sun, or solar eclipse. If part of the sun is visible throughout the eclipse, it is called a partial eclipse.

A total eclipse occurs when the sun is completely blocked from view as the region of full shadow passes over Earth as

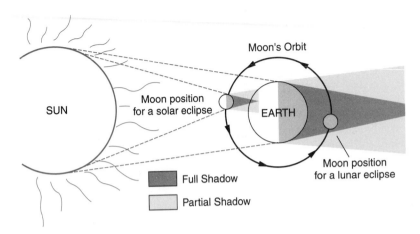

Moon's Orbit

SUN

Moon position
for a solar eclipse

EARTH

Full Shadow

Partial Shadow

Moon position
for a lunar eclipse

Figure 26-8 A solar eclipse occurs when the moon casts a shadow on Earth's surface. A lunar eclipse happens when the moon passes through Earth's shadow. (Like other diagrams in this chapter, neither object sizes nor distances are to scale.)

shown in Figure 26-8. In a total solar eclipse, the sky becomes much darker than usual and some stars may be visible briefly. A solar eclipse happens only in the new moon phase when the dark half of the moon faces Earth. If a solar eclipse occurs when the moon is relatively far from Earth, the moon may not be cover the entire sun's entire disk. In this type of solar eclipse, the sun may be visible as a ring of light surrounding the dark moon. (**CAUTION: Do not look directly at the sun.**)

 ## Lunar Eclipse

There are also eclipses of the moon, lunar eclipses. An eclipse of the moon occurs when the moon orbits to a position exactly opposite the sun. From Earth, the moon is seen to move into Earth's curved shadow. The moon becomes relatively dark and takes on a coppery-red color from a small amount of light that is refracted through Earth's atmosphere. The eclipse may continue for an hour or more until the moon moves out of Earth's shadow. A lunar eclipse can occur only at the full moon phase because the moon must be on the side of Earth opposite the sun. Figure 26-8 shows the relative position of the moon, Earth, and sun for both solar and lunar eclipses.

Predicting Eclipses

Precise observations of the moon and sun over hundreds of years enable astronomers to predict eclipses with great accuracy. Lunar and solar eclipses occur about once or twice a year. Lunar eclipses are easier to see because they are visible to everyone on the night side of Earth and they may last more than an hour.

Solar eclipses, however, are usually visible only to people located along a narrow path across Earth's surface. An eclipse of the sun usually lasts only a few minutes. If the moon orbits Earth each month, why do you not see eclipses of the sun and moon every month? The reason is that the moon's orbit is tilted at an angle of about 5° from the plane of Earth's orbit around the sun as shown in Figure 26-9. Eclipses occur only when the moon is in the part of its path where the two orbits intersect.

ACTIVITY 26-3 | THE NEXT ECLIPSES

Use the Internet to find when the next lunar and solar eclipses will occur. The time is often given in terms of Greenwich Mean Time. Convert the time of the next lunar eclipse into your local time. Where will the next solar eclipse be visible? When will the next solar eclipse be visible in your area?

Figure 26-9 The moon's orbit is inclined about 5° from the plane of Earth's orbit. Therefore, the new moon and the full moon are usually above or below the plane of the sun. This explains why eclipses are rare.

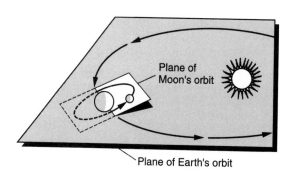

ACTIVITY 26-4 | **MODELING THE MOON'S PHASES**

Stand outside in direct sunlight holding a tennis ball. Consider your head to be Earth and the tennis ball to be the moon. Move the tennis ball to each of the eight positions representing the phases of the moon labeled in Figure 26-6.

Then move the tennis ball moon to the positions at which eclipses of the sun and moon would occur. If your head represents Earth, is the moon's size approximately to scale? Is its distance to scale?

 Apparent Size of the Sun and Moon

It is a coincidence that as viewed from Earth the moon and sun are each about $\frac{1}{2}°$ in angular diameter. That is, lines drawn from your eye to the sides of the sun or moon are about $\frac{1}{2}°$ of angle apart. In fact, the sun's true diameter is about 400 times the diameter of the moon. However, the moon is much closer to Earth. Consequently, the moon and sun look about the same size to us. The distance from Earth to each of them changes slightly. Sometimes the moon looks slightly larger and sometimes the sun looks slightly larger. Usually this change is not noticeable. However, if a solar eclipse occurs when the sun is relatively close and the moon is relatively far away, the sun can be seen as a ring of light surrounding the dark moon. This is called an annular solar eclipse.

TERMS TO KNOW

ellipse	gravity	phase	weight
focus	inertia	satellite	

CHAPTER REVIEW QUESTIONS

1. Astronauts have recovered basalt and other igneous rocks from the surface of the moon. What does this tell us about the history of the moon?

    (1) The moon was once a part of Earth.
    (2) At least part of the moon was once molten rock.
    (3) The moon has a weak magnetic field.
    (4) The age of the moon is greater than the age of Earth.

The table below provides information about the moon based on current scientific theories. Use this information to answer questions 2 and 3.

Information About the Moon

Subject	Current Scientific Theory
Origin of the moon	Formed from material thrown from a still-liquid Earth following the impact of a giant object 4.5 billion years ago
Craters	Largest craters resulted from an intense bombardment by rock objects around 3.9 billion years ago
Presence of water	Mostly dry, but water brought in by the impact of comets may be trapped in very cold places at the poles
Age of rocks in terrae highlands	Most are older than 4.1 billion years; highland anorthosites (igneous rocks composed almost totally of feldspar) are dated to 4.4 billion years
Age of rocks in maria plains	Varies widely from 2 billion to 4.3 billion years
Composition of terrae highlands	Wide variety of rock types, but all contain more aluminum than rocks of maria plains
Composition of maria plains	Wide variety of basalts
Composition of mantle	Varying amounts of mostly olivine and pyroxene

2. Which statement is best supported by information in the table above?

    (1) The moon was once a comet.
    (2) The moon once had saltwater oceans.
    (3) Earth is 4.5 billion years older than the moon.
    (4) Earth was molten rock when the moon was formed.

3. Which moon feature is an impact structure?

(1) crater

(3) terrae (highlands)

(2) maria (seas)

(4) mantle

The diagram below represents the orbit of a planet traveling around a star.

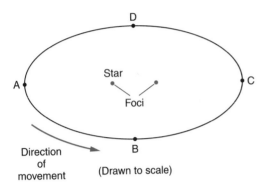

(Drawn to scale)

4. The calculated eccentricity of this orbit is approximately

(1) 0.01

(3) 5

(2) 0.2

(4) 12.8

5. At which position of the planet will the gravitational attraction between the star and the planet be greatest?

(1) A

(3) C

(2) B

(4) D

6. As the planet revolves in orbit from position A to position D, the orbital velocity will

(1) continually decrease.

(2) continually increase.

(3) decrease, then increase.

(4) increase, then decrease.

7. Because of the elliptical shape of the moon's orbit, the distance between Earth and the moon changes. Which of the following is a direct result of changes in the distance between Earth and the moon?

(1) the cycle of phases of the moon

(2) changes in the mass of the moon

(3) changes in the force of gravity between Earth and the moon

(4) the cycle of seasons that occur on Earth

8. Venus and Mercury are the only planets in our solar system that show a full range of phases, like Earth's moon. Where must Venus be located if we could see its surface in darkness like the new moon phase?

(1) Venus must be located outside Earth's orbit.

(2) Venus must be at right angles to the sun.

(3) Venus must be located between the Earth and sun.

(4) Venus must be on the side of the sun opposite Earth.

9. A cycle of moon phases can be seen from Earth because the

(1) moon's distance from Earth changes at a predictable rate.

(2) moon's axis is tilted.

(3) moon spins on its axis.

(4) moon revolves around Earth.

The diagram below shows the moon orbiting Earth as viewed from space above the North Pole. Use this diagram to answer questions 10, 11, and 12.

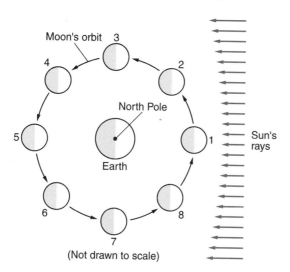

10. What is the approximate length of time for one complete cycle of the moon's phases?

(1) one day

(2) one week

(3) one month

(4) one year

11. An observer on Earth views the moon as the moon revolves from position 1 to 5 and back to position 1. How will the lighted portion of the moon's surface change *as she sees it* over this period?

(1) The lighted portion of the moon's surface will increase.
(2) The lighted portion of the moon's surface will decrease.
(3) The lighted portion of the moon's surface will increase and then decrease.
(4) The lighted portion of the moon's surface will decrease and then increase.

12. At which two positions of the moon is an eclipse of the sun or moon possible?

(1) 1 and 5   (3) 3 and 7
(2) 2 and 6   (4) 4 and 8

The graph below shows the maximum altitude of the moon, measured by an observer at latitude of 43° north during the month of June in a particular year. The names and appearances of four moon phases are shown at the top of the graph, directly above the date on which each phase occurred. Use this diagram to answer questions 13–15.

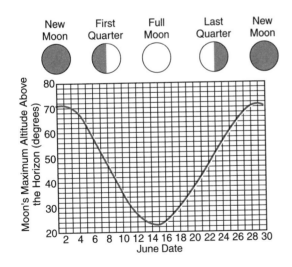

13. What is the maximum altitude of the moon on June 22?

(1) 40°   (3) 46°
(2) 43°   (4) 50°

14. Which diagram best represents the moon's phase on June 11?

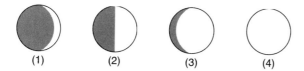

15. Which terms best describe both the changes in maximum altitude of the moon and changes in the moon's phases over a period of several years?

(1) cyclic and predictable

(2) cyclic and unpredictable

(3) noncyclic and predictable

(4) noncyclic and unpredictable

Open-Ended Questions

16. Unlike Earth, the moon has no atmosphere. Yet, the volcanic processes that released gases to make an atmosphere occurred on the moon as they did on Earth. Why did Earth keep its atmosphere, but the moon did not?

17. Earth is not at the center of the moon's orbit. Describe the exact position of the Earth in the orbit of the moon.

18. A person on the moon would weigh only about $\frac{1}{6}$ as much as his weight on Earth. Why is the weight of objects on the moon so much less than their weight on Earth?

19. The diagram below represents the sun and Earth viewed from space on a certain date. Please do not write in this book. On a copy of the diagram, draw a circle approximately 0.5 cm in diameter to show the position of the moon when it is in the full moon phase as observed from Earth.

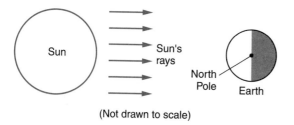

(Not drawn to scale)

The diagram below is an exaggerated model of Earth's orbit. Earth is closest to the sun at perihelion and farthest from the sun at aphelion.

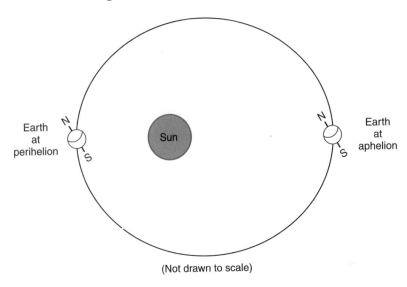

(Not drawn to scale)

20. *a.* State the actual geometric shape of Earth's orbit.

   b. Identify the season in the Northern Hemisphere when Earth is at perihelion.

   c. Describe the change that takes place in the apparent size of the sun, as viewed from Earth, as Earth moves from perihelion to aphelion.

   d. State the relationship between Earth's distance from the sun and Earth's orbital velocity.

Chapter 27

The Solar System

 COLONIZING SPACE

The idea of sending people to live away from Earth has been the subject of science fiction for many years. Astronauts have spent extended time in orbit around Earth and have spent several days on the moon. In January 2004, President Bush announced plans to return astronauts to the moon as early as 2015 to conduct extended lunar missions, with the goal of living and working there for increasingly extended period. Until that time there had never been a serious proposal to establish a colony far from Earth.

There are many reasons for considering such plans. People may wish to take advantage of resources, such as precious metals, found on other planets. Some manufacturing processes work better in reduced gravity and isolated from Earth's atmosphere. Governments may decide to establish a place to escape to in case Earth is endangered by a collision with a massive object from space. Nuclear war and bioterrorism could also cause a human disaster on this planet. In addition, a human colony could be used to support the exploration of other worlds.

Living Away from Earth

The life support requirements for people living outside the terrestrial, or Earthlike, environment are considerable. Things taken for granted, such as oxygen and liquid water, are absent in space and rare on other planets in our solar system. The cost of transporting these necessities would be prohibitive. Recycling these materials seems more practical. In a space environment near a star, there would be a steady supply of electromagnetic energy.

A project, called Biosphere II, attempted to satisfy the needs of its human and animal inhabitants by recycling as much as they could. In a special building in the Arizona desert, scientists constructed a small environment totally isolated from the outside world. The greatest challenge was keeping the proper balance of gases in the internal atmosphere. In spite of the scientific knowledge and technology expended in this project, the people running Biosphere II found it difficult to establish a proper equilibrium. Although scientists learned a more about how our environment sustains itself, they also learned that their current knowledge is insufficient to create an artificial environment for humans totally based on the natural recycling of materials.

Beyond the Solar System

Perhaps it is possible to find other planets in the universe with Earthlike environments. In the past few years, astronomers have detected planets that orbit other stars. However, these very large planets probably have gravity too strong to permit human habitation. Furthermore, they are so far away that it would take centuries to reach them with our present technology. If other planets like Earth exist, it seems logical that life and civilization might have developed on some of them. However, experiments to detect radio signals from other civilizations in space have not been successful. Therefore, there is no direct evidence that humans could exist anywhere outside Earth without complex artificial support systems.

These events have helped us understand how unique and how fragile our environment on Earth is. In spite of fears that a global disaster could make Earth uninhabitable, we have nowhere else to go. The best alternative is to preserve our current terrestrial environment in a way that allows all living things to prosper.

WHAT IS THE ORIGIN OF THE SOLAR SYSTEM?

Evidence from space indicated that the initial composition of the universe probably was nearly all hydrogen and helium. The early universe probably had a large portion of very massive stars that used up their hydrogen fuel relatively quickly. These stars ended as unstable objects that exploded and created clouds of debris with an abundance of heavy elements. In our solar system, there is an abundance of heavy elements, such as carbon, iron, silicon, and oxygen. This indicates that our solar system formed from the remains of a stellar explosion known as a supernova.

Nebular Contraction

A cloud of debris left over from this explosion may have come together under the influence of its own gravity. This theory is sometimes called nebular contraction theory. It suggests that the solar system began as a cloud of dust and gas in space, which is commonly called a nebula. More than 99 percent of the mass condensed to form the sun, which is at the center of the solar system. Debris that remained formed the planets, moons, and other objects orbiting the sun.

Just as an ice-skater spins faster when she brings in her arms, the collapse of this material produced the revolution of the planets in their orbits. It also caused the rotational motion of the sun and planets. All the planets revolve in the same direction that the sun rotates. Most of their moons also revolve in this direction. The spacing of the planets is remarkably ordered. Most planets are about twice as far from the sun as is their neighbor closer to the sun. In addition, the planets lie in nearly the same plane, orbiting the sun in a thin disk. This degree of order suggests that a single event formed the solar system.

From the study of radioactive elements within material that has fallen to Earth and rocks recovered from the older parts of the moon, scientists have inferred that these bodies are about 4.6 billion years old. Rocks from Earth are not as helpful because have been extensively recycled through the processes of the rock cycle and plate tectonics. Therefore, the oldest known terrestrial rocks are less than 4.6 billion years. Studies of the patterns of change in stars indicate that the sun is also about 5 billion years old. It therefore seems likely that the sun and the solar system were created in a single event that took place a little less than 5 billion years ago. This is only about one-third of the estimated age of the universe.

WHAT PROPERTIES DO THE PLANETS SHARE?

The definition of a planet has evolved through time. Early observers of the night skies thought of the planets as special stars that wander among the other stars. Later, with the use of telescopes, astronomers were able to see differences between these objects and stars. As astronomers shifted their thinking from the earth-centered (geocentric) hypothesis to the sun-centered (heliocentric) model, the idea of what a planet is changed. They began to think of planets not as wandering stars but as large objects that orbit the sun, reflecting the sun's light rather than creating their own light.

Like all satellites, the planets orbit the sun in ellipses with the primary, the sun, located at one focus. Their orbits have low eccentricities. That is, all the orbits are fairly close to the shape of a perfect circle. You learned in the last chapter that Earth moves a little faster in its orbit when it is closest to the sun in early January. The change in gravitational force between the planet and the sun causes the orbital velocity of Earth to change in a yearly cycle. All the planets move a little faster when they are closest to the sun.

The speed that planets move in their orbits also changes from planet to planet. As distance from the sun increases, the pull of the sun's gravity decreases. Therefore, orbital velocity is indirectly related to a planet's distance from the sun. That is, the greater the distance of a planet from the sun, the slower it travels in its orbit. The innermost planet, Mercury, moves about 10 times as fast along its orbit as the outermost planet, Pluto. Mercury, because it is closer to the sun, also has a shorter orbit than any other planet. Therefore, Mercury revolves around the sun in only 88 Earth days. Earth takes 1 year. Pluto takes nearly 250 Earth years to revolve around the sun. The difference in orbital period is so large because of the combined effects of the longer orbits and the slower speeds of the outer planets.

All the planets rotate on their axes. However, unlike the orbital properties of the planets, their rotational characteristics do not follow a regular pattern. Neither their size nor distance from the sun can be directly related to their period of rotation. The largest planet, Jupiter, takes only about 10 Earth hours to rotate 360°. That would be the length of a day on Jupiter. However, by far, the longest period of rotation is on Venus, about 8 Earth months. Furthermore, three of the planets (Venus, Uranus, and Pluto) do not rotate in the same direction as the other planets. Perhaps the rotations of these planets have been affected by collisions with other objects since the solar system formed nearly 5 billion years ago.

The number of moons orbiting the planets does show a general pattern with distance from the sun. The two innermost planets, Mercury and Venus, have no moons. Earth has one. Then the number of moons continuously increases as we move to the outer planets until the last two planets, Neptune and Pluto. It is difficult to distinguish between moons and other objects that orbit the outer planets. The question of how small an object can be and still be called a moon has no clear answer. For example, the rings of Saturn are made of millions of objects that orbit the planet in the same plane, but these objects are considered too small to be called moons. At this time, there is no known limit to the number of moons in the solar system. Table 27-1, based on the *Earth Science Reference Tables,* shows properties of the members of our solar system. You may notice that the *Earth Science Reference Tables* list a different number of moons for Jupiter, Saturn, Uranus, and Neptune. These planets have many small natural satellites that currently are being discovered. This table lists the number of confirmed moons as of January 2004. (*Note*: Time is expressed in units of Earth time. Thus, "days' are approximately 24 Earth hours long.)

TABLE 27-1 Solar System Data

Object	Mean Distance from Sun (millions of km)	Period of Revolution	Period of Rotation	Eccentricity of Orbit	Equatorial Diameter (km)	Mass (Earth = 1)	Density (g/cm³)	Number of Moons
Sun	—	—	27 days	—	1,392,000	333,000.00	1.4	—
Mercury	57.9	88 days	59 days	0.206	4,880	0.553	5.4	0
Venus	108.2	224.7 days	243 days	0.007	12,104	0.815	5.2	0
EARTH	149.6	365.26 days	23 h 56 min 4 s	0.017	12,756	1.00	5.5	1
Mars	277.9	687 days	24 h 37 min 23 s	0.093	6,787	0.1074	3.9	2
Jupiter	778.3	11.86 years	9 h 50 min 30 s	0.048	142,800	317.896	1.3	62
Saturn	1,427	29.46 years	10 h 14 min	0.056	120,000	95.185	0.7	31
Uranus	2,869	84.0 years	17 h 14 min	0.047	51,800	14.537	1.2	27
Neptune	4,496	164.8 years	16 h	0.009	49,500	17.151	1.7	13
Pluto	5,900	247.7 years	6 days 9 h	0.250	2,300	0.0025	2.0	1
Earth's moon	149.6 (0.386 from Earth)	27.3 days	27 days 8 h	0.055	3,476	0.0123	3.3	—

ACTIVITY 27-1 GRAPHING SOLAR SYSTEM DATA

Equally spaced, along the bottom axis of your graph paper, write "Sun" followed by the names of the nine planets. The vertical axis can be any one of the eight variables listed in Table 27-1. Each group can make a graph displaying different data. The graphs can be displayed to the class. What relationship does each graph show between the position of the planets and the characteristic graphed?

HOW ARE THE PLANETS GROUPED?

The planets are often separated into two groups. Mercury, Venus, Earth, and Mars are the closest planets to the sun. All four planets are solid objects surrounded by a relatively thin atmosphere or no atmosphere at all. Mars, with the largest orbit in the group, is less than one-third of the distance to the next planet, Jupiter. This group of planets is known as the **terrestrial planets** because they are similar to Earth in their rocky composition.

Four of the five outer planets are known as the Jovian planets because of their similarity to Jupiter. Jupiter is a large planet that has a relatively low density, since most of its volume is a thick shell of gas. Only the last planet, Pluto, does not fit this pattern. Figure 27-1 on page 692 compares the size of the sun and the planets.

 Terrestrial Planets

MERCURY The closest planet to the sun, Mercury, is a small planet. It therefore has weaker gravity than most of the other planets. Due to its location, lack of a magnetic field, and low gravitational force, electromagnetic radiation and charged particles given off by the sun (solar wind) have stripped Mer-

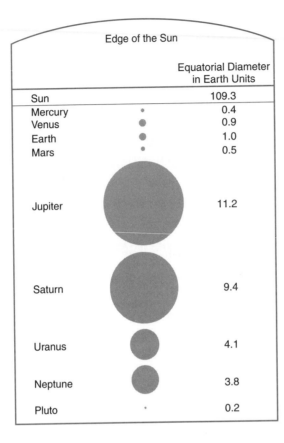

	Equatorial Diameter in Earth Units
Sun	109.3
Mercury	0.4
Venus	0.9
Earth	1.0
Mars	0.5
Jupiter	11.2
Saturn	9.4
Uranus	4.1
Neptune	3.8
Pluto	0.2

Edge of the Sun

Figure 27-1 The curved line at the top of this diagram represents the size of the sun. The dark circles represent the size of the nine planets to the same scale. The size of the sun accounts for the fact that it contains more than 99% of the total mass of the solar system.

cury of its atmosphere. Like Earth's moon, which is about the same size as Mercury, impact craters cover the surface of Mercury. Most craters probably formed from collisions with debris early in the history of the solar system. At that time there was more debris scattered through the solar system. With no atmosphere and no weather, these features are still prominent after billions of years. The slow rotation of Mercury and the lack of an atmosphere cause an extreme range in temperature on the surface, from 400°C during the day to −200°C at night. Mercury is visible from Earth only near sunset or sunrise.

VENUS   Like Mercury, Venus is visible from Earth only near sunset or sunrise. Venus is sometimes called Earth's twin because its diameter and mass are similar to Earth's. Venus therefore has about the same gravity as Earth. For many years astronomers wondered if Venus might have surface conditions similar to those on Earth. Thick clouds prevented direct observations of the solid surface. Radar images and mapping of the surface by artificial satellites revealed that Venus has volcanic features like Earth. However, its surface conditions are very different. Atmospheric pressure at the surface is 90 times the sea level pressure on Earth. The thick atmosphere of Venus is mostly carbon dioxide, which absorbs and reradiates infrared solar energy (greenhouse effect). Therefore, the surface temperature of Venus (about 500°C) is actually hotter than the daytime temperature on Mercury. In addition, the clouds of Venus are composed of droplets of sulfuric acid, a corrosive substance that is harmful to creatures living on Earth. The period of rotation of Venus is the longest of any planet. In fact, Venus takes longer to rotate on its axis than it takes to orbit the sun. (See Table 27-1 on page 690.) So much for our "twin planet."

Phases of Mercury and Venus As viewed from Earth, the portion of the lighted surface of Mercury and Venus that you see changes in a predictable cycle. Like the moon, these two planets show a full range of phases. Galileo was the first astronomer to document the phases of Venus as he observed them with his telescopes.

When Mercury and Venus pass between Earth and the sun, they, like the new moon, are not visible from Earth. Then, the lighted portion of each planet increases until it is fully lighted as the planets pass behind the sun. This change in phase is accompanied by a change in apparent size. Mercury and Venus appear largest at their closest approach to Earth in the crescent phase. As they move farther from Earth and the lighted portion increases, these planets appear smaller in angular diameter. The orbits of the other planets are outside Earth's orbit. The night side of Earth faces the lighted side of these planets. Therefore, they never show the dark, or new phase. They always appear mostly lighted as observed from Earth.

Earth Oceans that average about 4 km in depth cover about 70 percent of Earth's surface. In fact, Earth is the only planet scientists know of on which large quantities of water exist in

Figure 27-2 Astronauts on the Apollo 8 mission to the moon witnesses Earth rising.

all three states: ice, liquid water, and water vapor. Surface temperatures are moderate from a record low of $-89°C$ in Antarctica to $58°C$ in some desert regions.

The unique surface conditions on Earth permitted the evolution of life. In addition, living organisms have changed the planet. Earth's original atmosphere probably was mostly carbon dioxide like that of Venus. The carbon dioxide content of the atmosphere is now less than 1 percent (0.03 percent). Most of the original carbon content of the atmosphere has dissolved in the ocean or become part of plants and animals as well as surface deposits of organic sedimentary and metamorphic rock such as limestone and marble. Weathering and erosion coupled with plate tectonics constantly change Earth's surface. Figure 27-2 is Earth as seen from the moon.

MARS   Of the nine planets in our solar system, the conditions on Mars come closest to the favorable conditions for life found on Earth. Like Venus, the atmosphere of Mars is mostly carbon dioxide. However, the atmosphere of Mars is much thinner than that of Venus or Earth. Mars can warm up to a comfortable $20°C$ near the Martian equator. However, at night, about 12 hours later, it cools down to $-60°C$ because the thin atmos-

Figure 27-3 This image of the barren surface of Mars was taken by the Viking landing module and sent back to Earth by radio transmission. Mars is the next logical step for human exploration of space.

phere does not absorb and reradiate heat. Due to the tilt of the rotational axis of Mars, it has seasons like Earth. However, the polar regions become so cold that dry ice (solid carbon dioxide) forms on the surface.

Some Martian surface features look as if they were formed by large rivers, but there is no liquid water on the surface of Mars today. This has led some astronomers to speculate that most of the atmosphere of Mars has been lost and that the climate on Mars was warmer in the past. If the water is still there, it is probably in the form of ice under the planet's visible surface. The highest known surface feature in the solar system, Olympus Mons, is a Martian volcano four times as high as Earth's Mount Everest. Figure 27-3 is a photograph of the surface of Mars taken by a NASA landing module.

ACTIVITY 27-2   DESIGN A LANDING MODULE

NASA has hired your group to design an instrument package that will help scientists gather information about the surface of Mars. The task they have given your team is to design a module that will send back information about Mars after it reaches the planet. Your module may include only easily obtainable instruments that you can understand and operate. Make a drawing of the landing module and write an explanation of what each instrument is as well as what information it will gather.

NASA will send the module to Mars. Therefore, you should not include a design for a rocket. Neither the module nor any other object will be returned to Earth. Radio communication will be your only link. No people or animals will be sent to Mars on this mission.

The Outer Planets

The five remaining planets, the outer planets, are Jupiter, Saturn, Uranus, Neptune, and Pluto. The first four are known as **Jovian planets**. This name means that they are similar to Jupiter. These planets are larger and less dense than the terrestrial planets because they are composed mostly of gas. The last planet, Pluto, is an exception you will read about at the end of this section. Figure 27-4 is a scale model of the orbits of the planets.

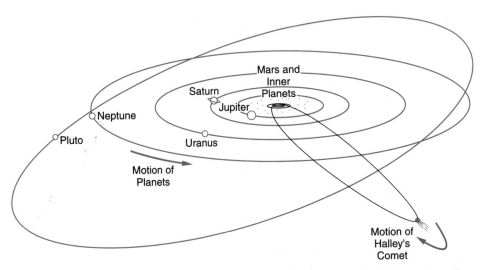

Figure 27-4 Most planets are roughly twice the distance from the sun as their nearest inner neighbor. The dotted circle shows the asteroids. The orbits of Mercury, Venus, and Earth are all inside the orbit of Mars. The orbit of Pluto is noticeably eccentric, but the eccentricity of Halley's comet is much greater.

The Jovian planets consist of a small, rocky core surrounded by a liquid mantle and a thick gaseous shell. Rather than the heavier elements of the terrestrial planets (iron, silicon, and oxygen), these planets are composed primarily of hydrogen and helium. However, they are still the most massive planets because of their immense size. Each also has a system of rings surrounding it. These rings are made of small particles that orbit independently in the plane of the planet's equator.

JUPITER   The first of the outer planets, Jupiter is the largest planet in the solar system. Jupiter alone accounts for two-thirds of the total mass of the nine planets. Hydrogen and helium make up more than 99 percent of the mass of Jupiter. This is more like the composition of a star than the composition of the terrestrial planets. Although the pressure at the center of Jupiter is about 10 times the pressure within Earth, it is not enough pressure to support nuclear fusion, which powers the sun and other stars.

The rapid rotation of Jupiter gives it a noticeable bulge at its equator. One of the most noticeable features on Jupiter is the Great Red Spot visible on its surface. Astronomers speculate that this may be either a raging, permanent storm or an area of calm within the turbulent atmosphere. Of the many moons of Jupiter, four can be seen from Earth with binoculars or a small telescope. They sometimes look like a line of stars that runs through the equator of Jupiter. These four satellites are known as the Galilean moons, named after the their discoverer, Galileo.

SATURN   The rings Saturn make it one of the most unusual objects in the night sky. These rings are sometimes visible through binoculars from Earth. (See Figure 27-5 on page 698.) Rings of the other Jovian planets are less dense and therefore more difficult to see. Second in size only to Jupiter, Saturn is also composed primarily of hydrogen and helium. It is the least dense planet with an average density less than that of water.

URANUS AND NEPTUNE   Because Uranus and Neptune are not as bright as the other planets, they were not known to the an-

Figure 27-5 Saturn has seven major rings separated by gaps caused by the gravitational effects of its many moons.

cient astronomers. Uranus, the brighter of the two, is barely visible without binoculars or a telescope. They are dim because they are so far from the sun and because they are not as large as Jupiter and Saturn. The rotational axis of Uranus is tilted nearly 90°, so it seems to spin on its side. Neptune is very similar in size and composition to Uranus, but nearly twice as far from the sun.

PLUTO   The most distant planet from the sun, Pluto has the coldest surface temperature of any planet. However, in other ways, Pluto is the exception to the rules. Unlike other outer planets, it is very small, in fact the smallest planet. The density of Pluto is greater than that of the Jovian planets, but it is still less than the terrestrial planets.

Pluto is the only planet that is not about twice as far from the sun as its inner planetary neighbor. The orbit of Pluto has a higher eccentricity than the orbit of any other planet. This allows Pluto in part of its orbit to be closer to the sun than Neptune. Pluto's orbit is tilted about 17° from the plane of most of the other planets. (See Figure 24-7.) Unlike the other outer planets, Pluto has only one moon, which orbits nearly perpendicular to the plane of the planetary orbits.

Why is Pluto so different from the other planets? Some astronomers believe that Pluto was a moon of Uranus that was knocked into an orbit around the sun by a passing object. Others suggest it is one of many smaller objects beyond the other planets and should not be listed as a planet. It may be many years before astronomers reach a consensus about the origin of Pluto because it is so small and so far away.

THE SOLAR SYSTEM TO SCALE

Use the data in the *Earth Science Reference Tables* to construct a scale model of the solar system. If you use a single scale for the size of the planets and their distance from the sun, you will probably need to work outside. To make the smallest planets visible as dots on a sheet of paper, the distance to Pluto may need to be greater than the size of your school building.

PLANETARY TRAVEL AGENCY

For this activity, each group will select a different planet or other object in the solar system and prepare a presentation designed to attract others to visit their planet/object. Exaggeration and humorous content are welcome. However, highlighted features must be based on fact.

WHAT OTHER OBJECTS ORBIT THE SUN?

The solar system contains other objects that orbit the sun. Sometimes these objects, such as comets and meteors, are visible in the night sky.

Asteroids

A ring of planetary debris known as the asteroid belt separates the terrestrial and Jovian planets. **Asteroids** are irregularly shaped rocky masses that are smaller than planets.

They orbit the sun in the same direction as the planets. Perhaps the asteroids are debris left over from the formation of the solar system that never came together as a single planet. However, evidence from asteroids that have been pulled to Earth by Earth's gravity indicate that they may be the remains of a planet that was struck by another object and broke into many fragments.

The spacing between Mars and Jupiter is larger than the pattern observed among other planets. This supports the idea of a missing planet. Some asteroids orbit outside the main part of the asteroid belt and have been known to approach or even fall to Earth.

 **Comets**

Occasionally, **comets** are visible in the night sky as a small spot of light with a long tail that points away from the sun. The highly eccentric shape of comets' orbits means that they spend most of their time in the outer parts of the solar system. Many comets orbit the sun beyond the orbit of Pluto. Some comets seen in the inner solar system were probably pulled into highly elliptical orbits around the sun by the gravity of a passing star or other object in space. When comets enter the inner part of the solar system near Earth and the sun, they move quickly under the influence of the sun's strong gravitational field. The brightest comets are usually visible for several weeks. Then, they return to the outer portion of their orbits where they spend most of their time.

Comets are sometimes described as dirty snowballs because they are made of ice and rocky fragments. Some of the water escapes from comets by sublimation when they come close to the sun, making the tails that are so distinctive. Only about 30 comets per century are bright enough to be clearly visible without a telescope. The best known is Halley's (HAL-ease) comet because it returns to the inner solar system about every 75 years and it usually becomes visible without a telescope. Halley's comet was visible in 1986, and will return in 2061.

 **Meteors**

On a clear night, it is often possible to see streaks of light that move across a portion of the sky in less than a second. They are sometimes called "shooting stars." These are solid bits of rock from outer space that enter the upper atmosphere at great speed. Friction with the atmosphere creates a streak of light known as a **meteor**. Table 27-2 lists the times of the year when meteors are especially numerous.

Table 27-2 Meteor Showers

Name/Source	Date of Maximum	Duration Above 25% of Maximum	Approximate Limits	Number per Hour at Maximum
Quadrantids	Jan. 4	1 day	Jan. 1–6	110
Aquarids	July 27–28	7 days	July 15–Aug. 15	35
Perseids	Aug. 12	5 days	July 25–Aug. 18	68
Orionids	Oct. 21	2 days	Oct. 16–26	30
Geminids	Dec. 14	3 days	Dec. 7–15	58

The objects themselves are called meteoroids as they move through space. Those that strike Earth's surface are known as meteorites. There are many craters caused by meteorite impacts. However, most meteoroids burn up before they reach the ground. Only the larger ones are able survive their fiery passage through the atmosphere. Collectors value meteorites for their unusual origin. Scientists study them for insights into the origin of the Earth and solar system, since nearly all of them are about 4.5 billion years old.

Meteorites can be separated into two groups. The most common are the stony meteorites, which are composed of minerals common in Earth's mantle layer. Other meteorites are mostly iron, which is thought to be similar to the composition of Earth's core.

One of the most remarkable recent meteor events happened in Peekskill, New York, in 1992. As 18-year-old Michelle Knapp was watching television at about eight o'clock in the evening, she heard a loud noise outside her home. She went outside to discover that a rock the size of a football had crashed through the trunk of her parked car. When she first saw it, the object was still hot. Before the object landed, it had been visible over several states for about 40 seconds as a bright green streak. Some observers reported that it was as bright as the full moon. Michelle sold the meteorite and the car. A private company, which lends them to museums for display, now owns both.

TERMS TO KNOW

asteroid	**Jovian planet**	**terrestrial planet**
comet	**meteor**	

CHAPTER REVIEW QUESTIONS

1. The Sun's position in space is best described as a place near the center of

    (1) a constellation.
    (2) the Andromeda Galaxy.
    (3) the Milky Way galaxy.
    (4) our solar system.

2. Radioactive dating of meteorites has been one of our most important methods to determine the age of the Earth and the solar system. What is the estimated age of the Earth and solar system?

    (1) 4.6 million years
    (2) 10 to 15 million years
    (3) 4.6 billion years
    (4) 10 to 15 billion years

Base your answers to questions 3 and 4 on the following information and Table 27-1.

Astronomers have discovered strong evidence for the existence of three large extrasolar (outside our solar system) planets that orbit *Upsilon Adromedae*, a star located 44 light-years from Earth. The planets are called planet B, planet C, and planet D. The diagram below compares part of our solar system with the *Upsilon Andromedae* planetary system. Planet distances from their respective star and the relative size of each planet are drawn to scale. (The scale for planet distances is not the same scale used for planet size.)

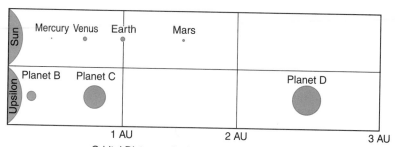

Orbital Distances in Astronomical Units (AU)
[1 AU = average distance of Earth from the Sun]

3. If our solar system had a planet located at the same distance from the sun as planet C is from *Upsilon Andromedae*, what would be its approximate period of revolution?

   (1) 100 Earth days
   (2) 300 Earth days

   (3) 1.5 Earth years
   (4) 10 Earth years

4. Planet D's diameter is 10 times greater than Earth's diameter. What planet in our solar system has a diameter closest in size to the diameter of planet D?

   (1) Venus
   (2) Jupiter

   (3) Saturn
   (4) Neptune

5. Compared with the average density of the terrestrial planets (Mercury, Venus, Earth, and Mars), the average density of the Jovian planets (Jupiter, Saturn, Uranus, and Neptune) is

   (1) less
   (2) greater

   (3) the same

6. The diagram below represents two planets in our solar system drawn to scale, Jupiter and Planet A.

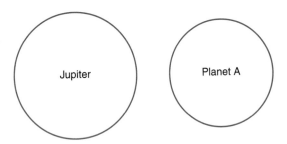

Planet A most likely represents

(1) Earth

(2) Venus

(3) Saturn

(4) Uranus

7. Some of the planets in our solar system are made primarily of dense, rocky material. Other planets are composed of a thick atmosphere that may have a small, solid core. Which planet is mostly material in the gaseous state?

(1) Venus

(2) Earth

(3) Mars

(4) Jupiter

8. Compared to Pluto, Mercury moves more rapidly in its orbit because Mercury

(1) is larger.

(2) is more dense.

(3) is closer to the sun.

(4) has a more elliptical orbit.

Base your answers to questions 9–13 on the information in Table 27-1.

9. Which planet takes more time to complete one rotation on its axis than to complete one revolution around the sun?

(1) Mercury

(2) Venus

(3) Mars

(4) Jupiter

10. Which planet is approximately 30 times farther from the sun than Earth is?

(1) Jupiter

(2) Saturn

(3) Uranus

(4) Neptune

11. Which planet has an orbit with an eccentricity most similar to the eccentricity of the moon's orbit around Earth?

(1) Earth

(2) Jupiter

(3) Pluto

(4) Saturn

12. Of these objects, which follows an orbit with the greatest eccentricity?

(1) Earth

(2) Pluto

(3) Earth's moon

(4) Halley's comet

13. A major belt of asteroids is located between Mars and Jupiter. What is the approximate average distance between the sun and this major asteroid belt?

(1) 110 million km

(2) 220 million km

(3) 390 million km

(4) 850 million km

14. Which graph best represents the change in gravitational attraction between the sun and a comet as the distance between them increases?

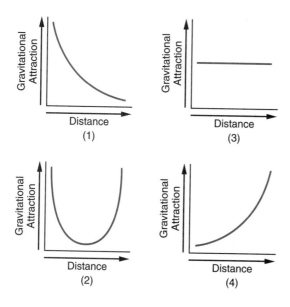

15. Which object does *not* move through the sky in a cyclic and predictable motion?

(1) the sun
(2) stars
(3) the planets
(4) meteorites

Open-Ended Questions

Base your answers to questions 16 and 17 on the data table below, which shows one cycle of equinoxes and solstices for the northern hemispheres of several planets in the solar system and the tilt of each planet's axis. Data for three planets are based on Earth's time system.

Data Table

Planet	Spring Equinox	Summer Solstice	Autumn Equinox	Winter Solstice	Tilt of Axis (degrees)
Venus	June 25	August 21	October 16	December 11	3.0
Earth	March 21	June 21	September 23	December 22	23.5
Jupiter	1997	2000	2003	2006	3.0
Saturn	1980	1987	1995	2002	26.8
Uranus	1922	1943	1964	1985	82.0
Neptune	1880	1921	1962	2003	28.5

16. *a.* State the length, in years, of the spring season on Uranus

   b. Describe the relationship between a planet's distance from the sun and the length of a season on that planet.

17. The illustration on page 707 shows an impact crater approximately 1-mile wide located near Canyon Diablo, Arizona. Describe the event that produced this crater.

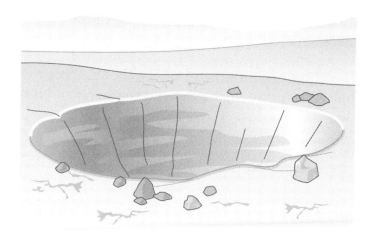

18. The diagram below represents the orbit of Mars around the sun. Make a copy of this diagram and base your answers to the next three questions on the diagram. Please do not write in this book.

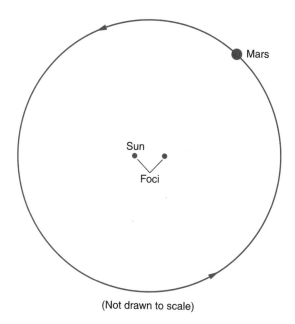

(Not drawn to scale)

On your copy of this diagram:

a. Draw and label the major axis of Mars's orbit.

b. Place an X on the orbit to show the location of Mars's greatest orbital velocity.

19. State the difference between the shape (not the size) of Earth's orbit and the shape of Mars's orbit.

20. This bar graph shows the equatorial diameter of Earth. On a copy of this diagram, make a bar that represents the equatorial diameter of Mars.

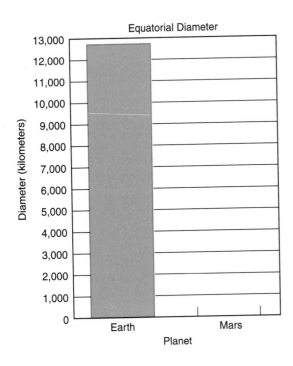

of energy. However, nuclear fusion can occur only under extreme conditions of heat and pressure. In the last chapter, you learned that Jupiter, the largest planet in our solar system, is too small to have enough internal pressure to support fusion.

ACTIVITY 28-2 | **MAKING LIGHT**

Make a list of the methods you can use to create light energy in an Earth science lab setting. This can be a competitive activity among lab groups with one point awarded for each method to create light energy and two points if you can safely demonstrate it. As in any other laboratory procedures, your teacher must approve all materials and methods you plan to use before you try them. Duplication such as burning two different substances counts as a single idea. Remember that you are looking for ways to create light energy and not methods to bring in light from another source like the sun.

 Energy Escapes from Stars

Once the energy is created deep in the sun, it moves to the sun's visible surface by radiation and convection. Convection is the same process of heat flow by density currents that distributes energy through Earth's atmosphere and oceans. Slow convection currents within Earth also carry heat energy from Earth's interior to the surface. From the solar surface, the energy escapes as electromagnetic radiation. The surface temperature of the star determines the kind of electromagnetic energy it radiates into space. The sun is a yellow star because its roughly 6000°C surface radiates most intensely as yellow light in the visible part of the spectrum.

Based on observations of other stars, astronomers predict that the sun will continue to radiate energy as it now does for approximately another 5 billion years. The next section will tell you about the evolution of stars such as the sun.

HOW ARE STARS CLASSIFIED?

In the early twentieth century, astronomers in Denmark and the United States discovered that they could classify stars on the basis of the amount of electromagnetic energy they generate and their temperature. The total energy output of a star is called its **luminosity**, or absolute brightness. Apparent brightness, or stellar magnitude, is how bright the star looks as seen from Earth. The closer a star is, the brighter it appears to us.

A good example is the sun. The sun is actually a smaller star, and gives off less light than most of the stars you see in the night sky. However, the sun is so close to Earth that during the day its light drowns out the light of the other stars. If we could see the sun at the same distance as the nighttime stars, it would be dimmer than most of them. Therefore, the brightness of a star depends on its absolute magnitude, or luminosity, and its distance from the observer.

You may have noticed that when you turn off an incandescent lightbulb the color of the hot wire briefly changes to red before it goes dark. Red is the coolest color of light visible to our eyes. If a material is heated beyond red-hot, it becomes white and then blue. Continued heating would push the radiation into the ultraviolet part of the spectrum and beyond. These forms of electromagnetic energy are not visible to us. However, they can affect us in other ways. Sunburn is caused primarily by ultraviolet light, which is a part of the spectrum of sunlight. Figure 28-2 compares the sun's spectrum with the spectrum of light radiated by hotter (blue) and cooler (red) stars.

 Hertzsprung-Russell Diagram

The graph used to classify stars is often called the Hertzsprung-Russell, or H-R, diagram in honor of the two men who developed it. This graph is printed below from the *Earth Sci-*

Figure 28-2 The solar spectrum illustrates why the sun is classified as a yellow star. It gives off its most intense radiation in the middle of the visible part of the spectrum. Blue stars are hotter and stronger in short-wave radiation. Red stars are cooler, and they radiate less energy per square meter of surface area.

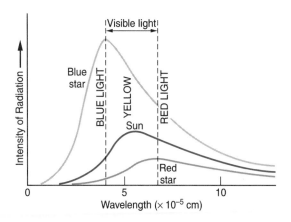

ence Reference Tables, where it is labeled "Luminosity and Temperature of Stars." (See Figure 28-3.) The graph is usually plotted with the temperatures *decreasing* to the right along the bottom axis. This is contrary to the way most graphs are made. (Usually, values increase to the right as well as upward on the vertical axis.) This graph is different because it is usually shown the way astronomers originally developed it.

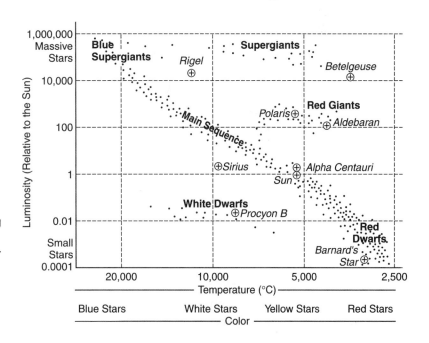

Figure 28-3 When stars are plotted on a graph according to their energy output (luminosity) and surface temperature (which determines the star's color), most stars fall into groups. Nine stars of special significance (⊕) are labeled by name.

Main Sequence Stars

When plotted on this graph, most stars fall into distinct groups. The greatest number of stars fall into an elongated group that runs across the luminosity and temperature diagram from the upper left to the lower right. This region of the graph is known as the main sequence. The position of a star along the main sequence is primarily a function of the mass of the star.

RED DWARF STARS The smallest stars, such as Barnard's Star, are red dwarf stars, which are barely large enough to support nuclear fusion. They are red in color because they are relatively cool. These stars are so dim that even the relatively close red dwarfs are difficult to see without a telescope. In fact, about 80 percent of the night stars closest to Earth are too dim to be visible to the unaided eye. This leads astronomers to infer that red dwarf stars are more numerous than all other groups of stars. However, we do not see them because they are so dim.

Small stars last longer than larger stars. The lower temperature and pressure in these stars allow them to conserve hydrogen fuel and continue nuclear fusion much longer than larger stars. The combination of small size and slow production of energy makes them very dim.

BLUE SUPERGIANT STARS At the other end of the end of the main sequence are the blue supergiants. These massive stars do not last as long as the smaller stars. The extreme conditions of temperature and pressure at the center of these stars cause rapid depletion of their large quantities of hydrogen. Some of them are a million times brighter than the sun. They are also much hotter than the sun, giving them a blue color. These largest stars are not nearly as common as the smaller stars, in part because they burn out quickly. The most massive stars last less than one-thousandth of the life of the sun. Rigel, a bright star in the winter constellation Orion, is 10,000 times as luminous as the sun. The blue color of Rigel is apparent if you compare it with Betelgeuse, another bright star in Orion. Betelgeuse is a red giant star on the opposite side of the same constellation.

Most other stars fall into one of the three groups on the temperature-luminosity chart. White dwarfs, red giants, and the supergiants are the most common star groups outside the main sequence.

HOW DO STARS EVOLVE?

Different sizes of stars have different life cycles. However, the evolution of stars can be illustrated by considering a star about the size of the sun.

Birth of a Star

Star formation begins when a cloud of gas and dust (mostly hydrogen) begins to draw together under the influence of gravity. There are two sources of this material. Some of it is hydrogen and helium left over from the formation of the universe about 14 billion years ago. The rest is the debris from the explosions of massive stars that formed earlier in the history of the universe. This initial phase takes place over a period on the order of 50 million years. (The process is faster for larger stars and slower for smaller stars.)

As the material draws together, heat from the collapse of the matter and from friction causes the temperature to increase until there is enough heat and pressure to support nuclear fusion. At this time, the star becomes easily visible since it produces and radiates great quantities of energy. The condensation process can be observed with binoculars or a small telescope in the constellation Orion. Several young stars below the belt of Orion can be seen shining through a giant cloud of gas that surrounds them.

Middle Age

The star becomes less luminous after it fully condenses, and it spends most of its life on the main sequence region of the luminosity and temperature chart. (See Figure 28-3 on page 715.) Gravitational pressure balanced by heat from nuclear fusion prevents the star from further shrinkage. This is the longest and most stable phase of stellar evolution.

 Death of an Average Star

After about 10 billion years, a star the size of the sun runs low on hydrogen. Fusion slows, and the core of helium collapses, causing the outer part of the star to expand quickly, becoming a red giant. Fusion of helium and other heavier elements replaces the hydrogen fusion process. The outer shell of gases expands and cools in the red giant stage, leaving behind a dense, hot core, which is a white dwarf star.

 Death of a Massive Star

Stars with more than about 10 times the mass of the sun end their period in the main sequence more violently. These stars create a variety of heavier elements before they collapse. The collapse process of larger stars generates so much energy that these stars end their life in an explosion known as a supernova. They briefly generate more energy than the billions of stars that make up the whole galaxy. Most of the mass of the star is blown into space. The core of the star may form an extremely dense object called a neutron star. Some stars are so massive that they form an object with gravity so strong that not even light can escape. This is called a black hole. Black holes cannot radiate energy, but they can be detected because energy is given off by matter that falls into the black hole. They can also be located by their gravitational effects on other objects.

HOW DO ASTRONOMERS STUDY STARS?

Stars are extremely hot and have no solid surface. Scientists can send instruments and cameras to land on Mars or other solid objects. However, these methods cannot be used to investigate stars. Any devices scientists build would melt and probably vaporize long before reaching the visible surface of a star. Furthermore, the night stars are too distant to reach with spacecraft. With our present technology, it would take tens or even hundreds of thousands of years for a spacecraft to reach even the nearest star beyond the sun. Therefore, most of the information astronomers have about stars comes from light and other electromagnetic energy they radiate into space.

 ## Optical Telescopes

Astronomers use telescopes to concentrate the light of stars. Telescopes allow them to observe objects that are too dim to be visible to unaided eyes. Some people think that the most important feature of a telescope is how much it magnifies. However, the stars are so distant that even the most powerful telescopes show nearly all of them as points of light. When an image is magnified, it will become dim, unclear, or fuzzy if the object is too far away.

Other factors are more important than magnification in telescope construction and use. The size, or diameter, of the front lens (or light-gathering mirror) of the telescope determines the dimmest object that can be observed. The farther astronomers look into space, the dimmer the objects become.

The second factor is the quality of optics of the telescope. If the lenses or mirrors that gather the light are not made with great precision, magnified images will not be sharp. Earth's atmosphere is also a limiting factor. This is why major observatories are built on high mountains, where the atmosphere is thin and has less effect on the light. Figure 28-4 on page 720 shows several buildings containing large

Figure 28-4 The telescopes of the Kitt Peak Observatory are located on a mountaintop to reduce problems associated with light passing through the atmosphere. The higher the observatory and the farther it is from atmospheric pollution and artificial lights, the better the quality of the observations and images.

telescopes on a mountaintop in Arizona. The Hubble Space Telescope, which orbits Earth above the distorting effects of Earth's atmosphere, was a major step forward in observational astronomy.

ACTIVITY 28-3   **MAKING A TELESCOPE**

You can construct a simple telescope using two convex lenses. A tube to hold the lenses at the proper distance and alignment is helpful but not essential. By using lenses with more or less curvature, you can change magnification. Moving the lens that is closest to your eye adjusts the focus.

 Radio Telescopes

Some telescopes gather long-wavelength radio energy rather than visible light. Radio telescopes, like those in Figure 28-5, are not blocked by clouds of dust and gas in space that block visible light. They are also useful in detecting objects that do not produce radiation in the visible part of the electromagnetic spectrum. Radio telescopes do not make sharp images, and it is difficult to tell the exact positions of a radio source. However, radio telescopes allow astronomers to make observations that would not be possible with telescopes that work

Figure 28-5 Objects that do not give off visible light can be investigated with radio telescopes. Radio signals penetrate clouds of dust and gas that block visible light. They have been especially useful in mapping the Milky Way Galaxy.

in the visible part of the spectrum.

Other Telescopes

Other kinds of telescopes allow astronomers to use electromagnetic wavelengths shorter than visible light, such as X rays and gamma rays. These instruments must be located in orbit above Earth's atmosphere, which filters out these forms of radiation.

Technology has changed the ways astronomers use telescopes. The first telescopes were used for direct observations. If astronomers wanted a permanent record of their observations, they had to draw by hand what they observed through the telescope. Chemical photography enabled astronomers to take pictures through their telescopes. Today, the more advanced telescopes use electronic sensors like those in digital cameras along with computers to create better quality images than ever before possible.

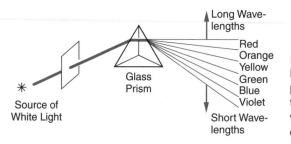

Figure 28-6 When white light passes through a glass prism, the light separates into the spectrum of colors, or wavelengths, of which it is composed.

 Spectroscope

The spectroscope is one of the most important tools that astronomers use. This instrument separates light into its component colors (wavelengths), like the glass prism shown in Figure 28-6. When starlight is passed through a spectroscope, dark lines appear in certain parts of the spectrum. These dark lines are produced when certain wavelengths of light are absorbed by gaseous elements within the outer parts of the star.

Each element has its own characteristic absorption lines. Since stars are composed primarily of hydrogen and helium, white light that passes through these elements shows dark lines in the orange, yellow, green, and blue colors that characterize hydrogen and helium. These spectral lines correspond to the energy that electrons absorb when they move to higher energy levels within the atoms. The atoms give off the same colors when the electrons fall to lower or inner energy levels. Each element has a unique set of energy levels. Therefore, these "spectral fingerprints" allow astronomers to identify the composition of distant stars.

ACTIVITY 28-4 MAKING A SPECTRUM

You can separate sunlight into its spectrum with a glass prism. This works best in a darkened room where windows face the sun. Close the shades so that a narrow slit of direct sunlight enters the room. Place the prism near the narrow opening that admits sunlight. The prism will bend the light beam and separate it into its colors. You may need to rotate the glass prism to project a visible spectrum. The spectrum can be projected onto a sheet of white paper. The stronger the light and the closer the paper is held to the prism, the brighter the spectrum will be. To increase the size of the spectrum, move the paper screen away from the prism. What two changes in the spectrum do you observe as the paper screen is moved away from the prism?

WHAT IS THE STRUCTURE OF THE UNIVERSE?

Early astronomers noticed fuzzy objects in the night sky. They called these objects nebulae (singular = nebula). The word nebula comes from the Latin word for cloud. Unlike the stars, these objects looked like dim fuzzy patches of light.

Nebulae and Galaxies

Telescopes revealed that some nebulae are regions of gas and dust where stars are forming. In addition, some nebulae were at greater distances than any known stars. Astronomers eventually realized that some nebulae are huge groups of stars held together by gravity. These objects are called **galaxies**. The whole Andromeda galaxy is visible as a small, faint patch of light high in the autumn sky. Powerful telescopes revealed that the Andromeda galaxy, like thousands of other galaxies, is a gigantic group of billions of stars. Galaxies are separated by vast distances that contain relatively few stars. Figure 28-7 on page 724 is a typical spiral galaxy.

The Milky Way

Astronomers realized that all the individual stars visible to us in the night sky are a part of the group called the **Milky Way Galaxy**. The sun and solar system are part of the Milky

Figure 28-7 Galaxy NGC 4414 is a typical spiral galaxy composed of billions of stars. Both the Milky Way Galaxy and its relatively nearby twin, the Andromeda Galaxy, are spiral galaxies.

Way Galaxy. This name came from observations of a faint, white band of light that can be seen stretching across the sky on very dark, moonless nights. (The Milky Way is not visible in urban areas where light pollution prevents the night sky from being dark enough to make it visible.) This broad band is actually made of thousands of stars.

Radio telescopes enabled astronomers to map the Milky Way Galaxy and estimate that it is composed of roughly 100 billion stars. Clouds of dust and gas that are also a part of our galaxy obscure most of them. The center of our galaxy is located in the direction of the summer constellation Sagittarius. The shape of the Milky Way Galaxy, like the Andromeda Galaxy, is a flattened spiral. The sun and solar system are located about two-thirds of the way from the center to the outer edge, as shown in Figure 28-8.

As stars orbit the core of the galaxy, inertia keeps gravity from drawing them together. Orbiting the core of the galaxy

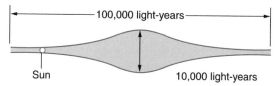

Figure 28-8 Earth and the solar system are located about two-thirds of the way from the galactic center to the outer edge of the Milky Way.

is an additional cyclic motion of Earth in space. Our planet rotates on its axis in a 24-hour cycle. It also revolves around the sun each year. The solar system revolves around the center of the Milky Way Galaxy in about 220 million years. Although this is a long time, the Milky Way Galaxy is so large that this motion is actually about 10 times faster than Earth's revolution in its orbit around the sun.

Clusters and Superclusters

The structure of the universe does not stop at galaxies. The Milky Way and Andromeda galaxies are two of about 30 galaxies known as the local group. Astronomers are now mapping superclusters of galaxies and even larger structures of matter. Why the matter of the universe is so unevenly distributed is one of the most important questions that astronomers are investigating today.

WHAT IS THE HISTORY OF THE UNIVERSE?

When you look at very distant objects in the universe, you are looking back in time. This is because light has a limited speed. You learned in an earlier chapter that you could estimate the distance to a lightning strike by counting the seconds between seeing the flash and hearing the thunder. In this procedure, you see the flash at essentially the same time it occurred. Light travels so fast that it could circle Earth about seven times in a single second.

Using Light as a Yardstick

Distances in space are so vast that light cannot reach Earth instantaneously. For example, to reach Earth, electromagnetic energy takes about 3 seconds to travel from the moon and 8 minutes from the sun. Light takes more than 4 years to arrive from the nearest night star, Proxima Centauri. The most distant object visible to the unaided eye is the Andromeda Galaxy. Light from the Andromeda galaxy takes about 2 million years to reach us.

In fact, light provides a good method to measure distances in the universe. A **light-year** is the distance that any form of electromagnetic energy can travel in 1 year: about 6 trillion miles, or 10 trillion km. Although the light year may sound like a measure of time, it is a measure of distance.

When astronomers look at distant objects in space, they see them as the objects were when the light started its long journey toward Earth. The farther away astronomers look into space, the farther back in time they see. At present, the most distance objects visible to astronomers are estimated to be about 13 billion light-years away. Astronomers can now look at the universe about a billion years after its origin, which is estimated to be 14 billion years ago.

 Redshift

Astronomer Edwin Hubble examined the spectra of distant galaxies in the early 1900s. He compared the dark absorption lines, or spectral lines, of light from these far away galaxies to the absorption lines of nearby stars. Nearby stars had spectral lines similar to those produced in the laboratory. The light from the distant galaxies did not show the dark lines in the same colors as the light from the nearby stars. However, the dark lines in the spectra of distant galaxies were shifted toward the red end of the spectrum. Hubble reasoned that the motion of distant galaxies away from Earth causes the **redshift** of spectral lines. The redshift of spectral lines is illustrated in Figure 28-9.

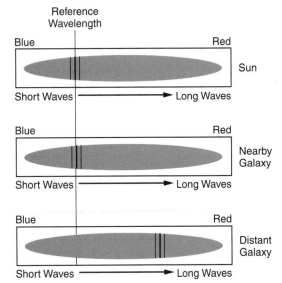

Figure 28-9 The sun and nearby galaxies show spectral lines similar to those produced in a laboratory. However, distant galaxies show these characteristic lines shifted toward the red and of the spectrum. Astronomers interpret this as evidence that the universe is expanding.

If the galaxies were moving toward Earth, the spectral lines would shift toward the blue end of the spectrum. The shift toward the red end of the spectrum indicates that the galaxies are moving away from Earth.

You can observe a similar change with sound. If you stand next to a racetrack, the high-pitched sound of the approaching car changes to a lower pitch as the car speeds past you. This apparent change in frequency and wavelength of energy that occurs when the source of a wave is moving relative to an observer is called the **Doppler effect**. It was named for Christian Johann Doppler, the scientist who explained it in 1842. The change in the frequency and wavelength of sound waves is similar to the changes that Hubble observed with light. The greater the redshift, the faster the object is moving away. Astronomers have found that the most distant galaxies are moving away the fastest.

ACTIVITY 28-5 DEMONSTRATING THE DOPPLER EFFECT

This procedure should be done only under teacher or adult supervision. This activity requires a noisemaker that can be tied to a strong cord, or string, and swung around your head. A noisemaker such as an alarm clock or a small battery-operated device from a science supplier works well. Swing the noisemaker in a circle around your head as people at a safe distance listen for changes in the pitch of the sound. Can the person swinging the device also hear the pitch of the sound change? For observers outside the circle, what part of the swing best represents redshift, and what part of the swing represents a "blueshift."

Two other factors supported Hubble's hypothesis of an expanding universe. The redshift of light of distant galaxies could be observed in all directions. In addition, the dimmer galaxies, which are thought to be dim because they are farther from Earth, showed greater redshift. Hubble explained that the redshift is caused by motion of distant galaxies away from Earth. He also reasoned that this kind of motion is a characteristic of an explosion.

You might think that the motion of distant galaxies away from Earth in all directions means that we are at the center of the expansion. However, from any position within the expanding matter of an explosion, matter is moving away in all directions.

In the 1960s, Arno Penzias and Robert Wilson were working on long-distance radio communications for the Bell Telephone Company. They constructed a special outdoor receiver to detect weak radio signals. However, the device picked up annoying radio noise that they were not able to eliminate. As they investigated the source of these radio waves, they realized that the energy they were picking up was billions of years old. They were actually listening to the origin of the universe. These radio signals, known as **cosmic background radiation**, are weak electromagnetic radiation left over from the formation of the universe.

The Big Bang

The outward motions of distant galaxies and the cosmic background radiation are evidence that the universe began as an event now called the **big bang**. The name was first proposed as a joke to make fun of the theory, but the name stuck. This theory proposes that at the time of its origin, the universe was a concentration of matter so dense that the laws of nature as we know them today did not apply. This matter expanded explosively, forming the universe. Even the most extreme conditions that exist within the largest stars could not compare with the beginning of the universe.

Experiments and the theories of physics show that the greatest possible velocity for matter or energy is the speed of light: about 300 million meters per second. Like the temperature of absolute zero (0 K), this is one of the absolute limits known to science. Astronomers reason that the universe is expanding at this rate.

By working backward, astronomers estimate that the universe began in the big bang 14 billion years ago. Expanding outward in all directions, the universe today could be as much as 28 billion light years across. This is so gigantic that it is nearly impossible for humans to comprehend how large this is. Earth, the solar system, and even the huge Milky Way Galaxy are incredibly small compared with the size of the universe.

ACTIVITY 28-6 | **A MODEL OF THE BIG BANG**

Inflate a round balloon into a small ball. Draw several small dots on the balloon's surface. Notice that as the balloon is inflated more, the dots always move apart. If observers were located anywhere on the surface of the balloon or even inside the balloon, as the balloon is inflated, they would see the dots moving away in all directions. No matter what location is chosen, it would appear that the observer is at the center of the expansion. Therefore, there is no way that astronomers can find the center of the universe.

WHAT IS THE FUTURE OF THE UNIVERSE?

From an astronomical point of view, Earth is in a reasonably unchanging state. Earth's orbit around the sun is stable and scientists expect the sun to continue its energy production on the main sequence for billions of years. Collisions with large objects from space, such as those thought to mark the ends of past geologic eras, are possible. Fortunately, these events are becoming less likely as the age of the solar system increases. However, the very-long-term future of the universe is not clear.

Three Possible Futures

The universe seems to have three possible futures. Some scientists propose that the expansion of the universe may be slowing due to gravity. However, it is possible that there is not enough gravity to stop the expansion. In this case, the universe will continue to expand without limit. Other scientists propose that there may be enough gravity to just stop the expansion, leading to a steady state. If the universe has enough gravity to reverse its expanding phase, it could fall back together in a very distant event that some astronomers call the "big crunch."

A good way to understand this is to consider a baseball thrown straight up. Gravity brings the ball back to the ground. However, if you could propel the ball fast enough, it would continue upward into space and never return to Earth.

In recent years, astronomers have found evidence that the universe is not only expanding, but that it is expanding at an increasing rate. What force could work against the force of gravity to cause this? It is as if you threw a ball up into the air and it did not fall back to Earth. In fact, it is as if the ball flew upward faster and faster with time. This would be surprising, indeed. Astronomers find these observations just as surprising.

Astronomers have named the mysterious cause of this accelerating expansion "dark energy." However, they cannot explain it. Nor can they explain the source of gravitational force that holds the rapidly spinning galaxies from breaking apart. This force is attributed to the gravitational attraction of "dark matter," which astronomers think makes up about 90 percent of the matter in the universe. Dark matter and dark energy, the mysteries of science just keep coming.

Therefore, the ultimate future of the universe depends upon the balance between the expansion of the big bang, gravity, and dark energy. To date, astronomers have not been able to determine which process will dominate. This remains one of many questions that guide scientific investigation.

TERMS TO KNOW

big bang	**luminosity**
cosmic background radiation	**Milky Way Galaxy**
Doppler effect	**nuclear fusion**
galaxy	**redshift**
light-year	**star**

CHAPTER REVIEW QUESTIONS

Base your answers to questions 1–4 on the *Earth Science Reference Tables* or Figure 28-3.

1. Which star has about the same surface temperature as the sun?

   (1) Betelgeuse   (3) Sirius
   (2) Polaris   (4) Procyon B

2. Which star is cooler, yet many times brighter than Earth's sun?

   (1) Barnard's Star   (3) Rigel
   (2) Betelgeuse   (4) Sirius

3. According to the "Luminosity and Temperature of Stars" graph in the *Earth Science Reference Tables*, the sun is classified as

   (1) a main sequence star.   (3) a blue supergiant.
   (2) a white dwarf.   (4) a red giant.

4. What is the color of a main sequence star that gives off about 100 times as much light as the sun?

   (1) blue   (3) yellow
   (2) white   (4) red

5. How do stars like the sun create energy that is later radiated away into space?

(1) nuclear fusion changing hydrogen into helium
(2) burning of carbon fuels
(3) changes in state such as melting and evaporation
(4) absorbing electromagnetic radiation from space

6. What instrument uses long-wave electromagnetic radiation to help astronomers make celestial observations?

(1) radio telescopes   (3) X-ray telescopes
(2) optical telescopes   (4) binoculars

7. According to the *Earth Science Reference Tables*, in what property do ultraviolet, visible, and infrared radiation differ?

(1) half-life   (3) wavelength
(2) atomic mass   (4) wave velocity

8. The Milky Way Galaxy is best described as

(1) a type of solar system.
(2) a constellation visible to everyone on Earth.
(3) a region of space between the orbits of Mars and Jupiter.
(4) a spiral-shaped formation composed of billions of stars.

9. In which list are celestial features correctly shown in order of increasing size?

(1) galaxy→solar system→universe→planet
(2) solar system→galaxy→planet→universe
(3) planet→solar system→galaxy→universe
(4) universe→galaxy→solar system→planet

10. What causes the spectral lines of light from distant galaxies to be shifted toward the red end of the spectrum?

(1) the gravitational field of Earth
(2) the gravitational field of the sun
(3) motion of the galaxies toward us
(4) motion of the galaxies away from us

11. The diagram below illustrates three stages of a current theory of the formation of the universe.

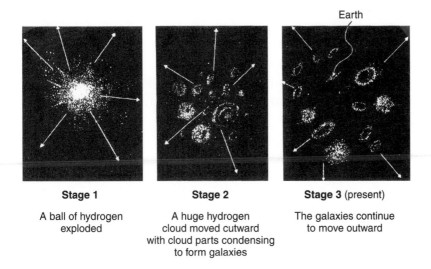

Stage 1
A ball of hydrogen
exploded

Stage 2
A huge hydrogen
cloud moved outward
with cloud parts condensing
to form galaxies

Stage 3 (present)
The galaxies continue
to move outward

A major piece of scientific evidence supporting this theory is the fact that wavelengths of light from galaxies moving away from Earth in stage 3 are observed to be

(1) shorter than normal (a redshift).
(2) shorter than normal (a blueshift).
(3) longer than normal (a redshift).
(4) longer than normal (a blueshift).

12. In the diagram below, the spectral lines of hydrogen gas from three galaxies, A, B, and C, are compared to the spectral lines of hydrogen gas observed in a laboratory.

What is the best inference that can be made concerning the movement of galaxies A, B, and C?

(1) Galaxy A is moving away from Earth, but galaxies B and C are moving toward Earth.
(2) Galaxy B is moving away from Earth, but galaxies A and C are moving toward Earth.
(3) Galaxies A, B, and C are all moving toward Earth.
(4) Galaxies A, B, and C are all moving away from Earth

13. Because of the Doppler redshift, the observed wavelengths of light from distant celestial objects appear closer to the red end of the spectrum than light from nearby celestial objects. The explanation for the redshift is that the universe is presently

(1) contracting, only.
(2) expanding, only.
(3) remaining constant in size.
(4) alternating between contracting and expanding.

14. How can we best describe the general pattern of motion that we observe for distant galaxies in the universe?

(1) Most galaxies are moving toward the Milky Way Galaxy, and the closer galaxies are generally approaching faster.
(2) Most galaxies are moving toward the Milky Way Galaxy, and the more distant galaxies are generally approaching faster.
(3) Most galaxies are moving away from the Milky Way Galaxy, and the closer galaxies are generally moving faster.
(4) Most galaxies are moving away from the Milky Way Galaxy, and the more distant galaxies are generally moving faster.

15. What could cause the expansion of the universe to slow?

(1) energy production by nuclear fusion
(2) energy production by nuclear fission
(3) gravitational force
(4) electromagnetic radiation

Open-Ended Questions

The graph below shows the inferred stages of development of the sun. Use this graph to answer questions 16 and 17.

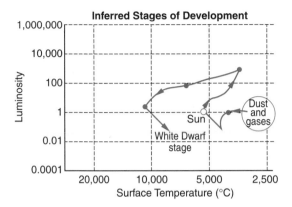

16. Describe the change in luminosity of the sun that will occur from its current Main Sequence stage to its final White Dwarf stage.

17. Which star shown on the "Luminosity and Temperature of Stars" graph in the *Earth Science Reference Tables* is currently in the sun's final predicted stage of development?

18. According to the "Luminosity and Temperature of Stars" graph in the *Earth Science Reference Tables*, what is the surface temperature of the sun?

19. According to the *Earth Science Reference Tables*, what kind of electromagnetic radiation has a wavelength of about 1 meter (100 cm)?

20. Name one characteristic that X rays, visible light, and radio waves have in common.

Appendices

LABORATORY SAFETY

Laboratory work is an important part of science. However, many materials, equipment and procedures that scientists use are potentially hazardous. Working in the science lab requires more attention to safety than do most other activities in the school setting. Work and conduct yourself with care in the laboratory. You should be familiar with the rules below, and be alert to any hazards.

1. Follow all safety precautions as directed by the teacher and stated in printed directions. If you do not understand the directions, ask your teacher to clearly explain them to you.

2. Notify the teacher at once if there is an accident or dangerous situation such as an electrical problem; an injury; or a spill of water, chemicals, or other substances. The teacher will supervise the clean up of spilled chemicals, broken glass, or broken equipment.

3. Never perform a laboratory procedure unless the teacher approves it.

4. Touch equipment and materials only as directed. If you do not understand how to perform a procedure, ask for help.

5. Wear laboratory safety goggles whenever you mix chemicals, heat liquids, perform procedures that could result in flying particles, or perform any tasks that could injure your eyes. When performing these procedures, everyone at the lab station must wear safety goggles.

6. When using a flame or a hot plate, keep combustible materials including paper, clothing, jewelry, and hair safely away from the heat source. If you are heating something in a container, be sure the container has an opening and point the open end away from people.

7. Hot plates, burners, glassware, and other materials remain hot for a long time. Use special caution during the cooling of these objects. You may need tongs or other special materials to handle these objects.

8. The use of acids, bases, and many other laboratory chemicals requires special precautions. You should handle these chemicals only as instructed, and their use must be supervised by your teacher.

9. Use sharp or pointed objects such a scissors, knives and drawing compasses in such a way as to avoid puncture injuries.

10. Dispose of materials as directed. Do not put solids into sinks or put hazardous materials into waste baskets.

11. Keep the work area neat and clean. At the end of the procedure, return equipment and materials as you are instructed. Unplug electrical devices and turn off gas and water unless instructed otherwise.

12. Know the location of and how to use safety equipment and materials such as running water and fire retarding devices. Also know how to notify another adult if the teacher is unable to respond to events.

Appendix B   A FORMAT FOR LABORATORY REPORTS

This lab report format is presented as a general guide to help students organize lab activities and written reports. Reports for laboratory activities are required for admission to the Regents examination. For some lab activities, your teacher may present labs in a written guided format. At other times, you may not be given special written instructions about how to conduct and report on a particular lab. Or, you may be asked to completely plan and conduct your own procedure. In those cases, this format should help you organize, conduct, and report on your laboratory activity. Be sure you understand whether each student must submit a report or if one report for the whole lab group is sufficient. Different lab groups should not work together or produce virtually identical lab reports. In general, it is best to keep lab reports neat, and brief, but complete.

Open-Ended Laboratory Report Format

Title: The name of the lab. Commonly a brief indication of what is being investigated.

Date: The date the lab procedure was performed.

Names: List all members of the lab group.

Objective(s): A clear statement of what your lab procedure is intended to explore.

Materials: Make a list of all the special equipment and supplies needed to conduct the laboratory procedure.

Procedure: In a sequence of numbered steps, tell how this lab was performed. These steps should be clear enough that a person who is not familiar with the activity can follow them and obtain good results. Please include any safety precautions or other warnings that may be necessary.

Observations/Data/Graphs: List the outcome of your procedure in a neat and organized format. This will include the data you collect, any mathematical steps taken with the data, and graph(s) to clarify your findings.

Conclusions or Discussion: The conclusion should relate your observations and results to the objectives at the beginning of the lab report.

Appendix C   THE INTERNATIONAL SYSTEM OF UNITS (S.I.)

S.I. units were based on European measures commonly known as the metric system They are used by the people of every major nation of the world except the United States. Scientific publications usually use S.I. or metric units. Most S.I. units are related to larger and smaller units by a factor of ten. The system of measures most commonly used in the United States is sometimes known as United States customary units.

Basic Units

Basic units (meter, liter, gram) are not composed of other units. Derived units are a combination of basic units. Examples of derived units include volume, which can be measured in meters cubed (distance^3); speed, which is measured in meters per second (distance/time); and density, which is measured in grams per cubic centimeter (mass/volume).

Quantity	S. I. Unit	Symbol	Approximate U.S. Equivalent
length	meter	m	39.37 inches (about 3 inches longer than one yard)
mass	kilogram	kg	2.204 pounds
time	second	s or sec	(Identical units)
temperature	kelvin*	K	(See conversion chart to degrees Fahrenheit in the *Earth Science Reference Tables)*

*(°Celsius is often used in metric measures)

S.I. and Metric Prefixes with Examples

Prefix	Symbol	Multiple	Example (length)
kilo	k	1000	kilometer (1 km = 1000 m)
centi	c	$\frac{1}{100}$	centimeter (1 cm = 0.01 m)
milli	m	$\frac{1}{1000}$	millimeter (1 mm = 0.001 m)

(Prefixes for other multiples of ten have been defined, but the three shown above are the most commonly used prefixes.)

Conversions Between S.I. (Metric) and U.S. Customary Units

(These conversion factors are limited to a convenient number of decimal places.)

Length

1 centimeter = 0.394 inch	(1 inch = 2.54 centimeters)
1 meter = 39.37 inches	(1 foot = 0.341 meter)
1 kilometer = 0.62 mile	(1 mile = 1.61 kilometers)

Volume and Capacity

1 liter = 1.06 quarts	(1 quart = 0.95 liter)
1 milliliter = 0.035 ounce	(1 ounce = 28.4 milliliters)

(The cubic centimeter is sometimes used as the equivalent of the milliliter.)

Mass

1 gram = 0.035 ounce	(1 ounce = 28.4 grams)
1 kilogram = 2.204 pounds	(1 pound = 0.4536 kg = 453.6 grams)

Appendix D   PHYSICAL CONSTANTS

Speed of light:	300,000,000 m/s (186,000 mi/s) [1 light-year is about 6 trillion (6×10^{15}) mi or 10 trillion km]
Lowest possible temperature:	0 K ($-373°$C or $-460°$F)
Density of water:	1 g/mL (at 3.98°C)
Mass of hydrogen atom:	1.67×10^{-27} kg

Appendix E   GRAPHS IN SCIENCE

It is often easier to understand data when it is presented in the form of a simple graph. Making and reading graphs are important skills to scientists and other informed adults. There are many kinds of graphs, but the two-dimensional line graph and the bar graph are the most commonly used forms. A line graph is used to show continuous change. A bar graph shows unconnected data values.

(Chapter 1 (pages 18 to 22) has a section on making graphs.)

If you need to make a graph, follow these rules to get the best results.

1. The purpose of a graph is to make data more meaningful. Keep the graph simple and neat, but include all necessary information.

2. Label the two axes with the quantity (such as mass, temperature, or whatever you are graphing) and the appropriate units (such as grams or °C).

3. Label the axes with numbers that are appropriate for the range of data. The graph line (or bars) should extend at least half way along each of the two axes. Graphs that simply represent general relationships may be shown without including numbers and units on the axes. It should be understood that values on graphs usually increase to the right along the x-axis and upward along the y-axis.

4. Usually, the independent variable is graphed along the bottom, or x-axis. For example, if your graph shows the amount of rock weathering that occurs though time, time should increase along the bottom axis. Because time influences the amount of weathering, rather than weathering affecting the passage of time, time is considered the independent variable. Values of the independent variable are usually selected to show how the dependent variable changes.

5. When the data have been plotted, a clear pattern may become visible. Sometimes the data points form a line that can be drawn with a straight edge. In other graphs, a smooth curve better fits the data. Draw the graph line accordingly.

6. Logically, some graph lines pass through the origin (0, 0), and others do not. Consider what you are graphing and draw the graph line accordingly.

7. All measurements include some error. Error can be made very small, but error cannot be eliminated. For this reason, it is possible that the most appropriate line will not pass through the center of each data point. It may be a straight line or a gentle curve that clearly shows the trend and estimated values of the data. This is sometimes called a "best fit" line.

8. A title, usually shown near the top of the graph, helps to identify the purpose of the graph. For example, the graph may be titled, "Average Monthly Temperatures at Buffalo, New York."

9. The guidelines above are not always followed. Graphs may violate one or more of these rules. As long as the graph makes the data more meaningful, it has fulfilled its principal function.

Glossary

This glossary can be used as a study aid. The terms that follow are helpful in learning Regents Earth science. Please be aware that the primary importance of these terms is their applications to understanding Earth systems. If you are not familiar with some of these terms, you may wish to use the index of this book to find them within their scientific context. Definitions alone are of limited use.

The entries for the terms are intended to help you understand how the words are applied in this specific course. Broader and more precise definitions can be found in textbooks that are more advanced and other reference materials.

The abbreviation ESRT is used to highlight terms used and sometimes clarified in the *Earth Science Reference Tables*. Many important terms that are not listed here, such as specific rock types and names of geologic ages, may be understood using charts in the *Reference Tables*.

abrasion: The grinding away of rock by friction with other rocks

absolute age: An age expressed as a specific amount of time, absolute age always includes a unit of time; numerical age (ESRT)

absolute humidity: The mass of water vapor in each cubic unit of air

acid precipitation: Precipitation (snow or rain) with corrosive (low pH) chemical properties, generally the result of pollution from the burning of fossil fuels

agents of erosion: Moving water, wind, or ice that cause the transport of weathered materials

air mass: A large body of air that is relatively uniform in temperature and humidity (ESRT)

altitude: The angular elevation of an object above the horizon

angle of insolation: The angle between Earth's surface and incoming rays of sunlight; angle of the sun above the horizon

Antarctic Circle: The latitude (66.5°S) south of which the sun does not rise on the Southern Hemisphere's winter solstice; the latitude (66.5°S) south of which the sun is in the sky for 24 hours on the Southern Hemisphere's the summer solstice

anticyclone: A region of relatively high atmospheric pressure

aquifer: An underground zone of porous material that contains useful quantities of groundwater

arctic air mass: A large body very of cold air that originated in the Arctic (ESRT)

Arctic Circle: The latitude (66.5°N) north of which the sun does not rise on the Northern Hemisphere's winter solstice; The latitude (66.5°N) north of which the sun is in the sky for 24 hours on the Northern Hemisphere's summer solstice

arid climate: A climate that has little rain and low humidity

asteroid: An irregularly shaped rocky mass that is smaller than a planet and occupies an orbit around the sun; most are found between the orbits of Mars and Jupiter

asthenosphere: The upper part of the mantle, capable of slow deformation and flow under heat and pressure (ESRT)

astronomy: The study of Earth's motions and the objects beyond Earth, such as planets and stars

atmosphere: The layer of gases that surrounds a celestial body (ESRT)

avalanche: The rapid, downslope movement of snow, similar to a landslide, that occurs on steep slopes

axis: An imaginary line that passes through Earth's North and South Poles

azimuth: The compass direction specified as an angle. Azimuth starts at 0° at due North and progresses through East (90°), South (180°), West (270°), and back to North (360°, or 0°).

banding: The light- and dark-colored bands of mineral that form parallel to foliation in metamorphic rocks (ESRT)

barometer: An instrument used to measure air pressure

barrier islands: Offshore features, similar to sandbars, that rise above sea level

bed load: The sediments that roll or bounce along the bottom of a stream

bedrock: The solid, or continuous, rock that extends into Earth's interior

big bang: The theory that the universe formed as a concentration of matter expanded explosively

bioclastic sedimentary rocks: Rocks composed of materials made from or by living organisms (ESRT)

biological activity: The actions of plants and animals that cause weathering

blizzard: A winter snowstorm that produces heavy snow and winds of 35 miles per hour (56 kilometers per hour) or greater

boiling: The change in state from liquid to gas (vapor) at the boiling point

caldera: A large, bowl-shaped depression formed when the top of a volcano collapses into the emptied magma chamber

calorie: The energy absorbed when the temperature of 1 gram of water increases 1 Celsius degree (ESRT)

capillarity: The tendency of a substance to pull water into tiny spaces, or pores, by adhesion

celestial objects: Things seen in the sky that are outside Earth's atmosphere

chemical change: A change that results in the formation of a new substance

chemical weathering: A natural process that occurs under conditions at Earth's surface, forming new compounds

classification: The organization of objects, ideas, or information according to their properties

clastic: Sedimentary rocks that are composed of the weathered remains of other rocks; fragmental (ESRT)

cleavage: The tendency of some minerals to break along smooth, flat surfaces (ESRT)

climate: The average weather conditions over a long time, including the range of conditions

cloud: A large body of tiny water droplets or ice crystals

cloud-base altitude: The height at which rising air begins to form clouds

comet: An object made of ice and rock fragments that revolves around the sun usually in a highly eccentric orbit; it may be visible periodically in the night sky as a small spot of light with a long tail

compounds: Substances made up of more than one kind of atom (element) combined into larger units called molecules

condensation: The process by which a substance changes from a gas to a liquid (ESRT)

condensation nuclei: Tiny particles of solids suspended in the air on which water condenses to form clouds

conduction: The movement of heat that occurs as heated molecules pass their vibrational energy to nearby molecules.

conservation: The careful use, protection, and restoration of our natural resources

contact metamorphism: The process in which an intrusion of hot, molten magma causes changes in the rock close to it (ESRT)

continental air mass: A large body of air that has relatively low humidity because it originated over land (ESRT)

continental climate: A climate characterized by large seasonal changes in temperature

continental glacier: A glacier that flows outward from a zone of accumulation to cover a large part of a continent

contour line: Lines on a map that connect places having the same elevation (height above or below sea level)

convection: A form of heat flow that moves matter and energy as density currents under the influence of gravity (ESRT)

convection cell: The pattern of circulation that involves vertical and horizontal flow

convergence: The act of moving together (ESRT)

convergent plate boundary: A place where lithospheric plates collide (ESRT)

coordinate system: A grid in which each location has a unique designation defined by the intersection of two lines (ESRT)

Coriolis effect: The apparent curvature of the path of winds and ocean currents as they travel long distances over Earth's surface; caused by Earth's rotation

correlation: Matching bedrock layers by rock type or by age

cosmic background radiation: Weak electromagnetic radiation (radio waves) left over from the formation of the universe (big bang)

crater: A bowl-shaped depression at the top of a volcano caused by an explosive eruption or the impact of an object from space.

crystalline sedimentary rocks: Sedimentary rocks that form by precipitation (ESRT)

cyclone: (1) A region of relatively low atmospheric pressure; (2) term applied to hurricanes in the Indian Ocean; (3) synonym for tornado

decay product: The stable, ending material of radioactive decay (ESRT)

decay-product ratio: A comparison of the amount of the original radioisotope with the amount of its decay product (ESRT)

deforestation: Cutting forests to clear the land for other uses

delta: A deposit of sediment built into a large body of water by deposition from a stream

density: The concentration of matter, or the mass per unit volume (ESRT)

deposition: The settling, or release, of sediments that have been carried by an agent of erosion (ESRT)

dew: Liquid water that forms by condensation on cold surfaces

dew point: The temperature at which air is saturated with water vapor (ESRT)

dew-point temperature: The temperature to which air must be cooled to become saturated with moisture (ESRT)

discharge: The amount of water flowing past a particular place in a specified time

divergence: The act of moving apart

divergent plate boundary: A place where lithospheric plates separate (ESRT)

Doppler effect: The apparent change in frequency and wavelength of energy radiated by a source as a result of the motion of the source or the observer

Doppler radar: A device that uses reflected radio waves to measure wind speed and direction at a distance

drainage divides: The high ridges, from which water drains in opposite directions, that separate one watershed from another

drainage pattern: The path of a stream, which is influenced by topography and geologic structures

drumlins: Streamlined hills of glacial origin aligned north-to-south that have steep sides, a blunt north slope, and a gentle slope to the south; made of till

dune: A hill or ridge of wind-blown sand

duration of insolation: The amount of time the sun is visible in the sky, or the number of hours between sunrise and sunset

dynamic equilibrium: The state in which opposing processes take place at the same rate; a state of balance of events

Earth science: A science that applies the tools of the physical sciences to study Earth; including the solid Earth, its oceans, atmosphere, and core, and surroundings in space

earthquake: A sudden movement of Earth's crust that releases energy (ESRT)

eccentricity: A measure of the elongation of an ellipse (ESRT)

eclipse: The partial or complete hiding of one celestial object by another. (An eclipse of the moon occurs when the moon orbits into Earth's shadow. An eclipse of the sun occurs when the moon's orbit takes it directly between Earth and the sun.)

ecology: The branch of science that is concerned with the relationships among organisms and their environment

El Niño: The periodic replacement of upwelling cold water by warm water along the western coast of South America

elements: The basic substances that are the building blocks of matter (ESRT)

ellipse: A closed curve formed around two fixed points such that the total distance from any point on the curve to both fixed points is constant

epicenter: The place on Earth's surface directly above an earthquake's focus (ESRT)

equator: An imaginary line that circles Earth halfway between the North and South Poles (ESRT)

equilibrium: A state of balance

equinox: One of the two days on which the sun rises due east and sets due west, on which the length of day and night are equal, on which the sun's vertical rays are at the equator; the first day of spring or fall

erosion: The transportation of sediments by water, air, glaciers, or by gravity acting alone. (See agents of erosion.) (ESRT)

erratics: Large rocks transported from one area to another by glaciers

escarpment: A steep slope or a cliff of resistant rock that marks the edge of a relatively flat area

evaporation: The process by which a substance changes from a liquid to a gas

evaporation: The change in state from liquid to gas when the temperature is below the boiling point

evolution: The gradual change in living organisms from generation to generation, over a long period of time

extinction: The death of every individual of a particular species (ESRT)

extrusion: The movement of magma onto Earth's surface (ESRT)

faults: Cracks in Earth's crust along which movement occurs

felsic: Describes light-colored minerals rich in aluminum or rocks made of these minerals (ESRT)

field: A region in which a force, temperature, land elevation, or another quantity can be measured at any location (ESRT)

floodplain: A flat region next to a stream or river that can be covered by water in times of flood

flotation: The method by which particles that are too large to be carried in solution or by suspension float on water

fluid: Any substance that can flow, usually a liquid or a gas

focus: (1) The place where rock begins to separate during an earthquake, usually located underground. (2) Either of the two fixed points that determine the shape of an ellipse (ESRT)

fog: Very low clouds that reach the ground (ESRT)

foliation: The alignment of mineral crystals, caused by metamorphism (ESRT)

fossils: A record of prehistoric life preserved in rock (ESRT)

fracture: The way minerals break along curved surfaces (ESRT)

fragmental: Describes sedimentary rocks that are composed of the weathered remains of other rocks; clastic (ESRT)

freezing: The change in state from liquid to solid

freezing rain: Rain that freezes on contact with Earth's surface

frequency: A measure of how many waves pass a given point in a given period of time

front: A boundary, or interface, between air masses (ESRT)

frost: Ice crystals that form when water vapor comes in contact with surfaces whose temperature is below 0°C

frost wedging: A form of physical weathering caused by repeated freezing and thawing of water within cracks in rocks

galaxy: A huge group of stars held together by gravity

geologists: Scientists who study the origin, history, and structure of Earth and how it changes

geology: The study of the rock portion of Earth, its interior, and surface processes

geosphere: The mass of solid and molten rock that extends more then 6000 kilometers from Earth's solid surface to its center

glacier: A large mass ice that flows over land due to gravity

global warming: A long-term increase in the average temperature of Earth's atmosphere, it is probably the result of the increased concentration of carbon dioxide and other greenhouse gases in the atmosphere

graded bedding: Within a layer of sediment, the gradual change in sediment size from bottom (large) to top (small) showing the order in which particles settled; vertical sorting

gradient: The change in field value per unit distance (ESRT)

gravity: The force of attraction between objects

greenhouse effect: The process by which carbon dioxide and water vapor absorb heat radiation, increasing the temperature of Earth's atmosphere

Greenwich Mean Time: The basis of standard time throughout the world; based on measurements of the position of the sun in Greenwich, England

grooves: Furrows of glacial origin in bedrock that are deeper and wider than striations

groundwater: Water that enters the ground and occupies free space in soil and sediment as well as openings in bedrock, including cracks, and spaces between grains

hail: Pellets of ice, which grow larger as they repeatedly become coated with water, and are then blown higher into cold air where the coating of water freezes; eventually the ice pellets become heavy enough to fall to the ground. (Hail is most common during thunderstorms.) (ESRT)

half-life period: The time it takes for half of the atoms in a sample of radioactive element to decay (ESRT)

hardness: The resistance of a mineral to being scratched (ESRT)

hazard: An event that places people in danger of injury, loss of life, or property damage

horizontal sorting: A decrease in the size of sediment particles with distance from the shore, produced as a stream enters calm water

hot spot: A long-lived source of magma within the asthenosphere and below the moving lithospheric plates (ESRT)

humidity: The water-vapor content of air (ESRT)

hurricane: A large storm of tropical origin that has sustained winds in excess of 74 miles (120 kilometers) per hour (ESRT)

hydrologic cycle: A model that represents water movement and storage within Earth, on the surface, and within the atmosphere

hydrosphere: Earth's liquid water, including oceans, surface water, and groundwater

hygrometer: An instrument used to measure atmospheric humidity

igneous rocks: Rocks that form by the solidification of melted rock (ESRT)

inclusion: A fragment of one type of rock that is enclosed in another rock

index fossils: Fossils used to establish the age of rocks; they must be easy to recognize, found over a large geographic area, and they must have existed for a brief period of geologic time (ESRT)

inertia: The tendency of an object at rest to remain at rest or an object in motion to move at a constant speed in a straight line unless acted on by an unbalanced force

inference: A conclusion based on observations

infiltration: The process in which water soaks into the ground

insolation: Solar energy that reaches Earth (*in*coming *sol*ar radi*ation*)

intrusion: The movement of magma to a new position within Earth's crust. A body of rock that was injected into surrounding rock as magma

island arc: A curved line of volcanic islands that are the result of partial melting of a tectonic plate where it descends beneath another oceanic plate

isobars: Isolines (q.v.) that connect locations with the same atmospheric pressure on a weather map

isoline: A line on a field map that connects places having the same field quantity value

isotherm: A line on a field map that connects places having the same temperature

isotopes: Atoms of the same element that contain different numbers of neutrons in their nucleus (ESRT)

jet streams: Wandering currents of air far above Earth's surface that influence the path of weather systems(ESRT)

Jovian planet: A planet whose composition is similar to Jupiter's; also know as a gas giant (ESRT)

kettle: A small closed basin formed in a moraine

lake-effect storms: Precipitation events that occur downwind from large lakes as the result of moisture that enters the air over the lake; especially common as early winter snow events

land breezes: Light winds that blow from the land to the water; they usually develop at night as the air over the land becomes cooler than the air over the water

landform: A feature of a landscape

landscape: A region that has landforms that are related by similarities in shape, climate, and/or geologic setting; the general shape of a large area of the land surface, such as plains, plateau, or mountain (ESRT)

landslide: The rapid, downslope movement of rock and soil

latent heat: Energy absorbed or released when matter changes state (ESRT)

latitude: The angular distance north or south of the equator (ESRT)

lava: Melted rock coming from a volcano or such rock that has cooled and hardened

levees: High banks along a river of natural or human origin

lightning: Sudden electrical discharges within clouds, between clouds, and between clouds and the ground that are seen as flashes of light

light-year: The distance electromagnetic energy can travel in one year, approximately 6 trillion miles (10 trillion km)

liquefaction: The process in which strong shaking allows water to surround the particles of sediment, changing the sediments into a material with the properties of a thick fluid

lithosphere: The solid rock that covers Earth (ESRT)

lithospheric plate: A rigid section of Earth's crust, which includes the crust and the rigid upper mantle

logarithmic: A scale in which an increase of one unit translates to a 10-fold increase in the quantity measured.

longitude: The angular distance east or west of the prime meridian (ESRT)

longshore transport: The motion of sediment parallel to the shore caused by waves

luminosity: The total energy output of a star; absolute brightness (ESRT)

luster: The way light is reflected and/or absorbed by the surface of a mineral (ESRT)

mafic: Describes dark-colored minerals rich in magnesium (ESRT)

magma: Hot, liquid rock within Earth (ESRT)

major axis: The distance across an ellipse measured at it widest point

maritime air mass: A large body of air that has relatively high humidity because it originated over the ocean or other large body of water (ESRT)

maritime climate: A humid climate that occurs over the oceans and in coastal locations

mass movement: The motion of soil or rock down a slope without the influence of running water, wind, or glaciers

meander: A curve that develops in the path of a river when the river flows over relatively flat land

mechanical weathering: The breaking up of rock into smaller particles without a change in composition; physical weathering

melting: The change in state from solid to liquid (ESRT)

Mercalli scale: A scale for measuring earthquake intensity based on the reports of people who felt the quake and observed the damage it caused

mesosphere: The layer of Earth's atmosphere directly above the stratosphere, in which temperature decreases with increasing altitude (ESRT)

metamorphic rocks: Rocks that form as a result of heat and/or pressure on other rocks causing chemical (mineral) or physical changes (ESRT)

meteor: A streak of light produced as a meteoroid burns due to friction with Earth's atmosphere

meteorologist: A scientist who studies the weather

meteorology: The study of Earth's atmosphere and how it changes

mid-latitude cyclone: An area of low pressure or a storm system, such as those that usually move eastward across the United States

mid-ocean ridges: A system of underwater mountain ranges that circles Earth like the seams on a baseball (ESRT)

Milky Way Galaxy: The group of billions of stars that includes the sun and our solar system, it is visible as a faint band of light across the night sky

mineral: A natural inorganic, crystalline solid that has a specific range of composition and consistent physical properties (ESRT)

model: Anything that is used to represent something else

Moho: The boundary between Earth's crust and mantle (ESRT)

Mohs' scale: A special scale of hardness used to identify minerals (ESRT)

monsoons: Seasonal changes in the direction of the prevailing winds, causing changes in temperature and rainfall

moraine: A mass of till deposited by a glacier

mountain landscape: A rugged landscape that has great relief from the top of the highest peaks to deep valleys, commonly underlain by resistant rock types and distorted structures including folds and faults

natural resources: Any material from the environment that is used by people

neap tides: The smallest tidal range, which occurs when the sun and moon are at right angles to Earth

nonrenewable resources: Resources that exist in a fixed amount or for which the rate of regeneration is so slow that use of these resources will decrease their availability

nuclear fusion: The process by which the nuclei of light elements, such as hydrogen, under intense heat and pressure form the nuclei of heavier elements, such as helium

oblate: Slightly flattened at the poles

observations: Information gathered through the use of sight, touch, taste, smell, and hearing

ocean trench: A deep-ocean location where old lithosphere moves back into Earth's interior; also called a subduction zone or a convergent plate boundary (ESRT)

oceanography: The study of the oceans that cover most of Earth

ores: Rocks that are mined to obtain a substance they contain of economic value

origin: How something was formed

origin time: The time at which a fault shifted to produce an earthquake (ESRT)

original horizontality: The principle that no matter the present angle or orientation of sedimentary rock layers, the layers were originally horizontal and were tilted after deposition

outcrop: A place where bedrock is exposed at Earth's surface

outgassing: The process in which bubbles of hot gas escape from magma exposed to reduced pressure at Earth's surface

outwash: Sorted sediments deposited by water from a melting glacier

overland flow: The water from precipitation that flows downhill under the influence of gravity until it reaches a stream or seeps into the ground; runoff

paleontology: The study of fossils

paradigm: A coherent set of principles and understandings

percent error: A comparison of the size of an error with the size of the value being measured (ESRT)

permeability: The ability of soil or sediment to allow water to flow through it

phase: The observed shape of the lighted portion of a celestial object, for example, the moon or Venus

phases of matter: The states of matter—solid, liquid, and gas

physical weathering: The breaking up of rock into smaller particles without a change in composition; mechanical weathering

plains: Relatively flat landscapes, commonly at low elevation and usually underlain by flat-lying sedimentary rocks; the range of elevation is small (ESRT)

plastic: A material that is solid under short-term stress, but flow like a liquid when stress is applied over a long period of time

plate tectonics: A theory of crustal movements that combines sea-floor spreading with continental drift (ESRT)

plateau: A rolling landscape or elevated, comparatively flat region with modest topographic relief (ESRT)

plutonic: Describes igneous rocks that form deep underground (ESRT)

polar air mass: A large body of cold air that originated near one of Earth's poles (ESRT)

polarity: The direction of a magnetic field determined with an instrument such as a magnetic compass

pollution: A sufficient quantity of any material or form of energy in the environment that harms humans or the plants and animals on which they depend

porosity: The ability of a material to hold water in open spaces, or pores

precipitation: (1) The settling of solids from solution, often the result of the evaporation of seawater (ESRT). (2) Water that falls to Earth as rain, show, sleet, or hail (ESRT)

prevailing winds: The most common wind direction and speed at a particular location and time of year (ESRT)

primary waves (*P*-waves): Longitudinal earthquake waves that cause the ground to vibrate forward and back along the direction of travel; the earthquake waves that travel the fastest; *P*-waves can travel through solids, liquids, and gases (ESRT)

prime meridian: The north-south line through Greenwich, England, from which longitude is measured (ESRT)

profile: A cross section, or side view of an object

psychrometer: An instrument, made up of two thermometers mounted side-by-side on a narrow frame, that is used to determine the dew-point temperature and relative humidity; also known as a wet and dry bulb thermometer (ESRT)

radar: A method or device that uses reflected radio waves to locate or map distant objects or weather events; an acronym from *r*adio *d*etection *a*nd *r*anging

radiation: The transfer of energy in the form of electromagnetic waves

radioactive: Describes atoms that breakdown spontaneously, releasing energy and/or subatomic particles to become different elements

radioisotope: An unstable isotope that breaks down spontaneously at a predictable rate

rain: Liquid precipitation that falls quickly; precipitation droplets larger than drizzle. (ESRT)

rain showers: Periods of rain of short duration. (ESRT)

redshift: A displacement of the spectral lines of very distant stars and galaxies, an increase in the wavelength of starlight caused by rapid relative motion of the star away from the observer. (See Doppler effect)

reflection: The process by which light bounces off a surface or material

refraction: The bending of light and other energy waves as they enter a substance of different density

regional metamorphism: The process in which a large mass of rock experiences increased heat and pressure due to large-scale movement of Earth's crust (ESRT)

relative age: The age of one thing compared to the age of another

relative humidity: A comparison of the actual water-vapor content of the air with the maximum amount of water vapor the air can hold at a given temperature (ESRT)

relief: The difference in elevation from the highest point to the lowest point on the land surface in a specific region

renewable resources: Resources that can be replaced by natural processes at a rate will not decrease their availability

residual soil: Soil that formed in place and remains there

Richter scale: A scale for measuring earthquake magnitude based on measurements from seismographs

rock: A substance that is or was a natural part of the solid Earth, or lithosphere (ESRT)

runoff: The water from precipitation that flows downhill under the influence of gravity until it reaches a stream, or seeps into the ground; runoff may also include stream flow; overland flow

sandbar: A low ridge of sand deposited along the shore by currents

satellite: An object in space that revolves around another object as a result of gravity

saturated air: The condition in which air is holding as much moisture as it can at a particular temperature

scattering: The reflection of light in many different directions

science: A universal method of gathering, organizing, and using information about the environment

sea breezes: Light winds that blow from the water to the land that usually develop in the late morning or afternoon when the land warms; they continue into the evening until the land cools

sea-floor spreading: The process in which new lithosphere is made at the mid-ocean ridges, and adds on to older material that moves away from the ridges on both sides

secondary waves (S-waves): Transverse earthquake waves that cause the ground to vibrate side-to-side, perpendicular to the direction of travel; S-waves travel through solids, but not liquids or gases (ESRT)

sediment: The loose material created by the weathering of rock (ESRT)

sedimentary rocks: Rocks that form as a result of the compression and cementing of weathered rock fragments or shells of once-living animals (ESRT)

seismic moment: A scale for measuring the magnitude of an earthquake based on the total energy released by the earthquake

seismograph: An instrument that measures the magnitude of earthquakes

seismologists: Scientists who study earthquakes

seismology: A science that deals with earthquakes

silicate: a mineral that contains silicon and oxygen

sleet: A form of precipitation that consists of rain drops that freeze before they reach the ground; also known as ice pellets. Unlike hail, sleet does not require violent winds aloft (ESRT)

smog: A mixture of fog and air pollution particles, especially smoke from the burning of fossil fuels

snow showers: Periods of snowfall of short duration. (ESRT)

soil: A mixture of weathered rock and the remains of living organisms in which plants can grow

soil horizons: The layers of a mature soil

solar noon: The time at which the sun reaches its highest point in the sky

solar time: Time based on observations of when the sun reach its highest point and crosses a north-south line through the sky

solution: The method by which dissolved solids are carried in water

sorting: The separation of particles of sediment as a result of differences in their shape, density, or size

source region: The location in which an air mass originated

species: A group of organisms so similar that they can breed to produce fertile offspring

specific heat: The energy needed to raise the temperature of 1 gram of a substance 1 Celsius degree (ESRT)

spring: a place where groundwater flows onto the surface of the ground

spring tides: The largest tidal range, which occurs when Earth, the sun, and the moon are in a line with one another (not related to Earth's seasons)

stratosphere: The layer of Earth's atmosphere directly above the troposphere, in which the temperature increases with increasing altitude (ESRT)

streak: The color of the powdered form of a mineral (ESRT)

stream: Flowing water, such as a brook, river, or even an ocean current

stream system: All the streams that drain a particular geographic area

stress: Force that tends to distort rock, resulting in slow bending

striations: Parallel scratches in bedrock that were made by rocks transported by glaciers

subduction zone: A region in which Earth's crust is destroyed as it is pulled down into the mantle (ESRT)

summer solstice: The name generally applied to the day of the year with the longest period of sunlight. (For observers in the Northern Hemisphere, this occurs near June 21. The Northern Hemisphere summer solstice occurs when the vertical rays of the sun are at the Tropic of Cancer. In the Southern Hemisphere, the summer solstice occurs in December when the vertical rays of the sun are at the Tropic of Capricorn.)

superposition: The concept that, unless rock layers have been moved, each layer is older than the layer above it and younger than the layer below it

surf zone: An area on the shore that extends from where the waves' base touches the ocean bottom to the upper limit the waves reach on the beach

suspension: The method by which small particles that settle very slowly are carried by water

tectonics: Large-scale motions of Earth's crust that are responsible for uplift and mountain building (ESRT)

temperate climate: A climate that has large seasonal changes in temperature

temperature: A measure of the average kinetic energy of the molecules in a substance (ESRT)

terminal moraine: Irregular, hilly deposits of till formed where a glacier stopped advancing and began to melt back

terrestrial coordinates: Coordinates based on Earth's system of latitude and longitude

terrestrial planet: A planets whose composition is similar to Earth's (ESRT)

texture: The surface characteristics of a rock that are the result of size, shape, and arrangement of mineral grains (ESRT)

thermometer: An instrument used to measure temperature

thermosphere: The highest layer of Earth's atmosphere, located directly above the mesosphere, in which temperature rises with increasing altitude (ESRT)

thunderstorm: A rainstorm that produces thunder, lightning, strong winds and sometimes hail (ESRT)

tidal range: The difference between the lowest water level and the highest water level

tides: The twice- (or once-) daily cycle of change in sea level caused by the gravitational influence of the moon and sun on Earth's oceans

till: Unsorted sediments deposited by a glacier

topographic map: An isoline map on which the isolines, called contour lines, connect places having the same elevation

transform boundary: A place where two lithospheric plates move past each other without creating new lithosphere or destroying old lithosphere (ESRT)

transpiration: The process by which plants release water vapor to the atmosphere, largely through pores in their leaves

transported soil: Soil that formed in one location and was moved to another location

travel time: The time between the breaking of the rocks that causes an earthquake and when the event is detected at a given location. (ESRT)

tributary: A stream that flows into a larger stream

Tropic of Cancer: The greatest latitude north of the equator reached by the sun's vertical ray; 23.5°N

Tropic of Capricorn: The greatest latitude south of the equator reached by the sun's vertical ray; 23.5°S

tropical air mass: A large body of warm air that originated close to the equator (ESRT)

troposphere: The lowest layer of Earth's atmosphere, in which temperature decreases with increasing altitude (ESRT)

tsunami: A series of waves caused by an earthquake or underwater landslide that can cause damage and loss of lives in coastal locations

unconformity: A buried erosion surface that represents a gap in the record of Earth's history

uniformitarianism: The concept that the geological processes that took place in the past are similar to those that occur now

urbanization: The development of heavily populated areas

valley glaciers: Glaciers that begin in high mountain areas and flow through valleys to lower elevations

vaporization: The change in state from liquid to gas (vapor) at any temperature (ESRT)

velocity: Speed; change in distance divided by change in time; sometimes velocity is used to include both speed and direction.

vent: A place where lava comes to the surface

vertical rays: Sunlight that strikes Earth's surface at an angle of 90°

vertical sorting: Within a layer of sediment, the gradual change in sediment size from bottom (large) to top (small) showing the order in which particles settled; graded bedding

vesicular: Rocks that contain gas pockets, or vesicles (ESRT)

volcanic: Fine-grained, extrusive igneous rocks (ESRT)

volcano: An opening in Earth's surface through which molten magma (lava) erupts

water table: The upper limit of the underground zone of saturation or the top surface on an aquifer

watershed: The geographic area drained by a particular river or stream; drainage basin

weather: The short-term conditions of Earth's atmosphere at a given time and place (ESRT)

weathering: The change in rocks that occurs when they are exposed to conditions at Earth's surface

winter solstice: The name generally applied to the day of the year with the shortest period of sunlight. (For observers in the Northern Hemisphere, this occurs near December 22. The Northern Hemisphere winter solstice occurs when the vertical rays of the sun are at the Tropic of Capricorn. In the Southern Hemisphere, the winter solstice occurs in June when the vertical rays of the sun are at the Tropic of Cancer.)

zenith: The point in the sky directly over an observer's head

zone of aeration: The part of the rock and soil in which air fills most of the available spaces

zone of saturation: The part of the rock and soil where all available spaces are filled with water

Index

Photo Credits

All photographs, except those listed below were provided courtesy of the author, Thomas McGuire.